FUNDAMENTALS OF
PETROLEUM

THIRD EDITION

Edited by Mildred Gerding

Published by
PETROLEUM EXTENSION SERVICE
Division of Continuing Education
The University of Texas at Austin
Austin, Texas
in cooperation with
ASSOCIATION OF DESK AND DERRICK CLUBS
Tulsa, Oklahoma
1986

Contributing Authors

Annes McCann Baker
Ron Baker
Cinda Cyrus
Mildred Gerding
Richard House
Jeff Morris
Jean Trimble Pietrobono
Ione Stelzner
Martine Stemerick

Research Assistants

Gale Hathcock
Jennifer Snowden

© 1986 by The University of Texas at Austin
All rights reserved
First Edition published 1979. Third Edition 1986
Fourth Impression 1991
Printed in the United States of America

Catalog No. 1.00030
ISBN 0-88698-122-0

Contents

Foreword

Our nation cannot live without energy. Oil and natural gas provide about 70 percent of the energy used in the United States. Because of its importance, people directly related to and affected by this industry should know more about it. The better people understand the petroleum industry, the greater the chances are for making the proper decisions toward solutions to our energy problems.

The petroleum industry is a leader in developing and applying advanced technology. The Association of Desk and Derrick Clubs recognizes the need for a better understanding of this complex and diversified technology and endorses *Fundamentals of Petroleum*, which is dedicated to individuals seeking knowledge of the petroleum industry. Designed for both the professional and layman as a basic guide on the practical aspects of the petroleum industry, this book is intended to be used by the association as a primary text for training in house, in junior colleges, and through correspondence courses. It is also intended to serve as an educational tool for allied industries, as well as professional and governmental agencies. It provides under one cover a basic discussion of the petroleum industry from geology through exploration, drilling, production, transportation, refining and processing, marketing, and economics.

Fundamentals of Petroleum was developed from an idea proposed by Loretta Owens of the Desk and Derrick Club of Fort Worth, Texas, and further stimulated by the Desk and Derrick Club of Corpus Christi, Texas. Under the leadership of Ms. Owens, a committee, representing eight geographical locations within the association, developed and presented an outline to Petroleum Extension Service of The University of Texas at Austin. With these guidelines, the Petroleum Extension Service staff researched, developed, and published this training manual.

Desk and Derrick is a unique organization of approximately ten thousand members employed in the petroleum and allied industries who are dedicated to the proposition that *greater knowledge* of the petroleum industry will result in *greater service* through job performance. Nonshareholding, noncommercial, nonprofit, nonpartisan, and nonbargaining in its policies, the organization has very positive concepts on the value of education for women.

Association of Desk and Derrick Clubs
411 Thompson Building
Tulsa, Oklahoma 74103

Preface

Fundamentals of Petroleum is designed to give an overall view of the petroleum industry in terms that can be understood by the layman as well as the professional. This book does not cover all procedures and equipment used in the industry, nor does it give detailed descriptions of oilfield operations. Rather, the authors and editor, mindful of its purpose, tried to include general, useful information while minimizing technical details.

To help the reader understand industry terminology, *A Dictionary of Petroleum Terms*, also published by the Petroleum Extension Service, is highly recommended for use with this book.

Fundamentals of Petroleum is used as a text for the School of Petroleum Fundamentals conducted by the Petroleum Extension Service at its training centers, for courses sponsored by the Association of Desk and Derrick Clubs, and for numerous petroleum-related college courses.

The first edition was based on an outline from the Association of Desk and Derrick Clubs. With the help of the ADDC Education Committee, who furnished industry reviewers for the text, this third edition was compiled by members of the Petroleum Extension Service staff. Incorporated in the new edition are the many comments and suggestions made by students, teachers, and readers of the first and second editions.

It is our sincere hope that this effort will meet the needs of people outside the industry who are interested in petroleum, as well as those in the industry, especially members of the Desk and Derrick Clubs everywhere. Their desire for knowledge is the stimulus needed by our staff whose job is to develop training aids for the petroleum industry.

Mildred Gerding

Acknowledgments

The knowledge and talents of many people and the resources of many organizations were utilized to produce this book. In addition to the work of the major authors and the creative staff listed, the content support provided by the following individuals is gratefully appreciated:

William C. Bailey, Jr., Mid-Continent Oil and Gas Association, Baton Rouge, Louisiana

William P. Biggs, Petroleum Engineering, The University of Texas at Austin, Austin, Texas

Jim Caples, Heritage Title Company of Austin, Austin, Texas

Keith Carter, Petroleum Land Management, The University of Texas at Austin, Austin, Texas

Robert S. Farnsworth, Transportation Division, Railroad Commission of Texas, Austin, Texas

John J. France, Southern Union Refining Company, Hobbs, New Mexico

Jerell J. Fromme, Diamond Shamrock Refining and Marketing Company, Three Rivers, Texas

A. T. Gibbon and L. A. Markley, Exxon Company, U. S. A., Houston, Texas

Jerry Gilbert, MORANCO Drilling Inc., Hobbs, New Mexico

Gary A. Gough, Gough Tank Trucks, Inc., Graham, Texas

Gene Hodge and Daniel Martinez, Valero Refining Company, Corpus Christi, Texas

L. W. Ledgerwood, Jr., Houston, Texas

Jodie Leecraft, Mark A. Welsh, Jr., and Bruce R. Whalen, Petroleum Extension Service, The University of Texas at Austin, Austin, Texas

Dennis Steffy, Statewide Petroleum Extension Program, University of Alaska, Soldotna, Alaska

Clem L. Ware, Certified Petroleum Landman, Odessa, Texas

Harold S. Wright, Columbia Gas Transmission Corporation, Charleston, West Virginia

M. E. Wright, Signal Hill, California

Also, sincere appreciation is extended to the following companies, organizations, and institutions for their generous contributions to this effort:

Alyeska Pipeline Service Company
American Gas Association
American Petroleum Institute
Amoco Chemicals Corporation
Amoco Public and Government Affairs Department
Ana-Log, Inc.
ARCO Chemical Company

Association of Oil Pipe Lines
Brown & Root, Inc.
Bureau of Land Management
Canadian Petroleum Association
Chevron Oil Field Research Company
Cities Services Company
Coastal States Gas Corporation
Colonial Pipeline Company
Conoco Inc.
Continental Pipe Line Company
Diamond Shamrock Refining and Marketing Company
Dome Petroleum Limited
El Paso Natural Gas Company
Esso Resources Canada Limited
Evans Transportation Company, Railcar Division
Exxon Company, U. S. A.
Exxon Corporation
Exxon Pipeline Company
Federal Energy Regulatory Commission
General American Transportation Corporation
Geosource, Inc.
Gould, Inc., DeAnza Imaging & Graphics Division
Gulf Canada Resources Limited
Gulf Oil Corporation
Halliburton Services
Hughes Drilling Equipment
Interstate Natural Gas Association of America
Longhorn Title Company, Georgetown, Texas
Marathon Oil Company
Marathon Pipeline Company
Martin-Decker
McGraw-Hill Book Company
Mitchell Energy & Development Corporation
Mobil Oil Corporation
Natural Gas Pipeline Company of America
Nicklos Drilling Corporation
Panhandle Eastern Pipe Line Company
PennWell Publishing Company
Petroleum Publishing Company
Phillips Petroleum Company
Pound Printing and Stationery Company
Pressure Transport, Inc.
Railroad Commission of Texas
Rotoflow Corporation
Saber Energy Company
The Scott Petty Company

Sheehan Pipe Line Construction Company
Shell Oil Company
Sivalls, Inc.
Standard Oil Company of California
Tennessee Gas Pipeline Co., Division of Tenneco Inc.
Texaco Inc.
Texas Christian University, Department of Geology
Texas Eastern Transmission Corporation
Texas Gas Transmission Corporation
Texas LP-Gas Association
Texas State Highway & Public Transportation Commission
Transco
Union Tank Car Company
The University of Texas at Austin, Bureau of Economic Geology
The University of Texas at Austin, Humanities Research Center
Valero Refining Co.
Wärtsilä Arctic, Inc.

Petroleum I Geology

Geology is the science that deals with the history and structure of the earth and its life forms, especially as recorded in the rocks. It is a science essential to the petroleum industry, for it is used to predict where oil accumulations might occur. Familiarity with the basic principles of geology is desirable for anyone interested in the petroleum industry.

Geology is based on observation and the knowledge derived from many other sciences. A first principle is that *the present is the key to the past;* that is, the processes acting on the earth today are basically the same as those that operated in the past. Using knowledge of these processes and observing rocks and rock formations, the petroleum geologist reconstructs the geologic history of an area to determine whether the formations are likely to contain petroleum reservoirs.

To be of interest to the petroleum industry, a petroleum deposit must be commercially valuable. That is, there must be a reasonable chance of making a profit on the eventual sale of the minerals, considering the market price of oil and gas, the amount of recoverable petroleum, the expected production rate, and the cost of drilling and producing the well.

A petroleum reservoir is not a large underground cavern, as the common term *oil pool* might suggest. A reservoir is a formation of mostly solid rock with tiny openings, or *pores*, that can hold fluids. Reservoir rock usually contains salt water in addition to the many different kinds of hydrocarbons. These fluids are layered, with the lightest (gas) on top, then the oil, and the heaviest (water) on bottom. Petroleum companies prefer that a reservoir contain all three fluids; oil and gas are valuable minerals, and gas and water often supply the force to drive petroleum out of the producing formation to the surface, making pumping unnecessary.

To accumulate a commercially valuable deposit of petroleum, a reservoir must have the right shape or configuration to hold the oil or gas, and some kind of seal or trap to keep it from escaping. It must be large and thick—usually 10 feet thick or more. It must have a *porosity* of 10 percent or more—that is, at least 10 percent of the rock must be pore space, capable of containing petroleum (fig. 1.1). It must be *permeable,*—that is, the pores must be connected so oil, gas, and water can flow through it by moving from one pore space to another (fig. 1.2).

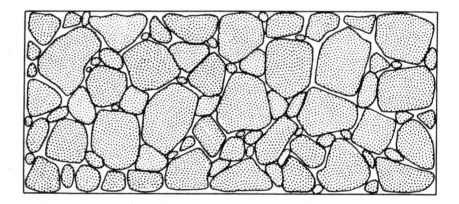

Fig. 1.1. A rock is *porous* when it has many tiny spaces, or pores.

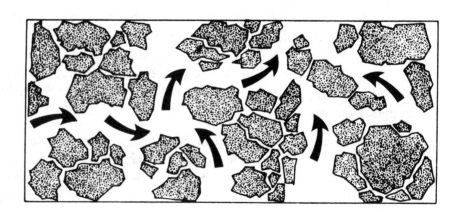

Fig. 1.2. A rock is *permeable* when the pores are connected.

BASIC CONCEPTS OF GEOLOGY

The earth is thought to have formed about 4.6 billion years ago out of a cloud of cosmic dust. As the planet was pulled together by its own gravity, the heat of compression and of its radioactive elements caused it to become molten. The heaviest components, mostly iron and nickel, sank to the earth's center and became the *core* (fig. 1.3). Lighter minerals formed a thick molten *mantle*, and certain minerals rich in aluminum, silicon, magnesium, and other light elements solidified into a thin, rocky *crust*.

The surface of the young planet was an inhospitable place. Molten rock (*magma*) erupted through thousands of fissures and volcanoes, as did the gases and water vapor that formed the early, oxygenless atmosphere. As the surface cooled, water vapor condensed and fell in torrential rains, and the first oceans began to form.

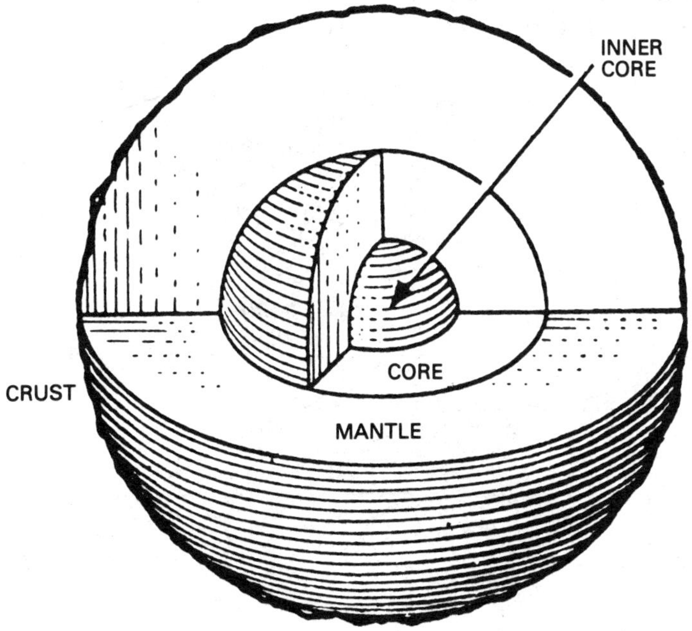

Early history of the earth

Fig. 1.3. A cross section of the earth would show its crust, mantle, core, and inner core.

Over time the earth's crust has grown thicker and more stable. Geologists today view the crust as an assemblage of plates. These plates fit together like a jigsaw puzzle. But unlike a jigsaw puzzle, the pieces of the earth's crust move and change shape. In some places they slide past one another; in others, they collide with or pull away from each other. The theory that explains these processes is called *plate tectonics.*

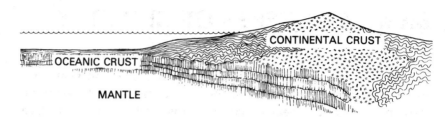

Fig. 1.4. Oceanic crust is thinner but heavier than continental crust.

Two basic kinds of crust are *oceanic* crust and *continental* crust (fig. 1.4). Oceanic crust is thin—about 5 to 7 miles—and made up of heavy *igneous* rock (rock that is formed from cooled magma). The rock of continental crust, however, is thick—10 to 30 miles—and relatively light. Because of these differences, continents tend to float like icebergs in a "sea" of heavier rock, rising high above sea level where they are thickest—in the mountains.

These continental heights are gradually worn away by running water and other agents. Particles of rock are carried into the sea, where they are deposited in thick sedimentary beds along the edges of the continents. Cemented together by minerals in the water and by the pressure of more sediments deposited on top of them, these beds become layers of *sedimentary* rock.

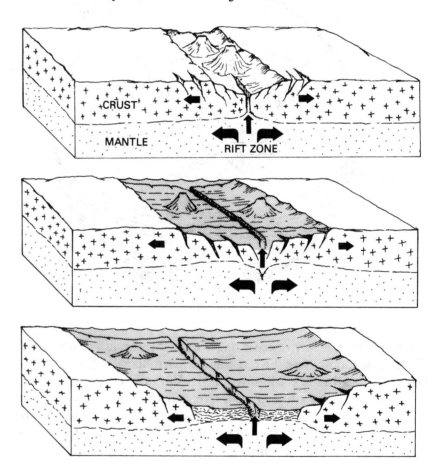

Fig. 1.5. Oceanic crust forms in the rift between two diverging plates.

Sometimes a crustal plate breaks into two or more smaller plates that begin moving away from one another, forming an ocean basin (fig. 1.5). As the two parts of the continent *diverge*, or separate, magma rises from the mantle and solidifies in the rift, forming a *midocean ridge*. New crust, thinner than the continents but denser, spreads outward between the two new continents. The Atlantic Ocean was born just this way about 200 million years ago when North and South America split away from Europe and Africa (fig. 1.6).

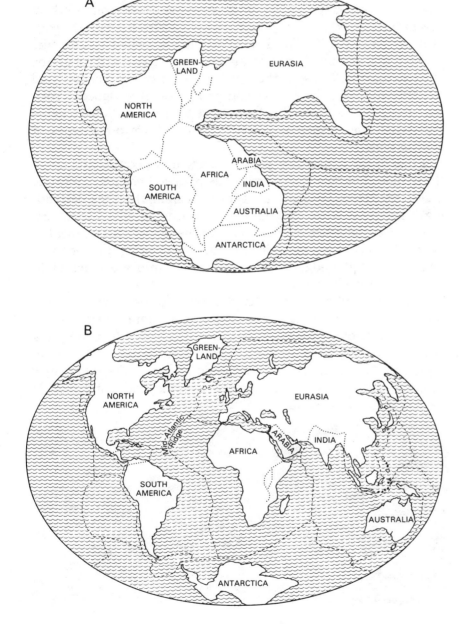

Fig. 1.6. The relative positions of the continents are shown 200 million years ago (*A*) and today (*B*).

6

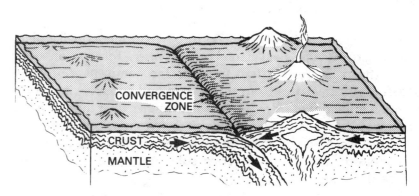

Fig. 1.7. When two oceanic plates converge, one dives beneath the other.

At its leading edge, a moving plate meets other plates head on. Several things can happen in this *convergence zone.* Where oceanic crust meets oceanic crust, one plate slips beneath the edge of the other (fig. 1.7). The descending plate is melted as it plunges into the hot mantle. The minerals that melt first rise through the crust as magma. Some of these rising bodies of magma cool and solidify within the crust, forming granite or other igneous rock that may later be exposed at the surface by erosion. Some magma may reach the surface as volcanic lava to form other types of igneous rock.

If one of the converging plates is made up of continental crust, the thinner, but heavier, oceanic plate plunges beneath it, forming a deep-sea trench. When this happens, magma from the descending plate may erupt in continental volcanoes like Mount St. Helens. On the other hand, if both of the plates are continental, the collision buckles and folds the rocks—including the sedimentary rocks at the edges of the continents—into great mountain chains like the Himalayas.

About 1.5 billion years after the earth formed, simple living organisms appeared in the oceans. However, more complex forms (fig. 1.8) did not appear in abundance until about 2.5 billion years later—at the beginning of the Cambrian period, only 550 million years ago. Not until the Devonian period, about 350 million years ago, did vegetation become widespread on the land. Land animals were scarce until even later.

Fig. 1.8. Abundant sea life helped form petroleum beneath the ocean floor. (*Courtesy of Continental Oil Company and American Petroleum Institute*)

Because life has evolved continuously from Precambrian times, the fossil remains of animals (*fauna*) and plants (*flora*) succeed one another in a definite, known order. Geologists have classified rocks in groups based upon this succession, as shown in table 1.1. The durations of the eras, periods, and epochs have been estimated from studies of radioactive minerals. The presence of life is essential to the petroleum story because organic matter is one of the necessary ingredients in the formation of oil.

TABLE 1.1
GEOLOGIC TIME SCALE

Era	Period	Epoch	Duration (millions of years)	Dates (millions of years)
Cenozoic	Quaternary	Recent	0.01	0.00
				0.01
		Pleistocene	1	
				1
	Tertiary	Pliocene	10	
				11
		Miocene	14	
				25
		Oligocene	15	
				40
		Eocene	20	
				60
		Paleocene	10	
				70 ± 2
Mesozoic	Cretaceous		65	
				135 ± 5
	Jurassic		30	
				165 ± 10
	Triassic		35	
				200 ± 20
Paleozoic	Permian		35	
				235 ± 30
	Pennsylvanian		30	
				265 ± 35
	Mississippian		35	
				300 ± 40
	Devonian		50	
				350 ± 40
	Silurian		40	
				380 ± 40
	Ordovician		70	
				460 ± 40
	Cambrian		90	
				550 ± 50
Precambrian			4,500 ±	

(After R. M. Sneider)

8

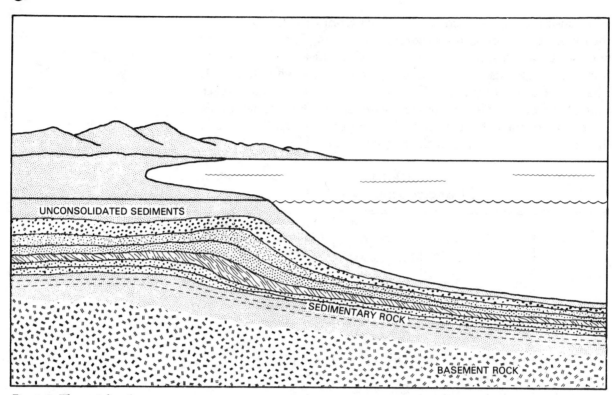

UNCONSOLIDATED SEDIMENTS

SEDIMENTARY ROCK

BASEMENT ROCK

Fig. 1.9. The weight of overlying layers compacts sediments into rock.

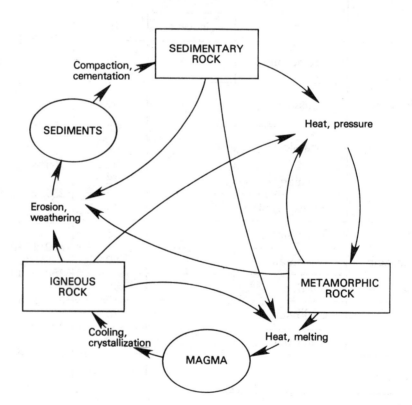

Fig. 1.10. The rock cycle shows how various processes change rocks from one type to another.

The rocks of the earth's crust are changed from one form to another by several geologic processes, including erosion, sedimentation, cementation, compression, and melting. Erosion is usually the result of flowing water, but it can result from the action of wind, freezing water, moving ice, and waves. Particles eroding from the continental highlands are deposited, in essentially horizontal layers, in the lowlands or in offshore shallows. As sediments continue to be deposited on sediments, the earlier unconsolidated deposits are compacted by the weight of the overlying sediments (fig. 1.9). Cemented together by water-borne minerals, these deposits are transformed into *sedimentary* rocks. Some of these rocks erode to produce more sediments; others are buried deep and subjected to tremendous heat and pressure, forming *metamorphic* rocks. Further increases in temperature and pressure melt some of the minerals in the rocks, forming magma. When the magma cools and recrystallizes, *igneous* rock is formed. Both metamorphic and newly formed igneous rocks in turn erode into new sediments, forming new sedimentary rocks. Thus, the cycle of erosion, sedimentation, and metamorphism continues to produce the various rock types (fig. 1.10).

The sedimentary rocks — those formed from sediments — are of most interest to the petroleum geologist. Most oil and gas accumulations occur in them; igneous and metamorphic rocks rarely contain oil or gas. Sedimentary rocks may be simply classified as clastic or nonclastic (table 1.2), or they may be categorized in some other system, depending on the purpose of the classification. *Clastic* rocks consist of individual grains derived from the erosion of preexisting rocks and are held together by crystalline or amorphous cement; they are generally classified by grain size. *Nonclastic* rocks are derived from other types of sediments, such as chemical precipitates and organic detritus, and are variable in texture. Some rocks, such as limestone, are found under several classifications because they are formed by a combination of processes or occur in different types.

The rock cycle

TABLE 1.2
SIMPLIFIED CLASSIFICATION OF SEDIMENTARY ROCKS

Clastic	Nonclastic			
	Carbonate	Evaporite	Organic	Other
Conglomerate	Limestone	Gypsum	Peat	Chert
Limestone	Dolomite	Anhydrite	Coal	
Sandstone		Salt	Diatomite	
Siltstone		Potash	Limestone	
Shale				

10

Earth movements

If some other process did not compensate for erosion, the land would be reduced to plains near the level of the sea. However, the land now stands about as high above the sea as it ever did. Obviously, the surface must have been uplifted to compensate for the wearing down of the mountains.

Both vertical and horizontal movements of the crust have occurred continually since the crust solidified. The earthquakes and volcanic eruptions that occur every year demonstrate that the crust is active even today. Most rocks near the surface are fractured by both internal and external stresses. If the rock layers on one side of a fracture have moved in relation to the other side, the fracture is called a *fault*. Displacement along a fault may range from only a few millimetres to hundreds of miles, as along the San Andreas fault in California, a boundary between two of the earth's largest crustal plates (fig. 1.11). The ground next to parts of the San Andreas fault moved 21 feet horizontally during the great San Francisco earthquake of 1906.

The remains of marine organisms found in some of the highest mountains and in the deepest oilwells prove that the rocks were deposited in ancient shallow seas and then rose or subsided to their present positions. Repeated movements of only a few inches at a time will gradually raise or lower the earth's surface enough to account for thousands of feet of displacement over millions of years. The Himalayas, formed by the collision of the Indian subcontinent with Asia 50 million years ago, are building by measurable amounts each year out of what used to be ocean floor.

Fig. 1.11. This view of the San Andreas fault in the Carrizo Plain, California, shows two stream channels that were offset about 70 feet by a right-lateral slip. (*Photo by John S. Shelton*)

Sedimentary rocks are deposited in horizontal layers called *strata,* or *beds.* However, these layers are often deformed by geological processes. Under intense pressure, even the hardest rock will bend or break. A common type of deformation is the buckling of the layers into folds (fig. 1.12). Folds are the most common structures in mountain chains, ranging in size from wrinkles of less than an inch to great arches and troughs many miles across. The upwarps or arches are called *anticlines;* the downwarps or troughs are *synclines* (fig. 1.13).

Geological structures

Fig. 1.12. This photo shows deformation of the earth's crust by the buckling of layers into folds.

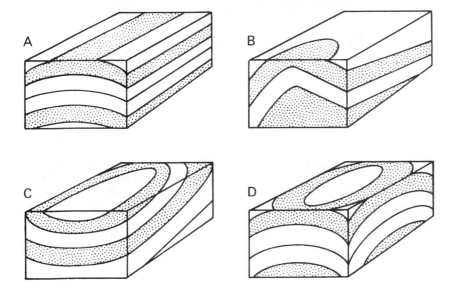

Fig. 1.13. Types of folds include symmetrical anticline (A), plunging asymmetrical anticline (B), plunging syncline (C), and dome (D).

The top of an anticline or syncline is rarely level; it tends to drop or *plunge* from one end to the other. A short anticline with its crest plunging in both directions from a high point is called a *dome*. Many domes are almost perfectly circular, plunging in all directions. Domes often have an intrusive core that lifts them, such as the salt domes along the U. S. Gulf Coast. Anticlines and domes are important to petroleum geologists because they often contain hydrocarbons.

Faults are important to the petroleum geologist because they affect the locations of oil and gas accumulations (fig. 1.14). Movement is mostly vertical in normal and reverse faults but horizontal in overthrust and lateral faults. Combinations of vertical and horizontal movement are also possible, as in growth faults.

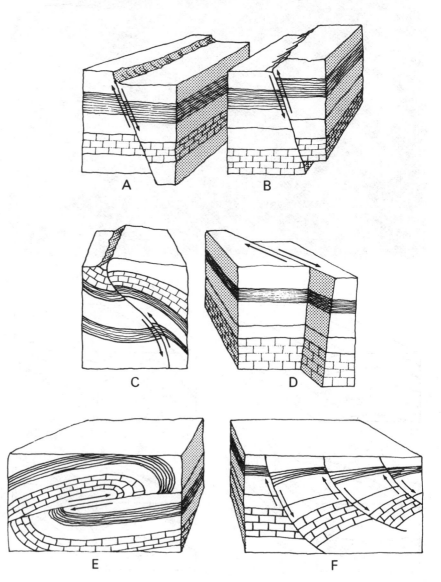

Fig. 1.14. Several common types of faults are normal dip-slip (*A*), reverse or thrust dip-slip (*B* and *C*), lateral (*D*), overthrust (*E*), and growth (*F*) faults.

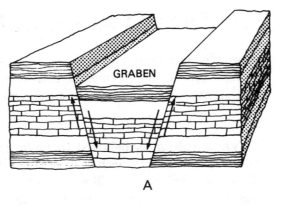

A

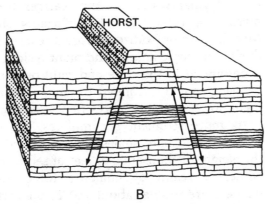

B

Fig. 1.15. Two fault-formed landscape features are the down-dropped graben (*A*) and the uplifted horst (*B*).

Faulting can produce certain recognizable surface features (fig. 1.15). A *graben* is a long, narrow block of crust that has subsided relative to the surrounding crust; a *horst*, on the other hand, is a similar block that has risen. In the North Sea, oil has accumulated in sediment-filled grabens beneath the ocean floor.

An eroded rock surface that has become buried under more recently formed rocks is called an *unconformity*. Two general kinds of unconformity are the *disconformity* and the *angular unconformity* (fig. 1.16). An unconformity may act as a barrier to trap accumulations of petroleum.

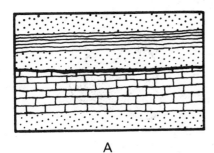

A   B

Fig. 1.16. Two general kinds of unconformities are a disconformity (*A*) and an angular unconformity (*B*).

PETROLEUM ACCUMULATIONS

For petroleum to accumulate, there must be (1) a source of oil and gas, (2) a porous and permeable bed of reservoir rock, and (3) a trap that acts as a barrier to fluid movement.

Origin of petroleum

Today's petroleum is believed to have originated from organic matter deposited along with rock particles during the formation of sedimentary rock many millions of years ago. This *organic theory* holds that oil and gas are of marine origin. In shallow seas and in the marginal waters of the warmer oceans, an abundance of marine life is eternally falling in a slow, steady rain to the bottom. Most is eaten or oxidized before it reaches bottom, but a portion of this microscopic animal and plant residue escapes destruction and is entombed in the ooze and mud on the seafloor. Bacteria take oxygen from the trapped organic residues and gradually break them down into substances rich in carbon and hydrogen.

As more sediment accumulates, the organically rich clay is squeezed into hard shales. Pressures and temperatures rise under the weight of thousands of feet of sediment. Within this deep, unwitnessed realm of immense force, oil is born. When the temperature reaches about 150°F, the carbon- and hydrogen-rich substances begin to recombine chemically to form hundreds of different kinds of hydrocarbon molecules. These hydrocarbons are well-organized chains of carbon atoms with hydrogen atoms attached (fig. 1.17). The conversion process reaches a maximum between 225° and 350°F. Above this temperature, the heavier long-chain molecules are broken into smaller, lighter ones, such as methane gas. However, at temperatures above 500°F, the organic material is carbonized and destroyed as source material. Source beds too deeply buried produce no hydrocarbons because of their extreme temperatures.

Petroleum is not formed in large concentrations; initially, it is as dispersed as the organic matter it originated from. Once formed,

Fig. 1.17. Hydrocarbon molecules contain hydrogen and carbon atoms. Heavier propane (*A*) has a chain of carbon atoms with hydrogen atoms attached. Methane (*B*), with only one carbon atom, is a smaller, lighter molecule.

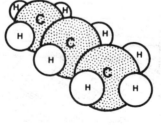

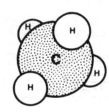

A   B

however, it can migrate through permeable rock. Formation pressures tend to squeeze it out of the relatively impermeable shale into fractures and open formations such as sandstone, where it can travel from pore to pore. Oil and gas tend to seek shallower levels. Unless they are trapped underground by geological formations, they may rise until they escape at the surface.

Reservoir rock

Reservoir rock is subsurface rock that is capable of containing gas, oil, water, or other fluids. To be a *productive petroleum reservoir*, the rock body must be large enough and porous enough to contain an appreciable volume of hydrocarbons, and it must be permeable enough to give up the contained fluids at a satisfactory rate when the reservoir is penetrated by a well. Sandstones and carbonates (such as limestone and dolomite) are generally the most porous of rocks and are also the most common reservoir rocks.

Close examination of reservoir rock with a powerful magnifying glass reveals thousands of tiny openings or *pores* (fig. 1.18). The greater the porosity of a formation, the more reservoir fluids it is able to hold. Porosity may vary from less than 5 percent in a tightly cemented sandstone or carbonate to more than 30 percent in unconsolidated sands. However, accurate determination of formation porosity is difficult.

Besides being porous, a reservoir rock must also be *permeable*. These connected pores allow petroleum to move from one pore to another so that petroleum has a way to flow out of the pores and into a well. The greater the *permeability*, the more easily petroleum can move or flow within the rock. The unit of measurement of permeability is the *darcy*. Most petroleum reservoirs have permeabilities so small that they are measured in thousandths of a darcy, or *millidarcys*. In a given formation, porosity and permeability are not necessarily related directly; however, highly porous formations are often highly permeable also.

Fig. 1.18. When reservoir rock is magnified, its porosity can be seen.

16

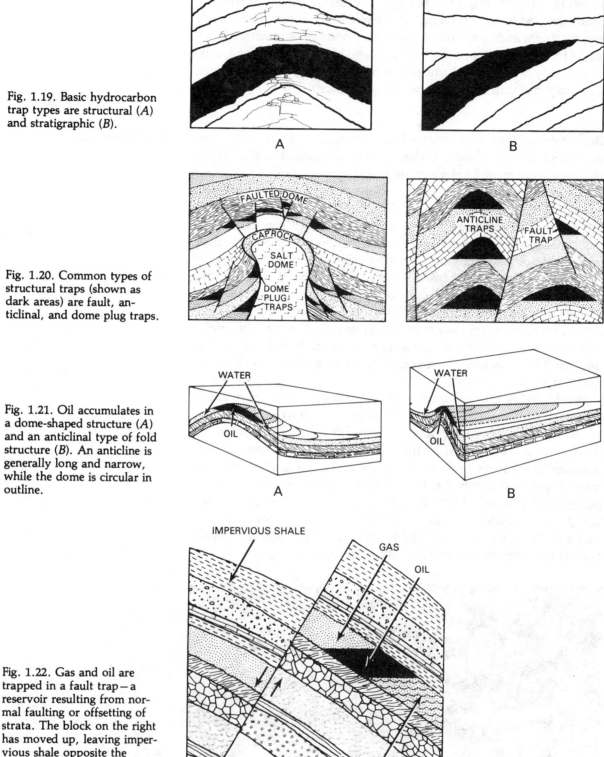

Fig. 1.19. Basic hydrocarbon trap types are structural (A) and stratigraphic (B).

A   B

Fig. 1.20. Common types of structural traps (shown as dark areas) are fault, anticlinal, and dome plug traps.

FAULTED DOME

CAPROCK

SALT DOME

DOME PLUG TRAPS

ANTICLINE TRAPS

FAULT TRAP

Fig. 1.21. Oil accumulates in a dome-shaped structure (A) and an anticlinal type of fold structure (B). An anticline is generally long and narrow, while the dome is circular in outline.

WATER   WATER

OIL   OIL

A   B

Fig. 1.22. Gas and oil are trapped in a fault trap—a reservoir resulting from normal faulting or offsetting of strata. The block on the right has moved up, leaving impervious shale opposite the hydrocarbon-bearing formation.

IMPERVIOUS SHALE

GAS

OIL

WATER

If there is a source of hydrocarbons, and if reservoir rock is sufficiently porous and permeable, then migration will occur. But if petroleum is to accumulate, there must be something to stop the migration. Otherwise, hydrocarbons will continue to move upward until they escape at the surface.

A geologic structure that prevents the escape of oil and gas from a reservoir rock is called a *trap*. Reservoir traps are of two general types: *structural* and *stratigraphic* (fig. 1.19). Structural traps are those formed by deformation of the reservoir formation; stratigraphic traps are those that result from an existing updip termination of porosity or permeability.

Structural traps. Structural traps vary widely in size and shape. Most are formed by the folding or faulting of reservoir rock (fig. 1.20). Some of the more common structural traps are anticlinal traps, fault traps, and dome plug traps.

Reservoirs formed by the folding of rock layers usually have the shape of anticlines or domes (fig. 1.21). These *anticlinal traps* are filled with petroleum that has moved in from a source below. In the anticline, further upward movement is arrested by the shape of the structure and by a seal or *caprock*—a layer of impermeable rock above the reservoir. Two examples of oilfields with anticlinal traps are the Santa Fe Springs field in California and the Agha Jari field in Iran.

Fault traps are formed by the shearing and offsetting of strata (fig. 1.22). The escape of oil from such a trap is prevented by nonporous rocks that have moved into a position opposite the porous petroleum-bearing formation. In traps of this type, oil is confined by the tilt of the rock layers and the faulting. A fault trap depends on the effectiveness of the seal at the fault. This seal may be formed by an impermeable bed that moved opposite the permeable reservoir bed, or by the impermeable material called *gouge* within the fault zone itself.

The simple fault trap may occur where structural contours provide closure against a single fault. However, in other structural configurations, two or even three faults may be required to form a trap (fig. 1.23). Fault trap accumulations tend to be elongated parallel to the fault trend; for example, the accumulations in the many oilfields along the Mexia-Talco fault zone extend from central to northeastern Texas.

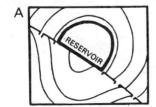 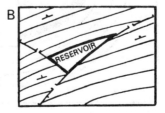

Fig. 1.23. Structure contour maps show simple (*A*) and compound (*B*) faulting.

18

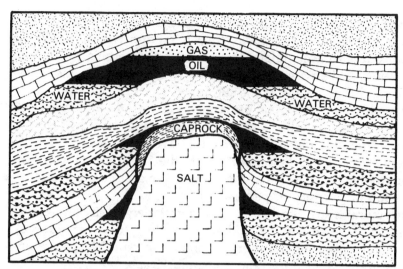

Fig. 1.24. A nonporous salt mass has formed dome-shaped traps in overlying and surrounding porous rocks.

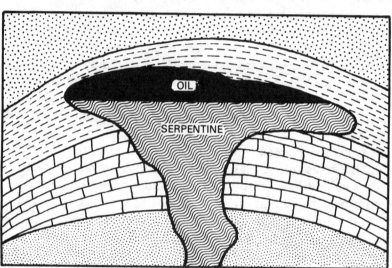

Fig. 1.25. A porous serpentine plug forms a reservoir within itself by intruding into nonporous surrounding formations.

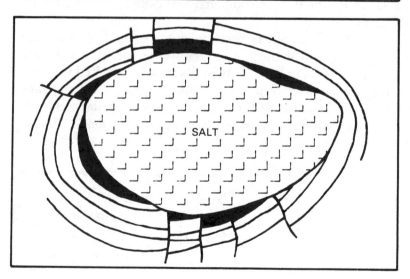

Fig. 1.26. Discontinuous peripheral traps form around a piercement salt dome.

Dome and *plug traps* are porous formations on or surrounding great plugs of salt or serpentine rock that have pierced, deformed, or lifted the overlying rock layers (figs. 1.24 and 1.25). Piercement may be nearly circular, as in a typical salt-dome oilfield on the U. S. Gulf Coast or in Germany, or it can be long and narrow, as in the Romanian oilfields. The salt and associated material form an efficient updip seal, blocking further migration of petroleum.

Hydrocarbon accumulations in the peripheral traps around a salt plug are usually not continuous but broken into separate segments by faulting (fig. 1.26). This discontinuity makes piercement traps difficult to drill successfully. The geologist knows the traps are there; he just cannot predict their precise locations accurately. As a result, many dry holes are drilled in attempts to tap the reservoirs.

Stratigraphic traps. A stratigraphic trap is caused either by a nonporous formation sealing off the top edge of a reservoir bed or by a change of porosity and permeability within the reservoir bed itself (fig. 1.27). One type of stratigraphic trap is the *unconformity*, in which part of a porous depositional sequence was eroded

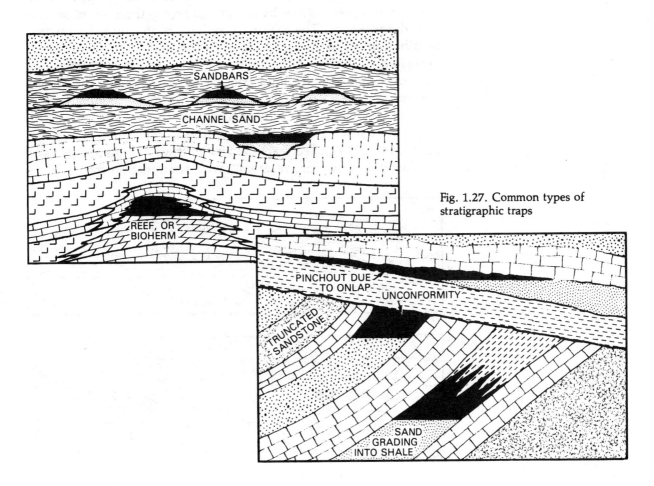

Fig. 1.27. Common types of stratigraphic traps

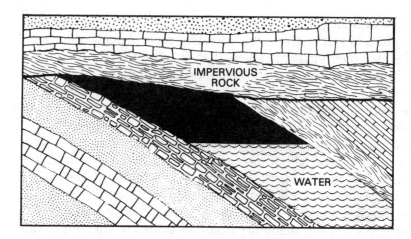

Fig. 1.28. Oil is trapped under an unconformity.

and then overlaid with impermeable caprock. In an *angular unconformity*, the layers above the unconformity are laid across the eroded edges of folded or tilted beds below, forming a trap (fig. 1.28). An example of a reservoir formed by an angular unconformity is the East Texas field.

Another type of stratigraphic trap is the *lenticular trap*, which is sealed in its upper regions by abrupt changes in the amount of connected pore space. The changes may have been caused by uneven distribution of sand and clay as the formation was being deposited, as in river delta sandbars. Oil is confined within porous parts of the rock by the nonporous parts surrounding it (fig. 1.29).

Combination traps. Many traps are formed by combinations of folding, faulting, porosity changes, and other conditions. These traps, both structural and stratigraphic in origin, are called *combination traps*. The Seeligson field in Southwest Texas and parts of the East Texas field are reservoirs with combination traps.

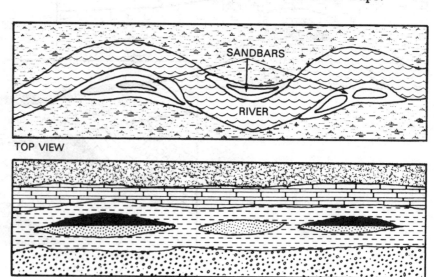

Fig. 1.29. Lenticular traps are often formed in buried river sandbars.

RESERVOIR FLUIDS

A fluid is defined as any substance that will flow. Three fluids are found in reservoirs: oil, water, and gas. Oil and water are liquids as well as fluids. Gas is a fluid but not a liquid in its natural state, although it can be liquefied by artificial means.

Water

Most oil reservoirs are composed of sediments that were deposited in or near the sea. These sedimentary beds were originally saturated with salt water. Part of this water was displaced when petroleum was formed. The salt water that remains in the formation is called connate interstitial water—*connate* from the Latin meaning "born with" and *interstitial* because the water is found in the interstices, or pores, of the formation. By common usage this term has been shortened to *connate water* and always means the water in the formation when the reservoir was being formed.

Connate water is distributed throughout the reservoir. However, nearly all petroleum reservoirs have additional water that accumulated along with the petroleum. It is this "free" water that supplies the energy for a water drive. *Bottom water* occurs beneath the oil accumulation; *edgewater* is found at the edge of the oil zone on the flanks of the structure.

Oil

Oil, which is lighter than water and will not readily mix with it, makes room for itself in the pores of reservoir rock by pushing the water downward. However, oil will not displace all the water. A film of water sticks to, or is adsorbed by, the solid rock material surrounding the pore spaces (fig. 1.30). This film is called *wetting water*. In other words, water is not only in the reservoir below the

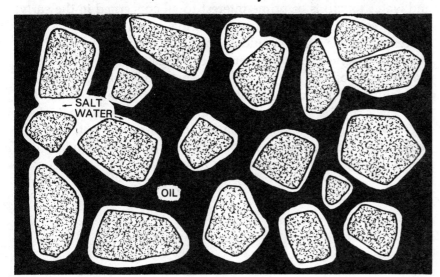

Fig. 1.30. Wetting water usually coats the grains of the reservoir rock.

oil accumulation but also within the pores along with the oil. The rare exceptions are oil-wet reservoirs, which have no film of water lining the pores but which may have an oil saturation of 100 percent of the available porosity.

Gas

Natural gas is usually associated with oil produced from a reservoir. The energy supplied by gas under pressure is probably the most valuable drive in the withdrawal of oil from reservoirs. The industry has come a long way since the day it was general practice to "blow" gas caps into the atmosphere so that a well in the gas zone of a reservoir could finally be induced to produce a little crude oil. Gas is associated with oil and water in reservoirs in two principal ways—as *solution gas* and as *free gas* in gas caps.

Given proper conditions, such as high pressure and low temperature, natural gas will stay in solution in oil in a reservoir. When the oil is brought to the surface and the pressure relieved, the gas comes out of solution, much as a bottle of soda water fizzes when the cap is removed. Gas in solution occupies space in a reservoir, and allowance has to be made for this space when the volume of oil in place is calculated.

Free gas—gas that is not dissolved in oil—tends to accumulate in the highest structural part of a reservoir, where it forms a gas cap. As long as there is free gas in a reservoir gas cap, the oil in the reservoir will remain saturated with gas in solution. Dissolved gas lowers the *viscosity* of the oil (its resistance to flow), making the oil easier to move to the wellbore.

Fluid distribution

The oil-water contact line (the level in the reservoir where the oil and water touch) is of prime interest to all concerned in the early development of a field because, to get maximum production from the reservoir, the water should not be produced with the oil. Practically all reservoirs have water in the lowest portions of the formation, with oil just above it. However, the oil-water contact line is usually neither sharp nor horizontal throughout a reservoir. Instead, it is a zone of part water and part oil from 10 to 15 feet thick. Much the same holds true for the gas-oil contact; but oil, being much heavier than gas, does not tend to rise as high into the gas zone as water does into the oil zone.

RESERVOIR PRESSURE

All reservoir fluids are under pressure. Pressure exists in a reservoir for the same reason that pressure exists at the bottom of the ocean. Imagine a swimmer in a large swimming pool, who decides to see whether he can touch bottom. Everything is going well except that his ears begin to hurt, and the deeper he dives, the more his ears hurt. The reason for the pain is that the pressure of the water is pressing against his eardrums. The deeper he goes, the greater the pressure.

Normal pressure

Just as there is water pressure in a swimming pool, there is fluid pressure in a reservoir. Under normal conditions, the only pressure in a reservoir is the pressure caused by the water in and above it. Contrary to what might seem logical, all the rocks that overlie a buried reservoir do not normally create pressure in the reservoir. Instead, it is as if there were no rocks there at all — at least as far as pressure is concerned.

Imagine leaving the swimming pool full of water but dumping a huge load of marbles into it. Naturally, the marbles would take up space formerly occupied by water, and the pool would overflow; water would slosh over the sides. However, a pressure measurement taken at the bottom of the pool after the marbles were dumped in would be exactly the same as one taken before. The pressure exerted by the water would remain the same.

The same thing happens in a reservoir. Usually, though, the reservoir's connection to the surface is much more circuitous; it may outcrop at the surface many miles away, or it may be connected to the surface through other porous beds that overlie it. In most cases, as long as the reservoir has some outlet to the surface, the pressure in it is caused only by the water and is considered to be normal pressure.

Abnormal pressure

Reservoirs that do not have a connection with the surface are usually surrounded by impermeable formations. In such cases, the overlying rock formations do have a bearing on reservoir pressure. What happens in this case is that the heavy weight of the overlying beds presses down and squeezes the reservoir. Since the water in the reservoir cannot escape to the surface, the reservoir pressure builds up to abnormally high levels. The situation is somewhat like blowing up a balloon, tying it off so that the air cannot escape, and then squeezing it. Since the air is confined, the pressure builds and builds until the balloon pops. Reservoirs do not pop, but the pressure builds.

Abnormally high pressure can also be produced by an artesian effect (fig. 1.31). In this case, the formation connects with, or outcrops at, the surface at an elevation much higher than the reservoir. A well drilled into this reservoir will encounter abnormal pressure, especially if the outcrop is on a higher plane than the derrick floor. As in an artesian water well, the well fluid seeks its own level.

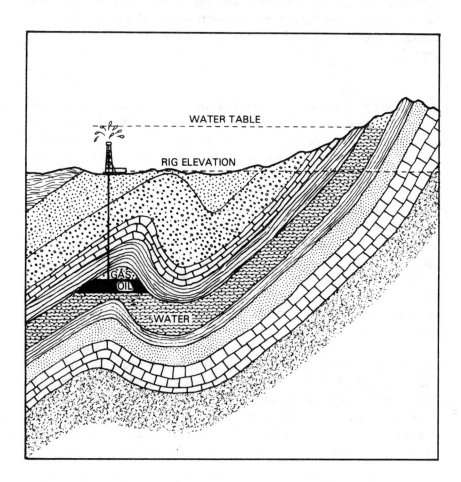

Fig. 1.31. Abnormal pressure can occur in formations outcropping higher than the rig elevation.

The petroleum geologist's job is twofold. He studies the geology of large areas to find likely locations for petroleum accumulations. When he finds one of these locations, he evaluates the reservoir to determine whether it has the potential to be commercially producible.

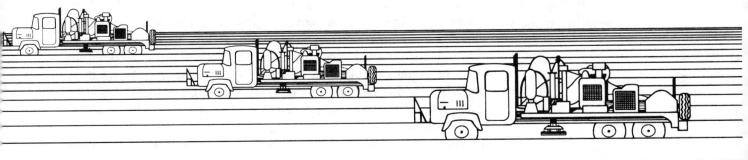

Petroleum Exploration

Exploring for petroleum was once a matter of good luck and guesswork; now it is based on many technical and scientific principles. The most successful oil-finding method in the early days of oil exploration was to drill in the vicinity of oil seeps, places where oil was actually present on the surface. Today, oil and gas discoveries are generally credited to surface and subsurface geology studies. Data gathered from aerial photographs, satellite images, and various geophysical instruments help determine where an exploratory well should be drilled. Then cuttings made by the bit as the well is drilled, core samples taken from the well, and logs that are generated by running special tools into the well as it is drilled all yield important subsurface information. Examining, correlating, and interpreting this information make it possible for the petroleum prospector to accurately locate subsurface structures that may contain hydrocarbon accumulations worth exploiting.

SURFACE GEOGRAPHICAL STUDIES

In a relatively unexplored area, the explorationist studies the surface topography and near-surface structures, as well as geographic features such as drainage and development. Sometimes the character of underground formations and structures can be deduced largely from what appears on the surface.

Aerial photographs and satellite images

Before a site for study is selected, geologists must contend with an unexplored area that may cover tens of thousands of square miles. In order to narrow this vast territory down to regions small enough for detailed surface and subsurface analyses, a combination of aerial photography and satellite imaging may be used. A series of landscape features that seem unrelated or insignificant to the ground observer may line up quite distinctly when viewed from the air.

Another type of remote sensing is imaging radar in which high-frequency radio waves are bounced off surface features to one or both sides of an airplane or satellite (Fig. 2.1). Return echoes are recorded to form a low-resolution relief image, which is useful in searching unexplored areas for potential oil-trapping structures and for discerning large-scale terrain features at a glance. Imaging radar used in airplanes is called side-looking airborne radar.

Fig. 2.1. This NASA Seasat radar image of Rio Lacantum, Mexico, shows prominent anticlines and pitted topography formed on limestone beds. (*Image processed by Jet Propulsion Laboratory and provided courtesy of Chevron Oil Field Research Co.*)

Fig. 2.2. Remote sensing system receives Landsat signals and modifies them by computer processing. (*Courtesy of Center for Remote Sensing and Energy Research, Texas Christian University*)

In the past, aerial photography was the sole tool used for this task even though it had some serious disadvantages. Besides the expense of flying, aerial exploration requires that a large number of photographs be taken from varying camera angles and distances. Because of the unequal quality of these pictures, large-scale geological interpretation is difficult.

Explorationists are now utilizing data gathered by the Landsat satellites, which were launched by the United States primarily for agricultural mapping and crop forecasting. Landsat 4, launched in 1982, collects data from two spectral bands in the infrared region that allow geologists to detect the presence of clays often associated with mineral deposits. After large areas are scanned by sensors on board the satellite, four separate images of each scene are sent back to earth simultaneously as electronic pulses. These signals are received at twelve ground stations situated throughout the world (fig. 2.2). Where the Landsat images are modified by computer processing, earth characteristics as small as 100 feet in diameter are enhanced (fig. 2.3). Companies using Landsat data, however, still need traditional exploration information to pinpoint the location of commercial deposits. Some members of the industry believe that Landsat's best value lies in ruling out large areas where mineral deposits are unlikely.

Satellite data may be purchased from the government or from companies that enhance the satellite images. If, however, an oil company invests in computer systems of its own to interpret raw digital data that are transmitted by the satellite, it can keep its discoveries secret. Such computer systems are quite expensive, but the price may be worth it. One industry expert estimates that satellite exploration costs less than $1 per square mile surveyed, as compared to $10 per square mile for aerial photographs and hundreds of thousands of dollars per square mile for seismic data.

28

TEXAS

RIO GRANDE

MEXICO

Fig. 2.3. This Landsat image of the West Texas area near Presidio shows the Rio Grande bordering Texas and Mexico at lower left. The original photograph is a false-color composite; thus dark areas are not discernible on this black and white copy. *(Image provided by the Center for Remote Sensing and Energy Research at TCU and printed by the Environmental Research Institute of Michigan)*

Oil and gas seeps are obvious signs of a subsurface petroleum source. Often, however, seepage is so slow that it is not easy to discern, since bacteria and weathering may decompose it as soon as it surfaces. Chemically testing the soil or water may reveal traces of hydrocarbons, leading to further exploration. Even plumes of gas rising from seeps on the ocean floor have led to off-shore exploration.

Many of the great oilfields of the world were discovered, in part at least, because of the presence of oil seeps at the surface. Oil seeps are located either updip or along fractures (fig. 2.4). Seeps at the outcrop of a reservoir bed may be active where oil or gas is still flowing out slowly, as at Mene Grande, Venezuela, only a mile up-dip from a large oilfield. In other cases, the sands near the surface are completely sealed, and the seep is no longer active, as at Coa-linga Field, California. The Athabasca "tar" sands in Canada appear to be a similar seep of the Cretaceous age. These sands were buried by later sediments and then exposed again by erosion. Seepage from fractures and faults is common and may be oil, gas, or mud (for example, the mud volcanoes of Trinidad and Russia).

The presence of oil seeps on anticlinal crests was observed as early as 1842. However, not until after the drilling of the famous Drake well in Pennsylvania in 1859 was it noted that newly discovered wells were being located on anticlines. Little practical use was made of this information until 1885, when I.C. White applied it in search of gas in Pennsylvania and nearby states. During the latter part of the nineteenth century, geologists searched for oil in the East Indies and Mexico. In 1897, geological departments were established by some U.S. oil companies. Many of the subsequent discoveries in the Midcontinent, Gulf Coast, California, and elsewhere were made by using geology to find petroleum reservoirs.

A

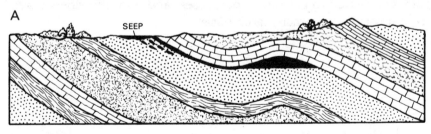

B

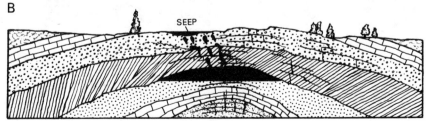

Fig. 2.4. Seeps are located either updip (A) or along fractures (B).

DATA COLLECTIONS

Much of the information that the explorationist works with comes from the files, libraries, and databases of the company he represents, public agencies, or privately operated service companies. Some of the data are proprietary and accessible only to those collecting it, while other information is available for a fee or is in the public domain.

Private company libraries

A private company or geological society may maintain a large collection of drilling and production data, maps, or well logs. Members pay dues for access to the information, which the libraries obtain both from public agencies and from members who contribute their own data. An oil company usually has a data library including all prior information collected on the areas that it has explored or developed. The company's geologist would use this data.

Public agency records

Agencies that regulate the oil and gas industry collect and file all types of data relating to drilling and production. These files are usually open to the public and accessible to anyone who knows the name of a well, the site of the well, who drilled it, and when it was drilled. Also, information may be obtained from a well log or a core sample library, often operated by a public university.

Databases

Both public and private organizations have established computer databases that offer access to a variety of information, generally regional and classified by field and reservoir. Most major oil companies subscribe to petroleum databases, and nonsubscribers may request service on a per-search basis.

GEOPHYSICAL SURVEYS

By 1920, anticlinal folding was only one of the geological factors known to control oil and gas accumulations. Surface mapping alone left much to be desired. Fortunately, geophysical methods of exploration came into existence about this time.

Geophysics deals with the composition and physical phenomena of the earth and its liquid and gaseous environments. The phenomena most commonly interpreted in petroleum exploration are earth magnetism, gravity, and especially seismic vibrations. Sensitive instruments are used to measure variations in a physical quality that may be related to subsurface conditions. These conditions, in turn, point to probable oil- or gas-bearing formations. Searching for oil and gas via geophysical means does not guarantee a successful find any more than other methods, but the combination of geophysical information and geological know-how reduces the chances of drilling a dry hole (fig. 2.5).

Geophysical exploration depends on a few fundamental variables in the earth's physical condition: gravitational change, magnetic field change, and electrical resistance change.

Fig. 2.5. Seismic exploration leads geophysicists from the deserts of Saudi Arabia (A), to Utah's canyon lands (B), and into the freezing plains of Alaska (C). (*Courtesy of American Petroleum Institute*)

Fig. 2.6. A surveyor helps determine site location for seismic operations. (*Courtesy of American Petroleum Institute*)

The first geophysical instruments used for oil and gas exploration were the refraction seismograph, the gravimeter, and the airborne magnetometer. The surveyor's instrument (fig. 2.6) continues to be a basic exploration tool.

Magnetic and electromagnetic surveys

The fact that the earth has a strong magnetic field has been turned to good use by geologists engaged in the search for minerals. Applying the principle that like rocks have like magnetic fields enables the geologist to compare slight differences in the magnetism generated by the minerals in the rocks.

Igneous rock, frequently found as the *basement* rock that lies under the sedimentary layers, often contains minerals that are magnetic. Basement rock seldom contains hydrocarbons, but it sometimes intrudes into the overlying sedimentary rock, creating arches and folds that could serve as hydrocarbon traps. Geophysicists can get a fairly good picture of the configuration of the geological formations by studying the *anomalies*, or irregularities, in the structures. A formation out of place with its surroundings may indicate the presence of a fault that may be associated with hydrocarbon deposits.

To measure the slight differences, two delicate instruments are used: a field balance and an airborne magnetometer. *A field balance* is used on the earth's surface and measures the differences in magnetism rather than the true or absolute magnetic variation of an area.

The *airborne magnetometer* records variations in the magnitude of the earth's total magnetic field. These variations are principally associated with the burial depth of magnetized rocks. Offshore, the magnetometer is trailed in the water behind a boat, and data are transmitted to a recording device on board, where they are translated onto paper or magnetic tape.

A development of airborne magnetics is the *micromagnetic* technique for oil exploration. Micromagnetic anomalies are detected from low-altitude surveys, normally flown at 300 feet above the ground. The magnetic data tapes from the aircraft are computer processed, producing a total magnetic field map, a flight line plot, and total field magnetic profiles. These profiles are then used to produce a fracture interpretation of the basement and characteristics of the overlying sediments. The net result is an efficient exploration method that defines relatively small areas for seismic surveys or drilling.

A remote sensing electrical method called *magnetotellurics* operates on the theory that rocks of differing composition have different electrical properties, namely, resistivities. The naturally occurring electromagnetic fields at the earth's surface are recorded, measured, analyzed, and interpreted to determine subsurface structures. This subsurface information, similar to that provided by electric logs run on an exploratory oilwell, helps to provide a complete database for making leasing or drilling decisions.

The magnetotellurics method is used primarily in reconnaissance (exploratory) surveying, although improved data processing techniques have made it increasingly useful for development surveys. The amount of detail is determined by the space between survey sites: closely spaced sites result in a detailed survey, which is helpful in locating step-out wells from areas of proven production. Geologic cross sections are constructed from regional surveys.

Although techniques continue to be improved, magnetic surveying does not assure the detection of all traps that contain hydrocarbons. It is useful, however, in giving the geologist a general idea of where oil-bearing rocks are most likely to be found.

Gravity surveys

Geophysicists also make use of the earth's gravitational field and the way it varies according to differences in mass distribution near the earth's surface. Simply explained, some rocks are denser than others, and, if very dense rocks are close to the surface, the gravitational force they exert is more powerful than that of a layer of very light rocks. The difference in mass for equal volumes of rock is due to variations in specific gravity. Geophysicists applied this knowledge, particularly in the early days of prospecting off the Gulf Coast. They could often locate salt domes by gravitational exploration because ordinary domal and anticlinal structures are associated with maximum gravity, whereas salt domes are usually associated with minimum gravity.

The *torsion balance*, first marketed commercially in 1922, was one of the earliest gravitational instruments invented. It, as well as another early instrument — the *pendulum* — was rather difficult to use. The most commonly employed instrument is the *gravimeter*, or gravity meter, a sensitive weighing instrument for measuring variations in the gravitational field of the earth (fig. 2.7).

Although the basic principle of the gravitational method remains the same, new technology and instruments continue to improve data collection. A small, portable, highly accurate gravity meter is now available for land work. Also, gravity data can be collected onboard a ship with a great deal of accuracy, and from the air as well, but with less resolution. Interpreters now use digital computers to relate gravity to geologic causes. Gravity maps and models help the geologist examine large areas of development and provide guidelines for planning a seismic exploration program.

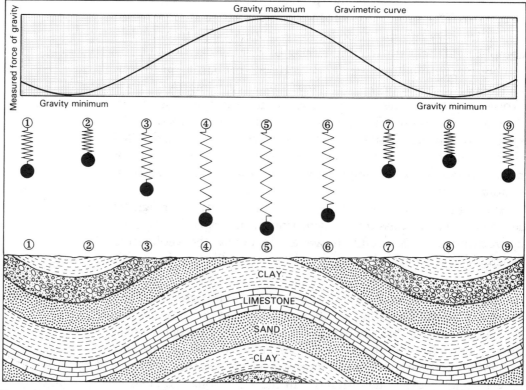

Fig. 2.7. Gravity meters are installed at observation stations (numbers 1–9) where the attraction of the gravitational force is measured. The measured force is then plotted in a curve (*top*).

A seismic survey is usually the last exploration step before a prospective site is actually drilled. Unlike gravity and magnetic surveys that provide general information, seismic surveys give the explorationist precise details on the structures and stratigraphy beneath the surface. Data is collected by creating vibrations, detecting them with a *seismometer*, recording them with a *seismograph*, and depicting them on a *seismogram*. Seismograms are used to generate a *seismic section*, which is much like a cross-sectional view of the subsurface (fig. 6.28).

Early methods. The first seismometer, as inventor David Milne called it, was used in 1841 to measure and record the vibrations of the ground during earthquakes. A few years later, Italian L. Palmieri set up a similar instrument, which he called a seismograph, on Mount Vesuvius. From these simple beginnings evolved seismic exploration.

Dr. L. Mintrop, a German scientist, developed one of the first practical uses of seismic data during World War I. He invented a portable seismograph machine for the German armies to use in locating the positions of the Allied guns. Mintrop set up three seismographs on the battlefield opposite Allied artillery, and, when a gun fired, he calculated the precise location so accurately that often the Germans could wipe out that position with one try.

Fig. 2.8. A seismic section indicates boundaries between formations.

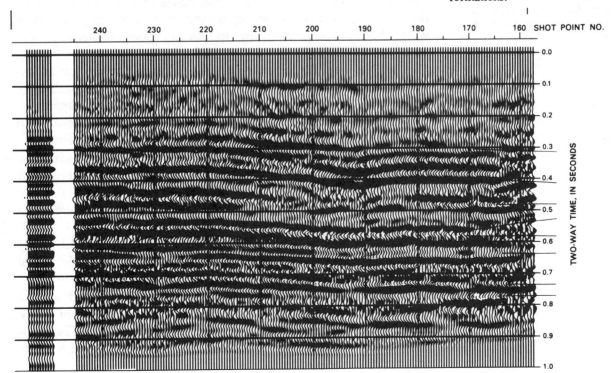

SHOT POINT NO.

TWO-WAY TIME, IN SECONDS

36

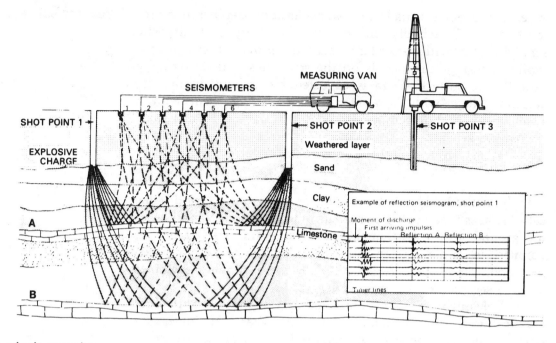

Fig. 2.9. In seismic surveying, an explosion at shot point 1 creates shock waves that are reflected by subsurface formations to seismometers and are recorded on equipment in the measuring van. An example reflection seismogram is shown in the inset.

The Germans perceived that miscalculations in distance were due to the variation of sound waves caused by geological formations through which the waves passed. Basic geological concepts were then applied, resulting in correct computations of the distance (fig. 2.9).

After the war, Mintrop reversed the process, setting off an explosion a known distance from the seismograph and measuring the return time of subsurface shock wave reflections to estimate the depths of formations. Finding his theories to be corroborated in the field, he put them into practice, forming the first seismic exploration company, Seismos. Soon after the company was started, the Gulf Production Company hired one of the Seismos crews and brought them to the Gulf Coast of Texas. With their arrival, news of seismic exploration spread, and soon rival companies were opening all over the state.

Two enterprising young brothers—Dabney E. and O. Scott Petty—decided to improve on Dr. Mintrop's methods. They resigned their jobs and spent a year developing a machine much more sensitive than Mintrop's. Their new seismograph used a vacuum tube sensitive enough to register the vibration of a "fly landing on a bar of steel," as O. Scott Petty explained it. The Petty Geophysical Engineering Company was established and became one of the early leaders in the field.

By the mid-1920s, the Pettys were "doodlebugging" (prospecting for oil) down by the Texas coast and in the Louisiana swamp. They hired an ingenious blacksmith named Pop Reichert to be their shot man—to set off the dynamite as the Pettys recorded the explosions on their instruments several miles away (fig. 2.10). He was a good mechanic and a fine fellow but had a bad habit of sitting on his dynamite. They warned him each morning that the dynamite was about 60 percent nitroglycerine, and that he would

Fig. 2.10. Pop Reichert poses with his first wooden shot-hole drilling rig. (*Courtesy of O. Scott Petty*)

38

be blown sky high if any of the flying debris from the blast fell on the box he was sitting on. One morning it happened. As usual, Pop was sitting on a box of explosives too near the shot, and the secondary explosion of hydrogen and oxygen that sometimes follows a blast almost buried him under flying dirt. After that, he was careful to keep his distance.

Modern land methods. Most of the methods used by these early geophysical pioneers have been replaced or supplemented by newer reflection methods. Various mechanical impactors and vibrators have been developed to create sound waves on the earth's surface that penetrate downward into the rock layers. Each formation reflects the sound waves back to the surface where sensitive instruments record and measure the intensity of the reflections. By interpreting the measurements, geophysicists are able to locate formations likely to contain hydrocarbons.

One of the first nondynamite sources of surface energy was the *Thumper*, developed by Petty-Ray Geophysical (fig. 2.11). This impactor drops a heavy steel slab from as high as 9 feet onto the surface, creating shock waves. Later, Sinclair Oil and Gas Company developed the *Dinoseis*, which uses a mixture of propane and oxygen in an expandable chamber to create an explosion. The explosion chamber is mounted under a truck and is lowered to the ground for use. Most commonly used today for land exploration is the *Vibroseis*, developed by Conoco (fig. 2.12). The Vibroseis generates continuous low-frequency sound waves whose reflections are picked up and changed into electrical impulses by instruments called *geophones*. These impulses are recorded on computer tape, processed on high-speed computers, and printed in the form of a seismic reflection profile. The profile is then analyzed to determine subsurface structures.

Fig. 2.11. The Thumper drops a 6,000-pound steel slab (surrounded by safety chains to warn personnel) 9 feet to strike the earth and create shock waves.

GEOPHONES ON CABLE

VIBRATOR

Fig. 2.12. The Vibroseis truck has a vibrator mounted underneath it (*inset*) that creates low-frequency sound waves. Geophones pick up the sonic reflections for recording. (*Courtesy of Trapp and Eskew Geophysical Consultants*)

Although geophones are usually used on the surface of the earth to gather seismic data, they are also used in the borehole of an existing well to generate a vertical seismic profile. In this method, geophones are run into a wellbore on wireline and attached to the wall of the hole at intervals of 20 to 100 feet. With the receivers deep in the earth, the signal from the sound source is less likely to be distorted by surface noise. Another advantage of the vertical seismic profile method is that it can be used to gather information about structures in the immediate vicinity of the hole.

40

Marine seismic methods. Methods for gathering seismic data offshore are similar in principle to those employed on land: echoes generated from a sound source are reflected by formations beneath the seafloor and received by *hydrophones*, the marine version of geophones. In the early days of marine seismic exploration, explosive charges suspended from floats were used to generate the necessary sound waves. Although used for many years, this method is now banned in many parts of the world because of environmental considerations.

Fig. 2.13. The *R/V Hollis Hedberg*, a geophysical vessel, has logged over 100,000 miles of geophysical data gathering since its christening in 1974. (*Courtesy of Gulf Oil Exploration and Production Co.*)

Fig. 2.14. Two computer systems (one is a backup) in the geophysical ship's laboratory can completely process all data through a full marine sequence as it is acquired. (*Courtesy of Gulf Oil Exploration and Production Co.*)

Today, offshore exploration centers around a seismic boat equipped to stay at sea several months if necessary. With a double crew, one for ship operations and one for seismic operations, the boat generates the sound waves using special equipment. After being reflected by subsea formations, the echoes are picked up by hydrophones located in a cable trailed behind the boat.

An air gun is one of the most common pieces of equipment used to generate the sound needed for offshore seismic work. Air guns contain chambers of compressed gas. As the gas is released under water, it makes a loud "pop," and this sound is reflected from the rock layers below the ocean floor. Both the "pops" and their reflected echoes are recorded on magnetic tape by instruments on the boat, and this tape is sent to a laboratory for analysis.

Another sound source used in offshore exploration is steam. After being manufactured in an on-deck boiler, the steam is suddenly released from a special chamber into the ocean. The sound is generated as the steam hits the cold sea and condenses rapidly into water. As with the air guns, reflected echoes from the steam are picked up by the hydrophones trailed behind the boat and recorded on magnetic tape for future analysis.

The state-of-the-art geophysical vessel is capable of complete onboard data gathering and handling (fig. 2.13). Its computer systems can process all data through a full marine sequence as it is acquired, eliminating the wait for onshore processing before a geological evaluation can be completed (fig. 2.14). The equipment and personnel aboard also permit basic geophysical data interpretation. For more sophisticated evaluation of offshore areas, processed data are taken, frequently by helicopter, directly to the operating company's offices.

Geochemical surveys

A geochemical survey provides direct, positive indications of subsurface petroleum accumulations. One such survey, known as the Petrex KV Fingerprint Technique®, was developed by Petrex International, Inc. It identifies regions that have hydrocarbon accumulations by collecting, measuring, and analyzing hydrocarbons from known subsurface accumulations. Volatile hydrocarbons are collected on activated charcoal samplers, analyzed, and evaluated by computer. Each oil type exhibits a somewhat different "fingerprint," reflecting its composition. Different oil types in the same section will have different fingerprints, and areas that have no significant hydrocarbon accumulations have very different fingerprints from those with accumulations. Therefore, samples with fingerprints that are similar to a known accumulation are judged to have a similar source. This information is then used to help develop the geologic model.

RESERVOIR DEVELOPMENT TOOLS

If the surface and subsurface data—collected, processed, and analyzed—indicate a strong possibility of hydrocarbons, an exploratory well or wells may be drilled. As drilling progresses, the geologist keeps a watchful eye on subsurface happenings via core samples, well logs, and test results.

All data, gathered and recorded as the well is drilled, are combined with the geologists' surface data and the subsurface findings of the geophysicists. This *synergistic* technique is especially useful for offshore exploration where developing, or exploiting, a reservoir is very costly. (*Synergy* is working together—combining actions or operations—so that the cooperative result is better than the sum of the actions or operations performed independently.) The quantity of oil or gas must be high enough to justify completing the exploratory well or drilling additional wells in the reservoir. Logs, test data, maps, models, computer analyses, and stratigraphic correlations all help the geologist further define the reservoir.

Driller's logs

A well log, in essence, is a systematic recording of data gathered during the drilling or production phase of a well. Several different kinds of logs are used, and each supplies the petroleum industry with specific information about the particular well and the formations encountered in it.

One of the most common well logs is the driller's log, which provides basic information to the geologist. It is a record kept by the driller of the depth, kind of rocks, fluids, and anything else of interest that he notices while drilling a well. Particularly when beds are alternating from soft to hard rock, a driller's log may be a very useful tool to geologists. It keeps track of exactly how long it took to drill through a particular bed, and that time record can be correlated with wells drilled later. An astute driller learns to recognize key beds as he drills in an area and reports them on the log.

Cores and cuttings

Core and cutting sample logs are used by geologists to identify lithologic sequence and key beds penetrated during drilling. Data for sample logs come from well cores and from cuttings that are brought to the surface in the drilling mud.

A core provides the most accurate information, since it is a slender column of rock that shows the sequence of rocks as they appear within the earth (fig. 2.15). Substituting a coring bit and barrel for the conventional drill bit allows the driller to obtain a

Fig. 2.15. Well cores are marked to maintain the sequence that they held in the earth. Then they are sent to a laboratory for analysis.

core ranging in length from 25 to 60 feet. Once this core is brought to the surface, it is carefully packaged and sent to a laboratory for analysis of many characteristics, including porosity, permeability, lithology, fluid content, and geological age. This information helps in determining the oil-bearing potential of the cored beds.

Since cuttings are fragments of rock and therefore do not form a continuous sample, they are not as useful as cores are to the geologist. Cuttings brought up with the drilling mud may not all come from the bottom of the hole but may include pieces of sloughing formations closer to the surface. Even with these limitations, however, cuttings can provide useful data and are regularly examined during drilling.

Wireline logs

Wireline well logging is the indirect analysis of downhole features by electronic methods. Numerous logs that gather data in many different ways under many different conditions are offered by the various wireline service companies; however, the procedure for each log is basically the same. An instrument called a *sonde* is lowered into the wellbore on conductor line, or electric wireline. A highly sophisticated electronic instrument, the sonde is designed to measure and record electrical, radioactive, or acoustic properties of formations. Then it transmits the signals up the conductor line while the sonde is being raised to the surface at a predetermined rate of speed. In the control cab of the wireline unit, these signals are translated—frequently, via computers—into graphs that can be interpreted by trained geologists and engineers.

The most useful and economical combination of logs must be chosen to provide the data needed to plan a good production program for a particular well. Some information may be taken from logs run on neighboring wells, while other logs are run routinely on every well. Among the most common wireline logs are spontaneous potential logs, resistivity logs, radioactivity logs, and acoustic logs.

SP logs. The *spontaneous potential (SP) log* records the weak, natural electrical currents that flow in formations adjacent to the wellbore. Most minerals are nonconductors when dry. Some, like salt, however, are excellent conductors when dissolved in water. When a layer of rock or mud cake separates two areas of differing salt content, a weak current will flow from the higher salt concentration to the lower. Drilling fluids usually contain less salt than formation fluids, which may be very salty. As freshwater filtrate invades a permeable formation, spontaneous potential causes weak current to flow from the uninvaded to the invaded zone. More importantly, current flows from the uninvaded rock into the wellbore through any impermeable formation, such as shale, above and below the permeable layer. Graphed against well depth, the SP log can be visually interpreted to show formation bed boundaries and thickness, as well as relative permeabilities of formation rocks. Because the SP log is so simple to obtain and provides such basic information, it is the most common and widely run log.

Resistivity logs. Resistivity logs record the resistance of a formation to the flow of electricity through it. Resistance depends on how much water the formation can hold, how freely the water can move, and how saturated the formation is with water rather than with hydrocarbons. High porosity, high permeability, and high water saturation each lower resistivity. On the other hand, oil and gas raise resistivity, since hydrocarbons are poor conductors. Common resistivity logs include the lateral focus log, the induction log, and the microresistivity log.

Radioactivity logs. Just as resistivity logs record natural and induced electrical currents, radioactivity logs record natural and induced radioactivity. Gamma ray logs record the emissions of natural radioactive elements in formation sediments (fig. 2.16). Since these elements are leached out of porous permeable rock, a gamma ray log is useful in identifying impermeable formations such as shale and clay-filled sands. Another type of radioactivity log is the neutron log, which records induced radiation in formation rock. As radiation emanating from the sonde bombards the rock around the wellbore, data about water and hydrocarbon saturations, salt content, rock types, and porosity are recorded.

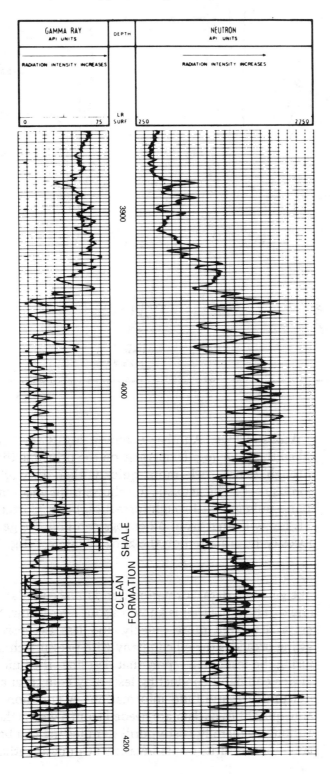

Fig. 2.16. Shale and clean formation points are marked on this combination gamma ray–neutron log. A wireline log such as this is transmitted from the sonde as it is raised in the wellbore at a predetermined rate. (*Courtesy of Dresser Atlas*)

46

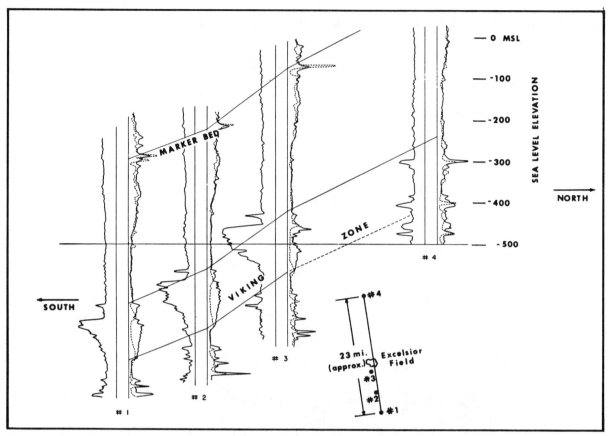

Fig. 2.17. Correlating data from electrical logs gives the geologist valuable porosity information.

Acoustic logs. Another common log is the *acoustic log*, also called the *sonic log*. It operates on the principle that sound travels through dense rock more quickly than through lighter, more porous rock. The acoustic log shows differences in travel times of sound pulses sent through formation beds. Correlating data from this log with that from other logs can give the geologist valuable porosity information on the zone of interest (fig. 2.17).

Formation test data

Like geologic data, formation test data accumulate over the life of a field. The more that is known about a reservoir, the easier it is to test a new well more quickly and less expensively. Formation testing provides the oil company with pressure charts or logs made during the test and a report that describes the fluids of the well. A service company conducting tests may collect, analyze, and interpret pressure and fluid information. A common test used when large quantities of data are needed from a well is the *drill stem test*. Although generally costly, drill stem testing can bring an oil company good returns on its investment in data gathering.

A useful test in reservoir development is the *stratigraphic test*, or *strat test*, as it is commonly called. A strat test involves drilling a hole primarily to obtain geological information. The borehole exposes complete sections of the formations penetrated, and fragments of the rock taken from the wellbore as it is drilled are analyzed by stratigraphers. (*Stratigraphy* is the study of the origin, composition, distribution, and succession of rock strata, which are distinct, generally parallel beds of rock.) Stratigraphers try to recognize and follow beds of rock from one well to another, going from a well or formation whose beds and lithologic (rock) sequence are known to an area that is unknown but assumed to be similar.

Strat tests

Stratigraphic correlation is the process of matching rock units of equivalent age. This process involves comparing fossils, rock hardness or softness, and electrical data from one well or outcrop with information from another well or outcrop (fig. 2.18). Wells that have information collected by drillers' logs, sample logs, and electrical logs enable the geologists to predict more precisely where similar rock formations will occur in other subsurface locations.

Stratigraphic correlation

Fig. 2.18. This stratigraphic cross section shows the effect of pinchout of sand 3.

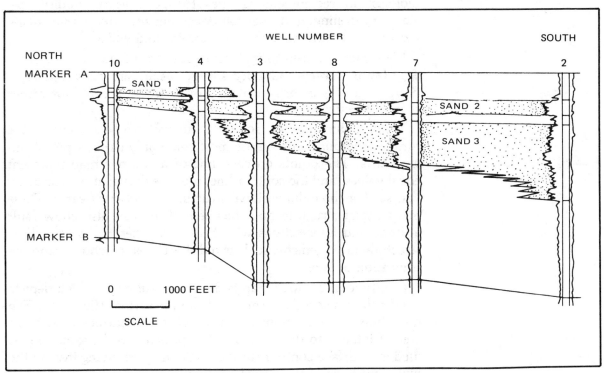

Subsurface correlation is based primarily on stratigraphic continuity, or the premise that formations maintain the same thickness from one well to another. A major change in thickness, rock type, or faunal content can be a geologic indicator that conditions forming the strata changed, or it may be a signal of an event that could have caused hydrocarbons to accumulate.

Formations, particularly marine ones, change rock texture and fossil characteristics very gradually. Thus, sudden changes in lithologic sequence are geologic indicators, and often particular fossils are used as markers for geologists trying to recognize the continuation of a formation in a new location. If the kind of rock is the same and the fossil markers are the same, the formation is probably a continuation of the one in the previous location.

The sequence of beds can also be used as a method of correlation. This method depends in part, as do the others, on drilling closely spaced wells so that the beds can be matched accurately as the stratigrapher attempts to correlate the data derived from several wells.

Maps

Maps of various kinds are used throughout the exploration procedure. *Base* maps that show existing wells, property lines, roads, buildings, and other surface geographic features are used by the geologist in recommending sites for geophysical studies, exploratory drilling, and reservoir development. *Topographic* maps show surface features such as mountains and valleys.

Maps are also used to represent subsurface data gathered from exploration surveys. Gravity survey results are displayed on a *Bouguer* gravity map, which is a geophysical representation rather than a structural one.

Topographic, gravity, and other surveys can be depicted in the form of *contour* maps—those most commonly used by petroleum geologists. The points on each line of a contour map represent equal values, and the contour lines must be drawn at regular intervals so that three-dimensional structures stand out clearly. These maps depict structure and thickness. They can also show fault attitude and intersections with beds and other faults as well as structural arrangements such as old shorelines, pinchouts, and the truncation of beds.

One type of contour map is the *structural* map, which depicts the depth of a specific formation from the surface (fig. 2.19). The map shows the elevations of the top of the subsurface layer of an area of interest to the geologist. The principle is the same as that used in a surface contour (topographic) map, showing instead the highs and lows of the buried stratum.

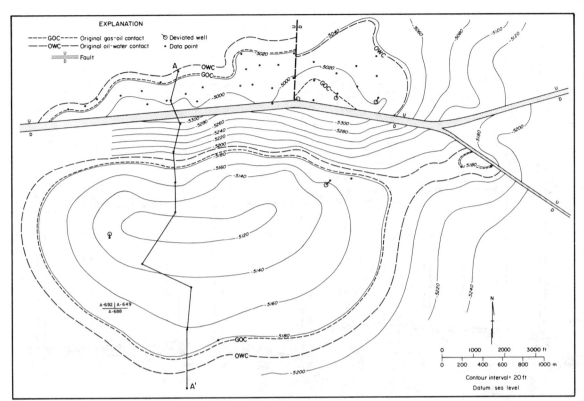

Fig. 2.19. A structure contour map shows the depth of a formation from the surface. *(Courtesy of Bureau of Economic Geology, The University of Texas at Austin)*

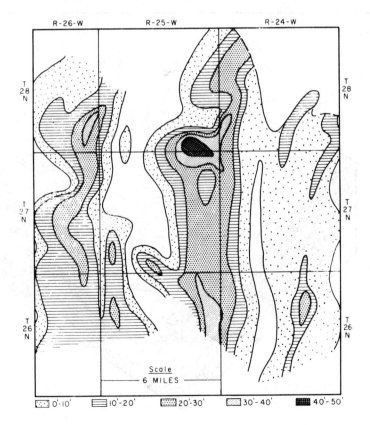

Fig. 2.20. This simplified isopach map shows the thickness of a porous formation based on microlog surveys in one subzone. Note the marked trend of the reservoir porosity, its length of some 15 to 20 miles, and its thickness of up to 50 feet.

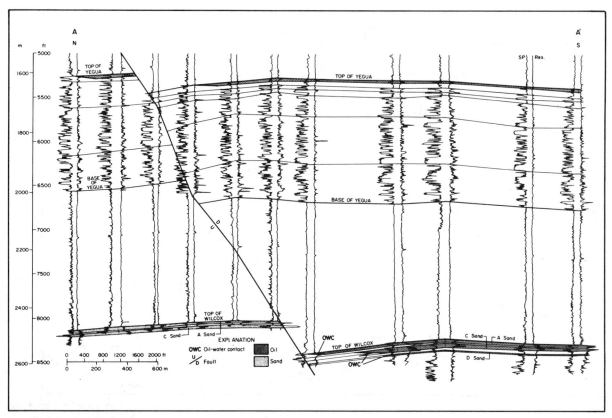

Fig. 2.21. This vertical cross section illustrates the fault pattern of the formation shown in figure 2.19. (*Courtesy of Bureau of Economic Geology, The University of Texas at Austin*)

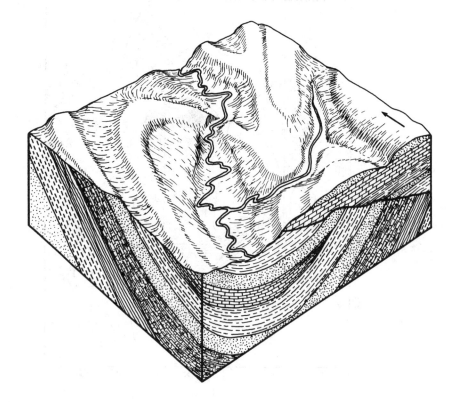

Fig. 2.22. A block diagram represents a section of the earth, showing subsurface strata and surface topography.

Another type of contour map is the *isopach* map (fig. 2.20). It shows the variations in thickness of a formation and is widely used in the calculation of reserves, in planning improved recovery projects, and as an aid in exploration work.

Similar to the isopach is the *isochore* map, which shows the thickness of a layer from top to bottom along a vertical line. To show the character of the rock itself and how it varies horizontally within the formation, a *lithofacies* map is used. This type of map has contours representing the variations in the proportion of sandstone, shale, and other kinds of rocks in the formation. Another type of facies map, the *biofacies* map, shows variations in the occurrence of fossil types.

Contour maps are frequently supplemented by vertical cross sections that show structure, fault pattern, and other pertinent details. A cross section represents a portion of the crust as though it were a slice of cake (fig. 2.21). Most cross sections show both structural and stratigraphic features together. They may show possible anticlinal and fault traps, or they may show only horizontal variations in lithology or thickness. The ability to show gravity anomalies in geologic contour form with cross sections has been an instrumental factor in spreading the use of gravity surveys in the exploration industry.

Vertical cross sections

One way to show geologic structures in three dimensions is by constructing a *peg model.* The peg model portrays the results of several stratigraphic tests. It uses pegs or rods for geologic columns, which are arranged on a map representing their actual locations. Contacts between formations occur at various depths on the columns; these points are connected with string to form an outline of the formation. Since the peg model is hard to construct and transport, it has largely been replaced by a transparent solid made of tinted acrylic plastic.

Models

The same information that was used to construct the peg model can be shown in a panel diagram where each formation type is represented by a continuous pattern. A panel diagram is useful in showing how formation structure and stratigraphic thickness vary in horizontal and vertical dimensions.

A block diagram is a perspective drawing of a section of the earth's crust as it would appear if cut out in a block (fig. 2.22). Two vertical sections whose faces are at right angles to each other and the top, either a subsurface view or the surface topography, are shown in a perspective view by a block diagram.

Panel and block diagrams

52

Fig. 2.23. By keyboarding parameters into a model program, the operator can transform statistics to Landsat imagery, usually in color but shown here in black and white. (*Courtesy of Gould Inc., DeAnza*)

Computers, which are economically feasible if large quantities of data must be analyzed, are being used more and more in exploration work. Gravity and magnetic data are frequently assessed by computer, and the results can then be applied to seismic data to produce a complex "three-dimensional" image of the geological features of the area of interest.

The skilled computer operator/programmer uses the computer and monitor screen to enhance Landsat, seismic, and other graphics (fig. 2.23). False or pseudo color can be added to highlight various features, which can be brought to the screen with a keystroke. High-resolution seismic "brightspots," for example, show specific trap definition, a technique used primarily in offshore exploration.

In addition to defining subsurface *structure*, seismic sections can now be used for *stratigraphic* studies as well. This new insight into the sedimentary processes is due largely to improved computer technology. Sophisticated data acquisition, processing, and graphic systems have been perfected to perform many heretofore time-consuming operations in seismic exploration. Mesh algorithms can be displayed over Landsat images to create three-dimensional images. Regional geology can be mapped onto a Landsat image to highlight surface features (fig. 2.24). Interactive interpretation will be practiced by petroleum explorationists as supercomputers and seismic workstations become the norm in seismic exploration.

Computer graphics and modeling

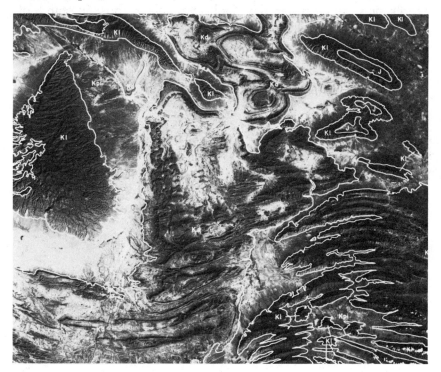

Fig. 2.24. Regional geology has been mapped onto this Landsat image of the Monterrey Region in Mexico. (*Image provided by the Center for Remote Sensing and Energy Research at TCU and printed by the Environmental Research Institute of Michigan*)

Reservoir modeling is another technique made practical by the computer. Information is manipulated by the use of a *model* — a computer program containing a set of parameters. The values of the parameters (such as the number of wells in a section, the production rates of the wells, etc.) are known for the present and for the production history of the reservoir. For predictions, new parameters are entered into the program in the form of mathematical "what if" questions; for example, "What if we add three wells in this area and increase the production rate by 20 bbl/day?" The computer plugs the new information into the model, performs all the calculations, and comes up with production predictions for a given period of time under a certain set of circumstances. This type of analysis requires a sophisticated computer system but is an invaluable aid to the decision maker.

While computer technology is not always cost-effective, its value in exploration and development work is unquestioned, and its use is becoming widespread. The mass of data that can be stored, recalled, and updated provides explorationists with information more readily than any other system, and the computerized analysis techniques are making reservoir modeling a major development tool.

Exploration for oil and gas is often difficult work, although technology has made the job of the petroleum geologist and geophysicist much easier. The tools available to those in search of oil are as basic as drillers' logs and as sophisticated as satellite probes of the earth's secrets. The continuing development of the industry depends on finding petroleum, and the challenge of doing so is one of the most exciting tasks of the business.

Aspects of Leasing

Finding petroleum is not the only requisite to production. Before a reservoir can be developed, the legal rights for such exploitation must be obtained. In most countries the mineral wealth, including petroleum, is owned by the state or national government. Companies that have the capital and expertise may negotiate a contract with representatives of the government claiming the reserves. Often, the host country retains controlling interest throughout exploration and development. Thus, extremely complex arrangements between the host country and consortia of petroleum companies, many of which are state or nationally owned, take on an air of international policy making.

Securing the rights to explore, drill, and produce differs from country to country (fig. 3.1). In the United States, three sources exist for the rights to petroleum: the private property owner, the state government, and the federal government. Each may hold title, or established ownership, to a given estate, and only the titleholder may grant the rights to develop the resources of the estate.

The instrument used to grant these rights is called a *lease*, traditionally an *oil, gas, and mineral lease*, or simply an *oil and gas lease*. For a lease to be valid, ownership of the minerals must be established, and provisions of the lease must be explicit and legally executed.

56

Fig. 3.1. In most oil-
producing nations outside the
United States, the mineral
rights are owned and con-
trolled by the government.

TYPES OF OWNERSHIP

Establishing ownership of the oil, gas, and mineral resources — called the *mineral estate* — is not as simple as it might seem. In the United States, the mineral estate is considered as part of the real estate. Real estate law is case law, or common law, and is determined by historical precedents established in court cases. Most of the states of the union have adopted the English common law of real property, except when contravened by statute. An exception is Louisiana, where civil law, derived from the Napoleonic code, is the rule as opposed to English common law. Since each state has its own court system in the form of a separate set of cases for the precedents, each state has in effect different laws regarding the ownership of petroleum.

States that follow the doctrine that oil and gas are owned in place, underground, are sometimes termed *absolute ownership*, or ownership-in-place, states. Texas is an example. However, if a reservoir lies under two properties, and one owner drills and produces but the neighbor does not, then the neighbor might well end up with no oil. The title has shifted to the owner who captured the oil and reduced it to personal property. Because an owner cannot protect a shifting title, courts in some states — Louisiana and Pennsylvania, for example — find that ownership in place is not an acceptable theory.

In states that follow a *nonownership-in-place* doctrine, no one owns the petroleum until it is captured. Title can be assumed upon production, and the oil and gas become personal property. In both types of jurisdictions, the lease is a conveyance, granting the exclusive right to explore, drill, and produce.

Regardless of the doctrine of ownership followed, two-thirds of onshore lands in the United States are in *private ownership*. Generally, the rights to the minerals, oil, and gas of these lands are also privately owned, although not necessarily by the same person who owns the surface. An individual or corporation may be a landowner in fee simple, a mineral estate owner, a surface estate owner, or an owner of a royalty interest in mineral production from the land. The remaining lands — both onshore and offshore — are public property, owned and administered by the individual *states* or by the *federal government*.

Preserved from the Middle Ages through English legal heritage, an estate of complete ownership in real property is a *fee*. An owner of *fee simple* property (henceforth called the *landowner*) owns the right to exploit what wealth the land might provide, whether above, on, or below the surface. Until the federal and state

Fee simple landowner

governments established certain restrictions to protect the rights of adjacent property owners and to establish conservation regulations protecting the overall hydrocarbon resources of the country, the landowner could do just about anything he wanted with his land.

Today, the landowner still has the exclusive right to search for and remove oil and gas from his property; however, the average landowner is not financially able to expend the enormous sums required to drill and produce. Hence, he might grant the right to exploit the resources to someone else for a given period of time through a leasing agreement.

The landowner can sell the mineral estate or a percentage of it to someone else by use of a *mineral deed*, or he can sell the surface and retain all or part of the mineral estate. (A mineral deed is distinguished from a lease in that a lessee will lose his rights to the oil and gas unless production is established within the time allowed by the lease.) In either case, fee simple ownership ends, and mineral estate and surface ownerships begin.

Mineral estate and surface owners

The rights of the mineral estate owner and the surface estate owner depend on (1) the state in which the property is located and (2) the minerals detailed in the sale agreement. Some states regard the mineral estate as a *possessory* estate — that is, as fee ownership of the minerals in place. Other states regard the mineral estate as a *servitude;* that is, it is subject to a specified use or enjoyment by one party, even though the surface is owned by another.

As a possessory estate, the mineral estate cannot be lost by mere failure to explore it. On the other hand, if the state regards the mineral estate as a servitude, the mineral estate owner has only the exclusive right to enter the land, explore, and produce the minerals, which are owned *only* after they have been reduced to possession. A servitude can be lost by failure to explore diligently. Louisiana, for example, has a ten-year statutory period within which mineral rights must be exercised to avoid their merger with the surface estate.

If the minerals in an agreement are oil and gas, the mineral estate is the dominant estate, and the surface is the servient estate. The surface owner may not prevent the mineral owner's access to and use of the mineral estate, nor may the surface owner lease the minerals. The mineral estate owner may use as much of the surface as is reasonably necessary to recover the oil and gas, with due regard to the rights of the surface owner. To avoid disputes, the lease usually contains specific agreements regarding surface use. The surface owner is normally entitled to compensation for property damage that results from exploration and development operations.

Like the fee simple owner, the mineral estate owner who cannot afford to exploit the petroleum resources himself may convey these rights to another by way of an oil and gas lease.

Another type of owner is one who holds a royalty interest in the mineral estate. A royalty interest is a share or percentage of the total or gross oil and gas production. A *participating* royalty owner is one who also owns all or part of the mineral estate and has the executive rights of a mineral estate owner. A *nonparticipating royalty interest* holder is one who owns no part of the mineral estate and therefore has no right to execute leases, enter the property, nor explore or produce any minerals on the property. He only receives his share of the profits from production.

A royalty interest can be transferred by sale or reserved totally or in part by a landowner who wishes to transfer only specified portions of his property. He might sell both the surface and mineral estates but reserve half of the royalty interest for himself and his heirs. On the other hand, an owner might keep his property and sell, by means of a *royalty deed*, some fraction of the royalty interest. He might even divide the royalty interest into several shares, retain a portion of the shares himself, and become a co-owner of the royalty interest along with several other owners. Depending on the laws of the state in which the land lies, these arrangements can be perpetual, for fixed terms, or for the lifetimes of the purchasers.

Royalty interest holder

Each state has a board or agency that governs the leasing of its lands. Again, state laws vary and are subject to change with legislation. Various state legislatures have provided for the leasing of some of their land to the petroleum industry. For example, the leasing of minerally classified lands in Texas (lands whose mineral estate, severed from the surface estate, is retained by the state) can be administered by the surface owner acting as agent for the state.

Offshore lands stretching from the shoreline seaward for 3 miles belong to the coastal states (fig. 3.2). Exceptions are Texas and Florida that own the area 3 leagues, or 10½ miles, from their Gulf Coast shores.

The state of Alaska has more federally owned land than it has state or private properties. Private land ownership increased with the passage of the Alaska Native Claims Settlement Act in 1971, when the federal government granted the natives title to 44 million acres. Native Corporations, special institutions created to handle this transfer of land and its associated bonus payments, are now

State ownership

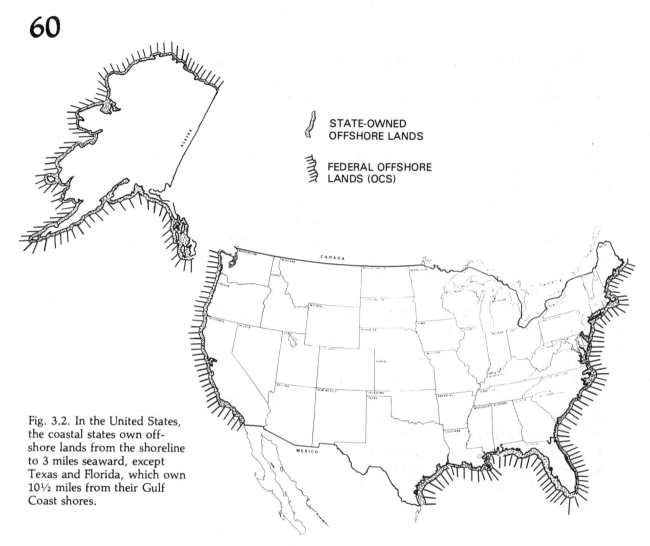

STATE-OWNED
OFFSHORE LANDS

FEDERAL OFFSHORE
LANDS (OCS)

Fig. 3.2. In the United States, the coastal states own offshore lands from the shoreline to 3 miles seaward, except Texas and Florida, which own 10½ miles from their Gulf Coast shores.

the largest private owners of land in Alaska. In spite of this Act and the fact that the state is still in the process of conveying acreage for patent to itself, the federal government still owns most of the onshore land with oil and gas potential.

Federal ownership

The federal government is a landholder of astonishing size; however, much of its land is unavailable for oil and gas production. Land set aside for military uses, national parks, wildlife refuges, and so on is not generally leased to the petroleum industry. The federal government *does* lease tracts of public domain land, acquired land, certain Indian land, and offshore land.

Both the mineral and the surface estates of land in the public domain belong to the nation. The federal government has also reserved the mineral rights to some property patented to individual citizens. *Public domain* land was originally owned by the United States and was not subsequently disposed of, while *acquired* federal lands were acquired by deed from earlier owners.

The federal government cannot lease *Indian tribal* land (land within a reservation), but it retains control over leases that affect *Indian-allotted* land (land that has been designated for use by a specific individual).

Federally controlled *offshore* areas—known as the Outer Continental Shelf—are the submerged lands seaward from state-owned areas 200 miles out or to a depth of 8,200 feet.

With some exceptions, each of the ten Canadian provinces—Newfoundland, Nova Scotia, New Brunswick, Prince Edward Island, Quebec, Ontario, Manitoba, Saskatchewan, Alberta, and British Columbia—owns and manages its own mineral resources. The oil and gas of the Northern Territories and offshore lands are owned and managed by the Canadian government.

Ownership in Canada

Some subsurface minerals are privately owned. In the four western provinces, a number of companies were granted land with attached mineral rights in exchange for their work in building railroads and thus opening up the provinces for settlement. Some early settlers of these provinces also acquired mineral rights with their homesteads, which are still in private ownership. Other mineral properties are owned by Canada's Indians and have separate regulations.

THE LEASE AND THE LAW

The most commonly used method for securing the rights to mineral production is the oil and gas lease. Because leasing is absolutely essential to the petroleum industry, a familiarity with lease laws, leasing practices, and regulatory agencies is necessary for a basic understanding of the field. The next step to understanding some of these legal aspects is to learn the language of leasing.

For *consideration*, usually money called *bonus*, and the further consideration of receiving a share of what is produced, known as *royalty*, the mineral estate or fee owner, called the *lessor*, gives exclusive rights to the oil and gas to a petroleum company or other party known as the *lessee*. In return, the lessee develops the lease in a timely and prudent fashion. Under most leases, the lessee pays a *delay rental* (money) each year to prevent automatic termination of the lease during the fixed or *primary term* (a period of time) in the absence of drilling. The law imposes certain *implied covenants*, or obligations, on the lessee. The lease contains specific or *expressed covenants* for development, and these supplant the implied covenants.

The language of leasing

The mineral interest is usually shared by the lessor and the lessee on a percentage basis. For example, if the royalty owned by the lessor is 25 percent, then the balance of 75 percent is the lessee's portion and is called the *working interest*. The party who ends up controlling the working interest and developing the resource is called the *operator*. The operator may be an individual, one of several in partnership, one or several corporations, or any combination agreed upon by all parties in a contract called the *operating agreement*. The lease then is a *negotiated agreement to create a working interest for purposes of exploiting a resource.*

The first leases

Before oil was discovered and commercially produced in Titusville, Pennsylvania, leasing agreements between prospectors and landowners were extremely simple. One lease made in 1853 gave the oil company the rights to "dig or make new springs," the cost of which would be deducted "out of the proceeds of the oil, and the balance, if any, to be equally divided" between the oil company and the landowner. "If profitable" is the concluding statement of the agreement, and the phrase reveals the uncertainty of the whole venture.

The length of the contract is also significant: both parties agreed that five years was sufficient time to discover and develop any resources that happened to be under the land. Prospecting was luck, not science; and, if petroleum was not discovered within a relatively short time span, it was universally believed that the oil and natural gas would "migrate" somewhere else, thus making further exploration unprofitable in any case.

Court rulings on oil migration

As scientists slowly discovered more about the natural laws that govern the behavior of oil and natural gas, some of the stranger theories concerning the migration of these resources were disproved. Landowners, who had thought their chances of profiting from an oil or gas lease were severely diminished if the company did not begin exploration and production immediately upon signing the lease, could not sue oil companies for forfeiture of the lease agreement on those grounds after 1875. Before that date, the courts had upheld landowners' claims that their lease rights included an "implied covenant" with the lessee that required the oil operator to develop or release the leasehold at once.

The courts, having heard endless litigation over the migration of oil and gas, finally decided that oil and gas are the property of the person who first captures the resource and reduces it to his control. The language of the hunt, "capturing" oil and "reducing it to the owner's control," comes from the old theory that oil and gas are migratory resources — they might wander onto one's property and then leave—almost like animals. The courts decided it was not their province to judge whose property the oil may have migrated from; the person who drilled a well and found oil was the legal owner. Although it seemed like a reasonable decision at the time, the consequences of the court rulings were not all fortunate.

Rule of capture. One consequence of the decision was the landowner's freedom from liability for drainage of a common reservoir even though part of the oil may have migrated from a neighbor's land. This so-called rule of capture, especially in states that have adopted the nonownership doctrine, allows the owner of a tract to drill as many wells as he can, provided he avoids drilling diagonally onto his neighbor's property. The rule does not give the landowner the right to draw a disproportionate amount from the common reservoir because government regulations prorate the amount of oil that can be produced and because the adjacent landowners can drill and produce wells from the same reservoir.

Offset drilling rule. The offset drilling rule, which is an outgrowth of the rule of capture, states that the landowner whose oil and gas reserves are being drained by his neighbor's wells cannot go to court and recover damages or stop the offending operator. The landowner's only recourse is to drill his own wells and produce as fast as he can. If the landowner leases the mineral interest, the lessee must assume the burden of the offset drilling rule. Production on an adjoining lease obligates the lessee to protect the lessor's interest by drilling a well nearby to offset the potential loss of oil.

Court decisions on mineral leases

In many of the older mineral leases, oil and gas were implied as the minerals (fig. 3.3); however, iron ore, lignite, and other minerals were mined under these leases to the detriment of the surface owner. As a result of several cases, the courts reasoned that the mineral estate owner did not have the right to "appropriate the soil." Thus, minerals substantially affecting the surface belong to the surface owner.

Producers 88 (12/79) Revised
With 320 Acres Pooling Provision

POUND PRINTING & STATIONERY COMPANY
2325 Fannin, Houston, Texas 77002 (713) 659-3159

OIL, GAS AND MINERAL LEASE

THIS AGREEMENT made this ____ day of ________________, 19____, between

Lessor (whether one or more), whose address is: ________________ Zip Code ________,

and ________________,

Lessee, (whether one or more), whose address is: ________________ Zip Code ________,

WITNESSETH:

1. Lessor in consideration of ________________ Dollars ($________), in hand paid, of the royalties herein provided, and of the agreements of Lessee herein contained, hereby grants, leases ~~and lets~~ exclusively unto Lessee for the purpose of investigating, exploring, prospecting, drilling and mining for and producing oil, gas and all other minerals, conducting exploration, geologic and geophysical surveys by seismograph, core test, gravity and magnetic methods, injecting gas, water ~~and other~~ fluids, and air into subsurface strata, laying pipe lines, building roads, tanks, power stations, telephone lines and other structures thereon and on, over and across lands owned or claimed by Lessor adjacent and contiguous thereto, to produce, save, take care of, treat, transport and own said products, and housing its employees, the following described land in ________________ County, Texas, to-wit:

Fig. 3.3. In an "Oil, Gas, and Mineral" lease, oil and gas are *implied* as the minerals as a result of past court decisions.

64

Government regulations

The combination of the rule of capture and the offset drilling rule led to excessive drilling and production from reservoirs. Not only was there waste from drilling unneeded wells, but the rapid production reduced reservoir pressures and altered drive mechanisms (the forces used to drive the oil and gas to the surface). The total amount of petroleum produced from a reservoir was often less than ideal.

When it proved impossible to provide efficient extraction of oil and gas and equitable distribution of the minerals among the owners under existing legislation, the state governments had to step in and form regulatory commissions. Conservation or regulatory agencies, eventually established in all oil-producing states, have the authority and responsibility to prescribe and enforce sound conservation rules such as well-spacing requirements and prorated production rates (fig. 3.4). State regulatory agencies also have the responsibility for protecting landowners' correlative rights. (*Correlative rights* are those rights afforded the owner of each property in a pool to produce without waste his just and equitable share of the oil and gas in such a pool.)

Like their counterparts to the south, the Canadian provinces enact new oil and gas legislation from time to time. The laws, administered by various government agencies, control mineral development and leasing practices. In addition to the Department of Energy, Mines and Resources in Ottawa, Ontario, each province has its own regulatory agency.

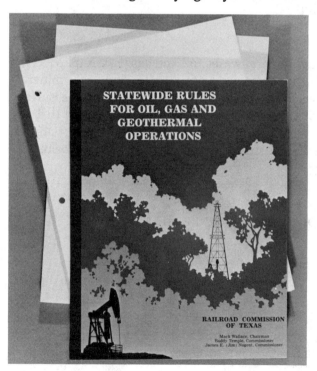

Fig. 3.4. Books of regulations are issued periodically by the Railroad Commission of Texas, the state agency that regulates the extraction and distribution of Texas oil and gas.

The U. S. government also involves itself in regulating the production of oil and gas through its Environmental Protection Agency, Bureau of Land Management, and Minerals Management Service of the Department of the Interior. Oil and gas production is now among the most heavily regulated of American industries.

PREPARATIONS FOR LEASING PRIVATELY OWNED LANDS

Once a decision is made by an operating company to lease privately owned land, the *landman*, or *leaseman*, comes on the scene. A landman (sometimes called an oil scout) is a person in the petroleum industry who negotiates with landowners for land options, oil drilling leases, and royalties; he also works with producers for the pooling of production in a field. Sometimes these preliminaries are handled by the owner and others on the staff, especially if the operating company is small.

If the operator is a large petroleum company, the company's exploration group will notify its own land department that a region is of interest, and a landman will investigate. The landman will find out who owns mineral rights in the area of the *play*, a term used in the industry to describe "geologically similar reservoirs or oilfields exhibiting the same source, reservoir, and trap characteristics."[1] If it is an "active" area, the landman will find out the prevailing terms of leasing, such as the amount of bonus and royalty. If the decision is made to pursue the play, the landman will contact the mineral estate owners and negotiate as the lessee, accumulating as much of the region as possible. Of course, other companies are aware of the action and are likewise investigating and leasing.

A landman can also operate independently of any one company. An independent landman is essentially a lease broker who may work with several operators or companies in the area in which he lives. If the area is a productive one, much bargaining and trading of the leases may take place.

Whether working for a company or as an independent, the landman's job is to acquire signed leases. Before leasing can take place, he must make sure that the legal ownership of the property has been established, that the owner has the capacity to legally sign a contract, and that the terms of the lease are agreed upon by both the lessor and lessee.

1. D. A. White, "Assessing Oil and Gas Plays in Facies-Cycle Wedges," *AAPG Bulletin*, Vol. 64, No. 8, 1980.

66

Determining ownership

The landman securing an oil and gas lease takes care to determine the ownership of the land and the minerals of that land in a preliminary check of the records (fig. 3.5). This research lets the landman know the names and addresses of the current mineral interest owners so that he can talk with them and eventually get their signatures on a lease. He also finds out who owns the surface of the land and whether the property is already leased. The landman tries to look at all the documents in the land's history of ownership and briefly describes these documents in a *run sheet* (fig. 3.6). This sheet will later be used to *clear the title* — that is, establish the full legal status of the land in question — before executing the lease.

Fig. 3.5. The landman checks deed records in the county or parish courthouse to establish ownership of a piece of property.

Fig. 3.6. With documents at hand, the landman prepares a run sheet.

Fig. 3.7. County records should show an unbroken chain of title from the patent of the state to the present owner.

Ideally, the recorded conveyances in the county records will show an unbroken chain of title extending from the patent of the state to the present owner (fig. 3.7). To minimize the risk of adverse claimants, the lease purchaser relies on his legal advisers to establish facts that confirm the chain of title. Otherwise, some heir or long lost relative of someone who once owned an interest in the land could show up demanding the whole well on the basis that it was leased from the wrong person.

Clearing the title

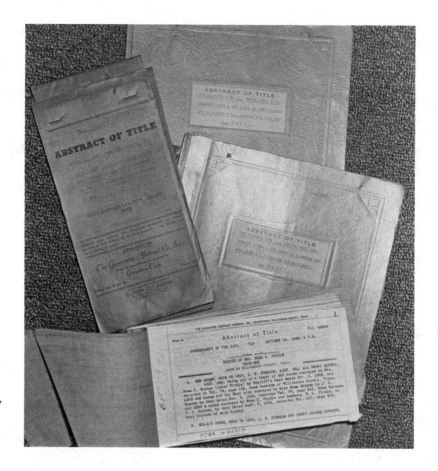

Fig. 3.8. A title examiner may use abstracts of the property to help in rendering a title opinion.

Fig. 3.9. A data bank hookup allows a title company to quickly retrieve title information with a computer display terminal and printer.

An old story illustrating the thoroughness expected in a title search concerns the Post Office Department in Washington, D.C., which was about to purchase a lot for a new post office in Louisiana. The department, dissatisfied because the title had been traced back only to 1803, requested further work on the case. The title attorney responded as follows:

> Please be advised that the government of the United States acquired the title of Louisiana, including the tract to which your inquiry applies, by purchase from the government of France in the year 1803. The government of France acquired title by conquest from the government of Spain. The government of Spain acquired title by discovery of Christopher Columbus, explorer and resident of Genoa, by agreement traveling under the sponsorship and patronage of her majesty, the Queen of Spain. The Queen of Spain received sanction of her title by consent of the Pope, a resident of Rome, and *ex officio* representative and vice-regent of Jesus Christ. Jesus Christ was the son and heir apparent of God. God made Louisiana.

Examining titles. One way to clear a title is by obtaining *title opinions* from a qualified title examiner—an attorney who examines titles to oil and gas properties. The title examiner, often with a landman's help, must investigate every aspect of a title before legal opinions can be given. Most companies require title opinions before paying bonuses and delay rentals. A common practice is to issue a post-dated draft to the lessor when the lease is signed. This gives the lessor time (usually around 15 to 20 days) to clear the title before payment is made. Opinions may depend on *abstracts* of the property (fig. 3.8) or a *takeoff*. An abstract is a collection of all the recorded instruments affecting title to a tract of land, usually presented in a shortened form. A takeoff is a brief description of the documents relevant to the title of a particular piece of property. A takeoff may be purchased from an abstract company. Also, local title companies usually carry complete records of real estate transactions and may be willing to share information for a fee. Many companies are now using data banks of title information and can quickly retrieve it with in-house computer terminals (fig. 3.9). In addition to *abstract-based opinions*, an examiner can render *stand-up opinions* based on the landman's run sheet and a careful examination of all documents in the appropriate county courthouse.

Curing titles. When the chain of title has gaps or other defects, the title examiner may obtain sworn statements or other documents that support the present occupant's title to the land, especially by possession and use. Some of these statements sworn before a notary public, known as *affidavits*, may already be recorded because of their need with regard to past title transfers. In all likelihood, additional investigation of facts will be required. The older affidavits may also leave some questions unanswered. Affidavits should, if possible, be secured from noninterested parties familiar with the facts, because a "self-serving" affidavit made by the present occupant is vulnerable to attack. Commonly used affidavits are those of *death and heirship; identity; nonproduction; use, occupancy, and possession;* and *adverse possession.*

In addition to affidavits, other documents and legal instruments that may assist the title examiner in curing a title are *tax receipts, rental receipts,* and *releases of old leases* and *mortgages* The landman or examiner may uncover a *deed of trust* to the property and, if still in effect, the holder of such an instrument—the creditor—has rights superior to any subsequent interest in the property. A *quitclaim deed* or *disclaimer* waives any rights to property; it is often used by the surface owner or tenant to affirm that he has no interest in the mineral estate. It should be remembered, however, that the title is not cleared by the affidavits and documents themselves; it is the existence of the appropriate *facts* that makes the title satisfactory.

Validating the owner's capacity to contract

The landman must also try to make sure the lessor has the capacity to contract. Simply stated, not all owners can grant valid leases. Thus, any person or company interested in obtaining lease rights to a piece of property has to investigate the legal status and powers of the owner very carefully before investing in a drilling venture. Special arrangements are necessary to lease from life tenants and remaindermen, cotenants, married persons, illiterates, owners as agents of the state, and fiduciaries.

Life tenants and remaindermen. A life tenant is a person who is entitled to exclusive possession of a property but cannot pass it on through inheritance. A remainderman is the person who has title to the property and will have full possession of the property after the death of the life tenant. Because they share interests in the land, both the life tenant and the remainderman should, under ordinary circumstances, execute an oil and gas lease together.

Cotenants. Cotenants are people who own land together. They are entitled to equal use of the land and *may* inherit one another's interest in the land. In the majority of states, one co-owner may execute a lease on jointly owned premises without the consent of the other cotenants, but the lease is binding only on the cotenant making the agreement, although all may profit from any production.

Married persons. The ability of married persons to make legally binding leases depends on the status of their land and the law of the state in which they reside. The estate may be *community property* (acquired after the marriage), *separate property* (owned separately by one spouse before marriage), or *homestead property* (occupied and used by the owner and his family). Some community property states (like California) and most noncommunity property states require both spouses to sign a lease.

Illiterates. The inability to read and write does not disqualify a person from making a legally binding contract. An illiterate can grant valid leases as long as the proper procedures concerning witnesses, executing the legal agreement, and making his mark are all fully complied with.

Owners as agents of the state. Some property in the state of Texas was given to its citizens who could enjoy all the rights of fee ownership except the right to lease and profit from its minerals. Later, the surface owner was given the right to grant an oil and gas lease as *an agent of the state*, with the owner and state splitting any benefits from that lease.

Fiduciaries. A fiduciary is a person who acts for another in a legal, financial, or other undertaking. Whether a fiduciary can execute a valid oil and gas lease depends on the laws of the state in which the property is located and the specific provisions of the empowering instrument. Among the more common fiduciaries are executors, administrators, guardians, trustees, persons given powers of attorney, and representatives of unknown or missing heirs.

The landman and the landowner do not always negotiate a lease over a cup of coffee. Many leases are acquired by telephone or letter, and many of those negotiated face-to-face are often sold or traded by the acquiring company. However a lease is acquired, it is always bargained for with each party trying to make the best deal he can. The landman must consider his company's resources and policies, be aware of the local situation, and know the prevailing prices in his area.

Negotiating the lease

The landowner will be concerned with making an informed evaluation of the landman's offer. He may contact neighbors, bankers, attorneys, accountants, or local lease brokers in order to evaluate the offer. The landman will probably not make his initial offer at the top of the scale of permissible bonuses or royalties. The landowner will probably bargain for a better price. The two usually compromise on primary term, bonus per acre, and royalty percentages. An astute landowner preparing to lease his land might also obtain a copy of the well-spacing and density regulations that apply to his area from his state regulatory agency.

Fig. 3.10. The landman and landowner "shake on it," but the written lease is the important part of this negotiation.

Since real estate law rather than contract law controls leasing, an oil and gas lease must be *written* (fig. 3.10). An oral agreement is not sufficient. Hence, in the course of negotiations, the landowner finds himself pondering a printed form that is several pages long and full of closely printed clauses filled with "hereins," "herebys," and "hereinafters." The instrument may be a lease form used by the landman or by the company he represents. Whether the lease is a preprinted form such as one of the well-known "Producers 88" forms, the landowner's own form (perhaps from an association), or one drafted by an oil and gas attorney, all of its provisions are subject to mutual agreement by both the lessor and the lessee.

Because of the language of the lease and past interpretations of the courts, the prudent landowner (lessor) will have a lease examined and explained to him (preferably by his attorney) before signing it. It is equally important for the oil and gas operator (lessee) to thoroughly understand any special clauses and amendments resulting from final negotiations between the landowner and the landman.

PROVISIONS OF THE LEASE

The provisions essential to a lease—conveyance, term, and royalty—are contained in standard lease clauses (fig. 3.11). Most leases also contain clauses that address the relative rights of both lessor and lessee when peculiar conditions exist. In addition to these clauses, a lease contains dates, names and signatures of parties involved, and the seal and signature of a notary public.

Producers 88 (12/79) Revised
With 320 Acres Pooling Provision

POUND PRINTING & STATIONERY COMPANY
2325 Fannin, Houston, Texas 77002 (713) 659-3159

OIL, GAS AND MINERAL LEASE

THIS AGREEMENT made this_____ day of_____, 19____, between

_____

Lessor (whether one or more), whose address is:_____

_____ Zip Code_____,

and_____,

Lessee, (whether one or more), whose address is: _____

_____ Zip Code_____,

WITNESSETH:

1. Lessor in consideration of_____   _____

Dollars ($_____), in hand paid, of the royalties herein provided, and of the agreements of Lessee herein contained, hereby grants, leases and lets exclusively unto Lessee for the purpose of investigating, exploring, prospecting, drilling and mining for and producing oil, gas and all other minerals, conducting exploration, geologic and geophysical surveys by seismograph, core test, gravity and magnetic methods, injecting gas, water and other fluids, and air into subsurface strata, laying pipe lines, building roads, tanks, power stations, telephone lines and other structures thereon and on, over and across lands owned or claimed by Lessor adjacent and contiguous thereto, to produce, save, take care of, treat, transport and

own said products, and housing its employees, the following described land in _____County, Texas, to-wit:

This lease also covers and includes all land owned or claimed by Lessor adjacent or contiguous to the land particularly described above, whether the same be in said survey or surveys or in adjacent surveys, although not included within the boundaries of the land particularly described

above. For all purposes of this lease, said land is estimated to comprise _____ acres, whether it actually comprises more or less.

2. Subject to the other provisions herein contained, this lease shall be for a term of _____ ( ) years from this date (called "primary term") and as long thereafter as oil, gas or other mineral is produced from said land or land with which said land is pooled hereunder.

3. The royalties to be paid by Lessee are:

(a) On oil, one-eighth of that produced and saved from said land, the same to be delivered at the well. If Lessor elects not to take delivery of the royalty oil, Lessee may from time to time sell the royalty oil in its possession, paying to Lessor therefor the net proceeds derived by Lessee from the sale of such royalty oil. Lessor's royalty interest in oil shall bear its proportionate part of the cost of treating the oil to render it marketable oil and, if there is no available pipeline, its proportionate part of the cost of all trucking charges.

(b) On gas, including all gases, liquid hydrocarbons and their respective constituent elements, casinghead gas or other gaseous substance, produced from said land and sold or used off the premises or for the extraction of gasoline or other product therefrom, the market value at the well on one-eighth of the gas so sold or used, provided that on gas sold at the well the royalty shall be one-eighth of the net proceeds derived from such sale. Lessor's royalty interest in gas, including all gases, liquid hydrocarbons and their respective constituent elements, casinghead gas or other gaseous substance, shall bear its proportionate part of the cost of all compressing, treating, dehydrating and transporting incurred in marketing the gas so sold at the wells.

(c) On all other minerals mined and marketed, one-tenth either in kind or value at the well or mine, at Lessee's election, except that on sulphur mined and marketed the royalty shall be fifty cents ($.50) per long ton.

(d) While there is a gas well on said land or on lands pooled therewith and if gas is not being sold or used off the premises for a period in excess of three full consecutive calendar months, and this lease is not then being maintained in force and effect under the other provisions hereof, Lessee shall tender or pay to Lessor annually at any time during the lease anniversary month of each year immediately succeeding any lease year in which a shut-in period occurred one-twelfth (1/12) of the sum of $1.00 per acre for the acreage then covered by this lease as shut-in royalty for each full calendar month in the preceding lease year that this lease was continued in force solely and exclusively by reason of the provisions of this paragraph. If such payment of shut-in royalty is so made or tendered by Lessee to Lessor, it shall be considered that this lease is producing gas in paying quantities and this lease shall not terminate, but remain in force and effect. The term "lease anniversary month" means that calendar month in which this lease is dated. The term "Lease year" means the calendar month in which the lease is dated, plus the eleven succeeding calendar months.

(e) If the price of any oil, gas, or other minerals produced hereunder is regulated by any governmental authority, the value of same for the purpose of computing the royalties hereunder shall not be in excess of the price permitted by such regulation. Should it ever be determined by any governmental authority, or any court of final jurisdiction, or otherwise, that the Lessee is required to make any refund on oil, gas, or other minerals produced or sold by Lessee hereunder, then the Lessor shall bear his proportionate part of the cost of any such refund to the extent that royalties paid to Lessor have exceeded the permitted price, plus any interest thereon ordered by the regulatory authority or court, or agreed to by Lessee. If Lessee advances funds to satisfy Lessor's proportionate part of such refund, Lessee shall be subrogated to the refund order or refund claim, with the right to enforce same for Lessor's proportionate contribution, and with the right to apply rentals and royalties accruing hereunder toward satisfying Lessor's refund obligations.

(f) Lessee shall have free use of oil, gas, coal, water from said land, except water from Lessor's wells, for all operations hereunder, and the royalty on oil, gas and coal shall be computed after deducting any so used.

CONVEYANCE

TERM

ROYALTY

Fig. 3.11. Conveyance, term, and royalty are contained in all standard leases.

Conveyance is the granting of interest in the petroleum to a person or company for the purposes of exploring, drilling, and producing. The conveyance part of the lease is addressed in the all-important *granting clause* and includes the consideration, legal land description, and usually a Mother Hubbard clause.

Term is the duration of the lease and is stated in the *habendum clause*. Royalty is a share of production provided for in several *royalty clauses* dealing with payments.

Dates

Any lease should be dated in order to avoid possible disputes as to which lease (if more than one exists) is the valid one. The controlling date for a lease is the one written in the instrument, not the date on which it was signed, notarized, or recorded.

Parties

Two parties are involved in a lease agreement: the lessor and the lessee. The *lessor* is the person or persons, usually the mineral estate owner or owners, granting the lease. The *lessee* is the person or company receiving the right to search for and produce oil or gas. A lease must name all persons involved in the agreement and, in many cases, give their addresses as well.

Granting clause

The granting clause conveys the described mineral interest from the lessor to the lessee for a consideration. By accepting the lease, the lessee accepts the covenants expressed and implied by the granting clause (fig. 3.12). The implied convenants vary with circumstances, but involve the need to exercise diligence and good faith in performance of what would be expected of an ordinarily prudent operator. Expressed covenants such as protecting and developing a lease might include drilling an offset well, drilling additional wells or recompleting wells to new depths, plugging abandoned wells, and treating and marketing the oil and gas. The granting clause is the place in which the minerals covered by the lease are clearly stated. Also, if the lessor is concerned about the restoration of his land after production has ceased, he may add a stipulation requiring the lessee to leave the property as he found it.

Consideration. The consideration is the benefit to the lessor and is required in some states to make the contract valid. Consideration is a term from contract law, often stated nominally as "$10 and other consideration." The $10 serves as the necessary consideration of the document, and no one else needs to know what other consideration, or *bonus*, the lessor assumed for signing the lease. Bonus money varies from $1 to $1,000 per acre, depending

CONSIDERATION

1. Lessor in consideration of _____ three thousand seven hundred and fifty _____ Dollars ($ 3,750 _____), in hand paid, of the royalties herein provided, and of the agreements of Lessee herein contained, hereby grants, leases and lets exclusively unto Lessee for the purpose of investigating, exploring, prospecting, drilling and mining for and producing oil, gas and all other minerals, conducting exploration, geologic and geophysical surveys by seismograph, core test, gravity and magnetic methods, injecting gas, water and other fluids, and air into subsurface strata, laying pipe lines, building roads, tanks, power stations, telephone lines and other structures thereon and on, over and across lands owned or claimed by Lessor adjacent and contiguous thereto, to produce, save, take care of, treat, transport and own said products, and housing its employees, the following described land in _____ LEE _____ County, Texas, to-wit:

LAND DESCRIPTION

50 acres, more or less, out of the James Joseph 1/3 League, Abstract 175, known as Lot No. 6, in the subdivision of N. F. Newhouse Estate, in Lee County, Texas, being particularly described by metes and bounds in the Deed from John Silver to Mark A. Brown, Jr., et ux dated June 25, 1965, of record in Vol. 162, Pg. 209, Deed Records of Lee County, Texas, to which said Deed and its record reference is hereby made for all pertinent purposes; SAVE AND EXCEPT 0.219 of an acre described in the Deed from C. B. Frieberg to the State of Texas, dated October 27, 1949, of record in Vol. 104, Pg. 68, Lee County Deed Records; and being the same property described in the Deed from Mark A. Brown, Jr., et ux to Robert E. Layman, et ux, dated March 1, 1971, of record in Vol. 194, Pg. 131, Lee County Deed Records.

MOTHER HUBBARD

This lease also covers and includes all land owned or claimed by Lessor adjacent or contiguous to the land particularly described above, whether the same be in said survey or surveys or in adjacent surveys, although not included within the boundaries of the land particularly described above. For all purposes of this lease, said land is estimated to comprise _____ fifty _____ acres, whether it actually comprises more or less.

Fig. 3.12. The granting clause conveys the mineral interest from the lessor to the lessee for a consideration.

on the area and circumstances. If desired, the total amount may be written into the consideration. In Texas, no consideration is necessary for the validity of the lease, because it is a conveyance of a *present interest* in land.

Land description. A legal description of the property involved is a necessary element in an oil and gas lease. If the lessor owns the entire undivided interest in the land being leased, the description will be straightforward. Otherwise, the lease may contain "lesser interest" or "proportionate reduction" clauses that allow lessees to reduce rents and royalties proportionately to the percentage of minerals owned. Whether the land is legally measured by *metes and bounds*, by the *rectangular survey system*, or as an *urban subdivision*, a clear and complete description of it is essential to an oil and gas lease.

Mother Hubbard clause. Primarily in the lease to protect the working interest, the Mother Hubbard clause allows the operator to regard, as part of the lease, lands not included in the land description. A Mother Hubbard (a term based on the loose, rather shapeless dress worn by the nursery-rhyme character) includes odd-shaped or small wedges of land inadvertently left out or incorrectly described. A lessor will limit this coverage to 10 percent of the described land so as not to include all of the land he owns but only that part intended to be included by both parties.

Habendum clause

The habendum clause fixes the duration of the lessee's interest (fig. 3.13). A typical habendum clause sets out the length of time for the initial (or primary) term of the lease, which states the amount of time the lessee has to begin to drill a well. If not met, the lease expires. The primary term, generally used as an exploration period, varies from one to ten years with three years a commonly used time period.

The lease will continue, however, in a secondary term "as long as oil, gas, or other mineral is produced." The lessee is not bound by the lease unless the well or wells produce in "paying quantities," a term held by the courts to mean sufficiently profitable to warrant production.

---

2. Subject to the other provisions herein contained, this lease shall be for a term of _____ one _____ (1) years from this date (called "primary term") and as long thereafter as oil, gas or other mineral is produced from said land or land with which said land is pooled hereunder.

Fig. 3.13. The habendum clause fixes the primary term of the lease.

Royalty is a share of production free of expenses, except for taxes **Royalty clauses**
and marketing expenses. The royalty clause (fig. 3.14) details the

---

3. The royalties to be paid by Lessee are:

(a) On oil, one-eighth of that produced and saved from said land, the same to be delivered at the well. If Lessor elects not to take delivery of the royalty oil, Lessee may from time to time sell the royalty oil in its possession, paying to Lessor therefor the net proceeds derived by Lessee from the sale of such royalty oil. Lessor's royalty interest in oil shall bear its proportionate part of the cost of treating the oil to render it marketable oil and, if there is no available pipeline, its proportionate part of the cost of all trucking charges.

(b) On gas, including all gases, liquid hydrocarbons and their respective constituent elements, casinghead gas or other gaseous substance, produced from said land and sold or used off the premises or for the extraction of gasoline or other product therefrom, the market value at the well on one-eighth of the gas so sold or used, provided that on gas sold at the well the royalty shall be one-eighth of the net proceeds derived from such sale. Lessor's royalty interest in gas, including all gases, liquid hydrocarbons and their respective constituent elements, casinghead gas or other gaseous substance, shall bear its proportionate part of the cost of all compressing, treating, dehydrating and transporting incurred in marketing the gas so sold at the wells.

(c) On all other minerals mined and marketed, one-tenth either in kind or value at the well or mine, at Lessee's election, except that on sulphur mined and marketed the royalty shall be fifty cents ($.50) per long ton.

(d) While there is a gas well on said land or on lands pooled therewith and if gas is not being sold or used off the premises for a period in excess of three full consecutive calendar months, and this lease is not then being maintained in force and effect under the other provisions hereof, Lessee shall tender or pay to Lessor annually at any time during the lease anniversary month of each year immediately succeeding any lease year in which a shut-in period occurred one-twelfth (1/12) of the sum of $1.00 per acre for the acreage then covered by this lease as shut-in royalty for each full calendar month in the preceding lease year that this lease was continued in force solely and exclusively by reason of the provisions of this paragraph, If such payment of shut-in royalty is so made or tendered by Lessee to Lessor, it shall be considered that this lease is producing gas in paying quantities and this lease shall not terminate, but remain in force and effect. The term "lease anniversary month" means that calendar month in which this lease is dated. The term "Lease year" means the calendar month in which the lease is dated, plus the eleven succeeding calendar months.

(e) If the price of any oil, gas, or other minerals produced hereunder is regulated by any governmental authority, the value of same for the purpose of computing the royalties hereunder shall not be in excess of the price permitted by such regulation. Should it ever be determined by any governmental authority, or any court of final jurisdiction, or otherwise, that the Lessee is required to make any refund on oil, gas, or other minerals produced or sold by Lessee hereunder, then the Lessor shall bear his proportionate part of the cost of any such refund to the extent that royalties paid to Lessor have exceeded the permitted price, plus any interest thereon ordered by the regulatory authority or court, or agreed to by Lessee. If Lessee advances funds to satisfy Lessor's proportionate part of such refund, Lessee shall be subrogated to the refund order or refund claim, with the right to enforce same for Lessor's proportionate contribution, and with the right to apply rentals and royalties accruing hereunder toward satisfying Lessor's refund obligations.

(f) Lessee shall have free use of oil, gas, coal, water from said land, except water from Lessor's wells, for all operations hereunder, and the royalty on oil, gas and coal shall be computed after deducting any so used.

Fig. 3.14. The royalty clause gives the percentage of production to be received by the lessor.

percentage of production given back to the lessor. Royalty varies, but one-eighth to one-fourth is common. From an oilwell, the lessor can receive royalty in *kind*—that is, in barrels of crude—or in *money*.

Gas royalty. Royalty for gas is traditionally payable only in money. Many leases for gas have royalty based on the value of the actual sale by the lessee if the gas is sold at the well. And if gas is sold off the premises of the lease, royalty is usually based on market value, expressed in the lease as "market price at the well." This market price is determined by comparable sales in the field, so that in some instances the lessee may have to sell for less than the market value of the gas, yet will have to pay royalty based on a higher value. Gas royalty is complicated by federal controls over marketing and field pricing through the Natural Gas Act of 1938 and the Natural Gas Policy Act of 1978.

Nonparticipating royalty. Royalty, or more often a fractional portion of the royalty, without other rights of the mineral interest, may be conveyed by the participating royalty owner to someone else for a definite term. Such an interest is termed *nonparticipating royalty*, and its owner cannot execute a lease or receive bonus or delay rental. He only receives a proportionate share of royalty payments. The *participating* royalty owner—one who also owns the mineral rights—can sign a lease and is said to have *executive rights*.

Shut-in royalty. The shut-in royalty clause for gas wells allows the lessee to maintain the lease in force by paying money to the lessor in lieu of actual production when a well is "shut in." The purpose of this provision is to allow an operator to hold the lease on a well that is capable of producing in commercial amounts but is not in production for lack of a market or an available transmitting pipeline.

Pooling and unitization clause

A pooling and unitization clause is a provision always used in the modern lease, authorizing the lessee to cross-convey interests in oil and gas (fig. 3.15). That is, two or more leases are combined to share interests in return for a proportionate sharing of royalty. For example, if John Smith had 100 acres and Mary Brown had 60 acres and the leases were pooled, a well drilling anywhere on the 160 acres would bring John $100/160$, or $5/8$, of the royalty, and Mary would receive $60/160$, or $3/8$, of the royalty.

The terms *pooling* and *unitization* are often used interchangeably but refer to different undertakings. Pooling is the combining of small or irregular tracts into a unit large enough to meet

5. (a) Lessee, at its option, is hereby given the right and power to pool, unitize or combine the acreage covered by this lease or any portion thereof as to oil and gas, or either of them, with any other land covered by this lease, and/or with any other land, lease or leases in the immediate vicinity thereof to the extent hereinafter stipulated, when in Lessee's judgment it is necessary or advisable to do so in order properly to explore, or to develop and operate said leased premises in compliance with the spacing rules of the Railroad Commission of Texas, or other lawful authority, or when to do so would, in the judgment of Lessee, promote the conservation of oil and gas in and under and that may be produced from said premises. Units pooled for oil hereunder shall not substantially exceed 40 acres each in area, plus a tolerance of ten percent (10%) thereof, and units pooled for gas hereunder shall not substantially exceed in area 320 acres each plus a tolerance of ten percent (10%) thereof, provided that should governmental authority having jurisdiction prescribe or permit the creation of units larger than those specified, for the drilling or operation of a well at a regular location or for obtaining maximum allowable from any well to be drilled, drilling or already drilled, units thereafter created may conform substantially in size with those prescribed or permitted by government regulations.

(b) Lessee under the provisions hereof may pool or combine acreage covered by this lease or any portion thereof as above provided as to oil in any one or more strata and as to gas in any one or more strata. The units formed by pooling as to any stratum or strata need not conform in size or area with the unit or units into which the lease is pooled or combined as to any other stratum or strata, and oil units need not conform as to area with gas units. The pooling in one or more instances shall not exhaust the rights of the Lessee hereunder to pool this lease or portions thereof into other units. Upon execution by Lessee of an instrument describing and designating the pooled acreage as a pooled unit, said unit shall be effective as to all parties hereto, their heirs, successors, and assigns, irrespective of whether or not the unit is likewise effective as to all other owners of surface, mineral, royalty, or other rights in land included in such unit. Within a reasonable time following the execution of said instrument so designating the pooled unit, Lessee shall file said instrument for record in the appropriate records of the county in which the leased premises are situated. Any unit so formed may be re-formed, increased, decreased, or changed in configuration, at the election of Lessee, at any time and from time to time after the original forming thereof, and Lessee may vacate any unit formed by it hereunder by instrument in writing filed for record in said county at any time when there is no unitized substance being produced from such unit.

(c) Lessee may at its election exercise its pooling option before or after commencing operations for or completing an oil or gas well on the leased premises, and the pooled unit may include, but it is not required to include, land or leases upon which a well capable of producing oil or gas in paying quantities has theretofore been completed or upon which operations for the drilling of a well for oil or gas have theretofore been commenced. In the event of operations for drilling on or production of oil or gas from any part of a pooled unit which includes all or a portion of the land covered by this lease, regardless of whether such operations for drilling were commenced or such production was secured before or after the execution of this instrument or the instrument designating the pooled unit such operations shall be considered as operations for drilling on or production of oil and gas from land covered by this lease whether or not the well or wells be located on the premises covered by this lease and in such event operations for drilling shall be deemed to have been commenced on said land within the meaning of paragraph 6 of this lease; and the entire acreage constituting such unit or units, as to oil and gas, or either of them, as herein provided, shall be treated for all purposes, except the payment of royalties on production from the pooled unit, as if the same were included in this lease.

Fig. 3.15. The pooling and unitization clause allows the lessee to combine leases to acquire sufficient acreage.

(d) For the purpose of computing the royalties to which owners of royalties and payments out of production and each of them shall be entitled on production of oil and gas, or either of them, from the pooled unit, there shall be allocated to the land covered by this lease and included in said unit (or to each separate tract within the unit if this lease covers separate tracts within the unit) a pro rata portion of the oil and gas, or either of them, produced from the pooled unit after deducting that used for operations on the pooled unit. Such allocation shall be on an acreage basis—that is to say, there shall be allocated to the acreage covered by this lease and included in the pooled unit (or to each separate tract within the unit if this lease covers separate tracts within the unit) that pro rata portion of the oil and gas, or either of them, produced from the pooled unit which the number of surface acres covered by this lease (or in each such separate tract) and included in the pooled unit bears to the total number of surface acres included in the pooled unit. Royalties hereunder shall be computed on the portion of such production, whether it be oil and gas, or either of them, so allocated to the land covered by this lease and included in the unit just as though such production were from such land. The production from an oil well will be considered as production from the lease or oil pooled unit which it is producing and not as production from a gas pooled unit; and production from a gas well will be considered as production from the lease or gas pooled unit from which it is producing and not from an oil pooled unit.

(e) The formation of any unit hereunder shall not have the effect of changing the ownership of any delay rental or shut-in production royalty which may become payable under this lease. If this lease now or hereafter covers separate tracts, no pooling or unitization of royalty interest as between any such separate tracts is intended or shall be implied or result merely from the inclusion of such separate tracts within this lease but Lessee shall nevertheless have the right to pool as provided above with consequent allocation of production as above provided. As used in this paragraph 5, the words "separate tract" mean any tract with royalty ownership differing, now or hereafter, either as to parties or amounts, from that as to any other part of the leased premises.

Fig. 3.15, *Continued*

state spacing regulations for drilling. Unitization is the combining of leased tracts on a fieldwide or reservoir-wide scale so that many tracts may be treated as one to facilitate operations such as enhanced recovery projects.

Pooling is a voluntary arrangement, although in Texas the Texas Mineral Interest Pooling Act allows the Railroad Commission of Texas to force pooling of interests in certain limited situations.

Drilling, delay rental, and related clauses

The *delay rental clause* provides the lessee with three choices (fig. 3.16). He may opt to (1) drill a well, (2) pay on an annual basis to delay drilling until later but within the primary term, or (3) terminate the lease by neither drilling nor paying delay rental. Most leases written today expire one year from the date on the instrument unless the lessee begins operations for drilling or makes timely payment of delay rental. Deferring drilling past the primary term of the lease generally voids the agreement. A *paid-up lease* provides for rental payment along with the cash bonus and thus requires no further action during the primary term.

A *dry hole clause* allows an operator to keep his lease in the event he drills a dry hole. He then has a specified period of time to begin drilling a second well or resume payment of delay rentals.

6. (a) If operations for drilling are not commenced on said land or on acreage pooled therewith as above provided on or before one year from this date, the lease shall then terminate as to both parties, unless on or before such anniversary date Lessee shall pay or tender (or shall make a bona fide attempt to pay or tender, as hereinafter stated) to Lessor or to the credit of Lessor in _____ Monroe State _____ Bank at _____ Lexington _____, Texas, (which bank and its successors are Lessor's agent and shall continue as the depository for all rentals payable hereunder regardless of change in ownership of said land or the rentals) the sum of _____ five hundred _____

Dollars ($ 500.00), (herein called rentals), which shall cover the privilege of deferring commencement of drilling operations for a period of twelve (12) months. In like manner and upon like payments or tenders annually, the commencement of drilling operations may be further deferred for successive periods of twelve (12) months each during the primary term. The payment or tender of rental under this paragraph and of royalty under paragraph 3 on any gas well from which gas is not being sold or used may be made by the check or draft of Lessee mailed or delivered to the parties entitled thereto or to said bank on or before the date of payment. If such bank (or any successor bank) should fail, liquidate or be succeeded by another bank, or for any reason fail or refuse to accept rental, Lessee shall not be held in default for failure to make such payment or tender of rental until thirty (30) days after Lessor shall deliver to Lessee a proper recordable instrument naming another bank as agent to receive such payments or tenders. If Lessee shall, on or before any anniversary date, make a bona fide attempt to pay or deposit rental to a Lessor entitled thereto according to Lessee's records or to a Lessor, who, prior to such attempted payment or deposit, has given Lessee notice, in accordance with subsequent provisions of this lease, of his right to receive rental, and if such payment or deposit shall be ineffective or erroneous in any regard, Lessee shall be unconditionally obligated to pay to such Lessor the rental properly payable for the rental period involved, and this lease shall not terminate but shall be maintained in the same manner as if such erroneous or ineffective rental payment of deposit had been properly made, provided that the erroneous or ineffective rental payment or deposit be corrected within 30 days after receipt by Lessee of written notice from such Lessor of such error accompanied by such instruments as are necessary to enable Lessee to make proper payment. The down cash payment is consideration for this lease according to its terms and shall not be allocated as a mere rental for a period. Lessee may at any time or times execute and deliver to Lessor or to the depository above named or place of record a release or releases of this lease as to all or any part of the above-described premises, or of any mineral or horizon under all or any part thereof, and thereby be relieved of all obligations as to the released land or interest. If this lease is released as to all minerals and horizon under a portion of the land covered by this lease, the rentals and other payments computed in accordance therewith shall thereupon be reduced in the proportion that the number of surface acres within such released portion bears to the total number of surface acres which was covered by this lease immediately prior to such release.

(b) Lessor hereby designates_____ Monroe State _____ Bank at _____ Lexington _____, Texas, and its successors as Lessor's agent to serve as the depository for any payment due with respect to any shut-in gas well. Payment of shut-in gas royalty may be made in the manner provided in paragraph 6(a) hereof for the payment or tender of rentals, including all terms with respect to the deposit of same in the designated depository bank, notwithstanding paragraph 6(a) being otherwise stricken or inoperative due to this lease having a primary term not exceeding one year, if such be the case.

Fig. 3.16. The delay rental clause allows extension or termination of the lease if a well is not drilled during the primary term.

9. The rights of either party hereunder may be assigned in whole or in part, and the provisions hereof shall extend to their heirs, successors and assigns; but no change or division in ownership of the land, rentals or royalties, however accomplished, shall operate to enlarge the obligations or diminish the rights of Lessee; and no change or division in such ownership shall be binding on Lessee until thirty (30) days after Lessee shall have been furnished by registered U.S. mail at Lessee's principal place of business with a certified copy of recorded instrument or instruments evidencing same. In the event of assignment hereof in whole or in part, liability for breach of any obligation hereunder shall rest exclusively upon the owner of this lease or of a portion thereof who commits such breach. In the event of the death of any person entitled to rentals, shut-in royalty or royalty hereunder, Lessee may pay or tender such rentals, shut-in royalty or royalty to the credit of the deceased or the estate of the deceased until such time as Lessee is furnished with proper evidence of the appointment and qualification of an executor or administrator of the estate, or if there be none, then until Lessee is furnished with evidence satisfactory to it as to the heirs or devisees of the deceased and that all debts of the estate have been paid. If at any time two or more persons be entitled to participate in the rental payable hereunder, Lessee may pay or tender said rental jointly to such persons or to their joint credit in the depository named herein, or, at Lessee's election, the proportionate part of said rentals to which each participant is entitled may be paid or tendered to him separately or to his separate credit in said depository; and payment or tender to any participant of his portion of the rentals hereunder shall maintain this lease as to such participant. In event of assignment of this lease as to a segregated portion of said land, the rentals payable hereunder shall be apportionable as between the several leasehold owners ratably according to the surface area of each, and default in rental payment by one shall not affect the rights of other leasehold owners hereunder. If six or more parties become entitled to royalty hereunder, Lessee may withhold payment thereof unless and until furnished with a recordable instrument executed by all such parties designating an agent to receive payments for all.

Fig. 3.17. The assignment clause allows the lessee to assign the lease rights to another party.

12. When drilling, production or other operations on said land or land pooled with such land, or any part thereof are prevented, delayed or interrupted by lack of water, labor or materials, or by fire, storm, flood, war, rebellion, insurrection, sabotage, riot, strike, difference with workers, or failure of carriers to transport or furnish facilities for transportation, or as a result of some law, order, rule, regulation or necessity of governmental authority, either State or Federal, or as a result of the filing of a suit in which Lessee's title may be affected, or as a result of any cause whatsoever beyond the reasonable control of Lessee, the lease shall nevertheless continue in full force and effect. If any such prevention, delay or interruption should commence during the primary term hereof, the time of such prevention, delay or interruption shall not be counted against Lessee and the running of the primary term shall be suspended during such time; if any such prevention, delay or interruption should commence after the primary term hereof Lessee shall have a period of ninety (90) days after the termination of such period of prevention, delay or interruption within which to commence or resume drilling, production or other operations hereunder, and this lease shall remain in force during such ninety (90) day period and thereafter in accordance with the other provisions of this lease. Lessee shall not be liable for breach of any express or implied covenants of this lease when drilling, production or other operations are so prevented, delayed or interrupted.

Fig. 3.18. The force majeure clause keeps the lease in force even though delays specified in the lease prevent the lessee from meeting his obligations.

If production stops, a *cessation of production statement* in the lease allows the operator a specified length of time to restore production (as in the case of a workover on a sluggish well), to drill a new well, or to return to paying delay rentals.

A *continuous development clause* is designed to keep drilling operations going steadily past the primary term. It requires the operator to develop leased land up to its allowable density.

Assignment clause

Interest in a lease can usually be transferred by either the lessor or the lessee to another party. In fact, leases frequently change hands a number of times before and sometimes after production begins. Although an assignment by one party to the lease does not materially affect the interests of the other party, both parties may be required to give notice of such transfers. The nature of this notice is provided for in the assignment clause (fig. 3.17).

Damage clause

Many operators voluntarily pay for damage to the surface, but most leases include a clause that makes the lessee liable for damages or losses suffered because of drilling or production. The damage clause is usually tailored to the requirements of the landowner.

Force majeure clause

The force majeure clause (fig. 3.18) allows the lease to continue in force while the lessee is prevented from meeting the conditions of the lease during delays caused by "acts of God" or other events beyond the control of the operator. These events do not include delays due to equipment failures or labor problems. This clause also contains a reminder that the lease is subject to state and federal laws and usually excuses delays caused by government interference that cannot be blamed on the lessee.

Warranty and proportionate reduction clauses

While the warranty clause seems to guarantee clear title, the proportionate reduction clause provides for the possibility that an owner owns less than his land description claims. If an owner's interest turns out to be less than he thought, the lessee can proportionately reduce the rentals and royalties he pays. The clauses also include the lessee's rights in case the lessor defaults on tax or mortgage payments.

Special provisions and amendments

Provisions that outline specific rights of the lessee or lessor concerning the physical operation of the leasehold are included in the final lease in the form of additions, deletions, or amendments. A

84

landowner may request specifics concerning the drilling of an off-set well. One reason for such a well may be to protect his property from drainage. Another landowner may want the right to approve the placement of equipment, storage facilities, pipelines, and roads, or he may require that cattle guards be installed to protect his animals.

The landowner may insist on a provision that wells be plugged according to state regulations and that the well site be restored as nearly as possible to its original condition.

In some parts of the country, owners may bargain for gas from a gas-producing well for their own use—either free or purchased at the wellhead price.

To prevent a lessee from holding a tract under lease by marginal production from a shallow well, the landowner may insist on a *Pugh clause* or *Freestone rider*. This provision also releases non-productive or untested zones, as well as lease acreage outside a producing pooled unit if drilling or exploration does not take place by the end of the specified time.

For the operating company, or lessee, the provisions might include rights to use gas, oil, or water produced on the leasehold for its drilling operations (except, of course, the landowner's water wells or stock tanks); the right to remove its equipment when production has ceased; and the right to know in advance about pending changes in ownership or forfeiture of mortgage for non-payment.

Lease-terminating provisions

The lessee's obligations concerning the termination of the lease agreement are usually covered in the delay rental and habendum clauses. Failure to comply with conditions set in these clauses can result in *forfeiture* of the property rights by the lessee. For example, the landowner can force, through the courts, the operator to relinquish his rights because of failure to pay delay rental or to drill within a given period of time.

On the other hand, the lessee can voluntarily give up his rights by neither drilling nor paying delay rental. Or, if the lessee ceases production (generally because the well is no longer profitable), he may terminate the lease by *abandonment*—that is, he may physically abandon, or vacate, the premises and allow the lease to expire.

If the lease contains a *surrender* clause, the lessee must notify the lessor of his intent to surrender the lease. In some states, the lessee is required to give written notice of his surrender of the lease so that the lessor has a clear title should he want to lease the property again. Failure to comply with surrender requirements could result in a fine or suit for damages.

EXECUTION OF THE LEASE

A properly executed oil and gas lease is a written instrument signed by the granting party, acknowledged by a notary public or other witnesses, and officially recorded in the appropriate county records. This step usually requires the help of an attorney who is licensed to practice law in the relevant state and preferably one well-versed in oil and gas law.

The lessor, or granting party or parties, must sign the lease to make it valid. The signature should conform to that used on any earlier document, such as a deed, by which the land was acquired. For example, a woman whose name has changed because of marriage since she acquired her property should show both names. People signing as a legal representative for someone else must identify their capacity and normally the party for whom they sign.

Signing the lease

Acknowledgment proves proper execution, or signing, of a lease. A notary public or other accepted officer witnesses the signing and pledges with a written certificate and an official signature that the grantor signed his lease freely (fig. 3.19). The pledge also assures the identity of the signing party.

Acknowledging the lease

Fig. 3.19. A lease is acknowledged when signed by the lessor in the presence of a notary public.

Recording the executed lease

Generally, a lease is recorded as soon as possible after it is executed and consideration has been paid. Recording the lease involves officially entering it into the appropriate county records and is ordinarily handled by the lessee or his representative. Either the original or a copy, depending on the state, is kept on file by the lessee, and a copy is provided for the lessor. This record shows other interested parties that a valid lease exists for that property and prevents possible title challenges on the grounds that the transaction has not been recorded.

TRANSACTIONS AFTER LEASING

Once he has signed a lease and received his bonus, the lessor can go about his business. The lessee may do further exploration, obtain the required drilling permit, arrange for drilling a well, or negotiate other agreements that may be necessary to expedite the exploration and development of the leased property. After production begins, division orders are executed—sometimes by the purchaser and sometimes by the producer, depending primarily on whether the product sold is oil or gas.

Well permits

When the title to a drill site is cleared by the examiner, the operating company in many states must obtain a well permit from the appropriate state agency. The permit stipulates spacing regulations, proposed depth of well, operator's name, and other pertinent information. Once the permit has been approved, drilling can commence.

Division orders

When a well is brought in and production begins, a division order is drafted, based on the terms of the lease, the title opinion, and any subsequent agreements affecting ownership of the oil or gas. The division order gives the names of all parties who have interests in the well (mineral owners, royalty owners, and working interest owners) and their proportionate shares of the payments. In addition to warranting title and guaranteeing correct percentages, the division order gives the purchaser certain rights in handling the oil or gas, accounting procedures, and establishing market values.

The purchaser of *oil* usually handles the division orders and payments to the operator, royalty owners, and all parties holding a guaranteed interest in the minerals. The title to all *gas* usually goes to the lessee when it is produced. When gas is sold, the lessee handles the division orders and payments. Whether prepared by the operator or the purchaser, all division orders must be executed, or signed, by the operator, the royalty owners, and anyone else having an interest in the production. The division order, then, serves as a contract of sale between the mineral owners (usually represented by the operator) and the purchaser.

Support can be offered in the form of money or an assigned interest in the leased property in exchange for drilling a well. A company seeking support money is probably planning to drill on one of its own leases. If it wants to drill a well on someone else's lease, it may suggest trading an assigned interest in the leased property in exchange for drilling. A lessee may farm out some of his leased acreage to a third party who wants to drill a well on it. An agreement that trades drilling obligations for an interest in the property is known as a *farmout* to the granting party (the *farmor*) and as a *farm-in* to the receiving party (the *farmee*) (fig. 3.20).

Support agreements

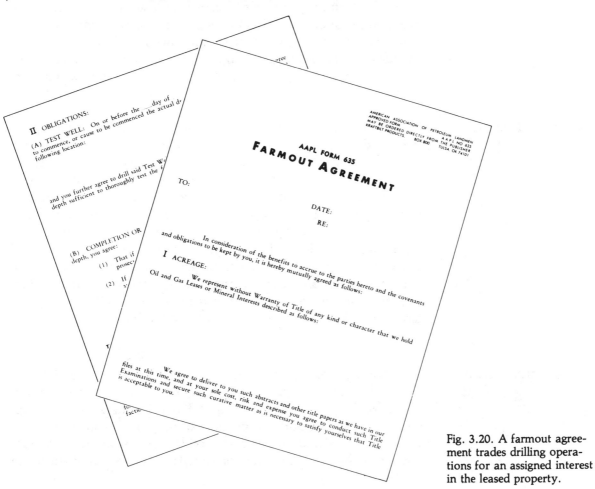

Fig. 3.20. A farmout agreement trades drilling operations for an assigned interest in the leased property.

88

Acreage acquisition agreements

The simplest way to acquire acreage is to purchase the lease. In a lease purchase agreement, one company buys a block of leases from another company that has already set up files and plats of the area. Another transaction common in the industry is the agreement for acquiring acreage with the option to explore and then lease. In order to carry out seismic exploration, a company will secure blocks of acreage from one or more landowners by paying a given price per acre. After gathering exploration data, the company then usually has the option to lease selected acreage that is deemed promising for an additional fee.

Joint operating agreements

Two or more co-owners of the operating rights in a tract of land probably share the costs of exploration and possible development by means of a joint operating agreement. Usually, one of the owners serves as the operator and manages the drilling and related costs. An agreement may take many forms and is usually a complex contract. Joint operating agreements often follow farmouts when the farmor and farmee become co-owners of the rights to drill and produce. Also, joint operating agreements make possible expensive explorations that few individuals could attempt alone. In addition to exploratory operations, pooling and fieldwide unitization require operating agreements of one kind or another.

Joint ventures

Unlike a joint operating agreement, where management is delegated and co-owners are *not* generally liable for certain operator actions, participants in a *joint venture* share liability for third-party claims. Of course, liability claims are settled in courts, and court decisions and legal interpretations vary from state to state and from one agreement to another. The company supporting a drilling venture may be an investment company that represents many individual investors. Most nonoperators are not liable for the operator's actions because clauses disclaiming joint liability are usually written into the agreements. Nonoperators who form a statutory *limited partnership* avoid joint liability but also relinquish any voice in operational matters.

Overriding royalty agreements

Overriding royalty is an expense-free share of the production and thus similar to the royalty received by the lessor, but it is paid out of the working interest rather than the royalty share. For example, a lessee might sell portions of the working interest in the lease to other operators and reserve an override. Because an override does not affect the lessor's interest or royalty, it is not provided for in the lease.

LEASING PUBLIC LANDS

Extensive private ownership of mineral resources is the exception, not the rule, among nations. Since much of the U.S. oil and gas is found on state and federal lands, and Canada's oil and gas largely belongs to the provinces, the lessee of such lands deals with a government leasing agency instead of individuals representing their own interests. These agencies can also provide information on how to lease from cities, counties, school districts, and other political units within the state or province.

Leasing state lands

To lease land owned by a state, certain procedures must be followed, depending upon the state and the region of interest. For example, Texas holds title to lands in various categories such as riverbeds, estuaries, Gulf Coast areas, public school lands, university lands, park lands, and so on. If an area is of interest to an operator or landman, a request is filed with the state, and the Texas General Land Office will determine whether it can be leased and what the minimum terms will be. The Land Office will periodically distribute a *notice for bids* offering certain tracts of the lands and describing procedures and limitations to the development. Bids are mailed in with payment and the Land Office will award the lease to the highest bidder (fig. 3.21).

Fig. 3.21. In some states, state-owned land is leased to the highest bidder on a mail-in basis.

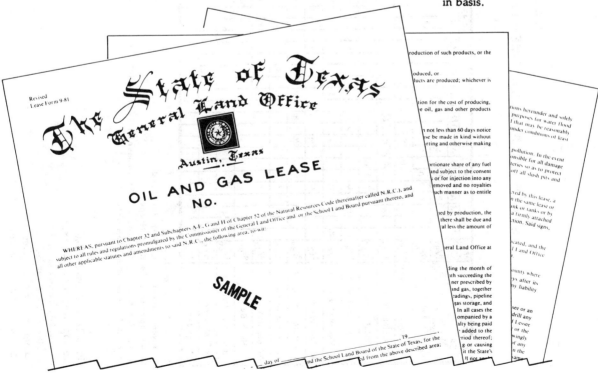

MILESTONE CHART

STATE OF ALASKA FIVE-YEAR OIL AND GAS LEASING PROGRAM

REVISED 6/15/82

Milestone chart with columns grouped by year (1982, 1983, 1984, 1985, 1986), each subdivided into months J F M A M J J A S O N D.

PROPOSED SALE AREA	PROPOSED DATE	Milestones (in sequence)
37 Middle Tanana & Copper River Basins	8/82	P, D, S
37-A Chokok River Exempt	8/82	P, D, S
34 Prudhoe Bay Uplands	9/82	P, D, S — CANCELED
38 Norton Basin	1/83	A H, P, D, S
39 Beaufort Sea	5/83	A H, P, D, S
40 Upper Cook Inlet	9/83	C, R, M, A H, P, D, S
42 Minchumina Basin	1/84	NC, R, M, A H, P, D, S
43 Beaufort Sea	5/84	C, R, M, A H, P, D, S
41 Bristol Bay Uplands	9/84	R, NC, A H, P, D, S
46 Holitna Basin	1/85	M, R, NC, A H, P, D, S
47 Kuparuk Uplands	5/85	M, R, A H, P, D, S
45 Hope Basin	9/85	M, NC, A H, P, D, S
48 Kuparuk Uplands	1/86	M, R, NC, A H, P, D, S
49 Cook Inlet	5/86	M, R, NC, A H, P, D, S
50 Camden Bay	9/86	M, R, A H, P, D, S

Legend:

C = Call For Comments Only
R = Nominations/Comments Received
M = Proposed Sale Area Mapped
A = Draft Social, Economic, and Environmental Analysis (SEEA)
NC = Call For Nominations / Comments

H = Public Meeting On (SEEA)
P = Preliminary Finding / Preliminary .345 (a)(3) Notice/ACMP/ Final SEEA
D = Departments Final Decision/Final .345 (a)(4) (Notice Of Sale And Terms)
S = Sale

Fig. 3.22. In Alaska, basins are approved for a five-year oil and gas leasing program, and leases are awarded by the Commissioner of the Department of Natural Resources.

By contrast, in Alaska where only 2 percent of the land is in private ownership, the state controls development through a leasing program. The Division of Mineral and Energy Management, Department of Natural Resources, administers the process that takes into account social, economic, and environmental analyses of the potential development. Stipulations are added to the lease to minimize adverse impacts. Offshore, for example, certain activities must stop during bowhead whale migration. In some areas, training employees in local customs and concerns may be needed. The basins within the region are identified, and specific tracts within the basins are approved for leasing (fig. 3.22). Potential lessees are informed of the stipulations and bidding procedures. The highest bid gets the lease, although the Commissioner of the Department of Natural Resources may still reject it as insufficient.

Procedures for leasing federal onshore lands for oil and gas were set up in the Mineral Lands Leasing Act of 1920 and the Acquired Lands Act of 1947. Coal, phosphates, oil shales, and certain other resources are leased under these Acts as well. Under the Mining Law of 1872, the government issues land patents for copper, gold, and certain other metallic resources. A lease from the federal government does not convey title but grants the rights to explore, drill, and produce.

Leasing federal onshore lands

The Department of Interior (DOI) is in charge of these resources. Indian lands are generally leased now from the tribal authorities, although the Bureau of Indian Affairs may intercede in certain circumstances.

Leasing of onshore federal lands can take one of three forms: (1) noncompetitive, over-the-counter leasing; (2) noncompetitive lottery, or simultaneous leasing; or (3) competitive bidding. Land that has never been leased before and is not known to cover favorable geologic structures is leased on a first-come, first-served basis. A prospective lessee of such land can simply file for the tract that interests him, pay the filing fee, and submit the necessary forms—over the counter or by mail.

Annual delay rental fees are $1 per acre for the first five years and $3 per acre thereafter. The lease has a ten-year primary term, with 12½ percent royalty.

Land for which a lease has expired, been cancelled, or returned must be reoffered through a *simultaneous leasing procedure*. The Bureau of Land Management (BLM), part of DOI, lists such tracts in a bimonthly publication. All applications received before the published deadline are considered to have been filed simultaneously. The lessee is chosen by *lottery*, or a drawing of all applications submitted. Leases are issued under the same terms as those under the over-the-counter method.

Land over a known, producible structure is leased in a *competitive procedure* (fig. 3.23). The BLM reviews nominations from interested parties and accepts sealed bids. The bidder offering the highest per-acre bonus receives the lease. If the BLM feels all of the bids are too low, it will not award the lease but may reoffer the tract in a later sale. Leases under the competitive bidding procedures have a five-year primary term with delay rental at $2 per acre. Royalty is on a scale of from 12½ to 25 percent, sliding up or down according to the volume of oil and gas produced.

	FORM APPROVED OMB NO. 1004—0074 Expires: April 30, 1985
UNITED STATES DEPARTMENT OF THE INTERIOR BUREAU OF LAND MANAGEMENT	Name of oil and gas field
COMPETITIVE OIL AND GAS AND **GEOTHERMAL RESOURCES LEASE BID** 30 U.S.C. 181 et. seq.; 30 U.S.C. 1001–1025	Known geothermal resources area
	State / Date of sale

The following bid is submitted for ☐ competitive oil and gas ☐ geothermal resources lease on the land identified below.

PARCEL NUMBER OR LAND DESCRIPTION	AMOUNT OF BID	
	TOTAL BID	DEPOSIT SUBMITTED WITH BID

1. Are you a citizen of the United States? ☐ Yes ☐ No

2. If a corporation or other legal entity, specify kind

3. Are you the sole party in interest in this lease? ☐ Yes ☐ No

CHECK APPROPRIATE BOX BELOW CONSISTANT WITH PURPOSE OF THIS BID

☐ I CERTIFY That my interests, direct and indirect, in oil and gas leases in the above State do not exceed 246,080 acres, including the average covered by this bid, of which not more than 200,000 acres are under options. If this bid is submitted for lands in Alaska, I further certify, as above, that by holdings in each of the Alaska leasing districts do not exceed 300,000 acres of which not more than 200,000 acres are under option in each said districts.

☐ I CERTIFY That I am qualified to hold any lease which may issue as a result of this sale under the Geothermal Steam Act of 1970 (84 Stat. 1566) and the regulations thereunder and that my interest in geothermal leases in the above State does not exceed 20,480 acres.

_____ (Signature of Bidder)   _____ (Address of Bidder)

_____ (Type or print name of Bidder)   _____ (City, State, and zip code)

Title 18 U.S.C. Section 1001, makes it a crime for any person knowingly and willfully to make to any department or agency of the United States any false, fictitious or fraudulent statements or representations as to any matter within its jurisdiction.

(Instructions on reverse)   Form 3000—2 (July 1984)

Fig. 3.23. The Bureau of Land Management uses this type of form as an invitation to bid for competitive oil and gas leases.

Most of the onshore leases are issued noncompetitively. Of the 13 million acres leased in fiscal year 1979, about 70,000 acres were leased competitively. Any one corporation or person is limited to leasing a total of 246,000 acres of federal land in any one state except Alaska, where 300,000 acres may be leased in the north and 300,000 more in the south. In all federal leases, once production ends or if the lease is otherwise surrendered, all rights revert to the government.

The federal government controls the area seaward from the states' coastal zones to 200 miles, or about 8,000 feet of water depth. This region is known as the Outer Continental Shelf. OCS leasing is guided by the OCS Lands Act of 1953 as amended. The DOI cites three goals for the leasing program: (1) orderly and timely resource development; (2) protection of the marine, coastal, and human environment; and (3) a fair market rental for the federal estate.

Leasing federal offshore tracts

The Minerals Management Service (MMS) of the U.S. Department of the Interior is now responsible for OCS leasing and production programs and royalty management for both onshore and offshore leases. The MMS was created after the Linowes Commission report of 1982 detailed problems of oilfield theft and irregularities in reporting, collecting, and auditing royalty payments for government and Indian leases.

A five-year schedule of lease sales is set out, with each sale following a series of procedural steps (fig. 3.24). The first step is the call for information. Suggestions are taken for which areas should or should not be leased. DOI then selects the areas for study, and environmental impact statements (EIS) are prepared and published. Public notice is given and affected states comment on the proposals. The tracts and bidding procedures selected are made known when DOI issues the final notice of sale.

Tracts are offered for sale by sealed competitive bids in which the variable item may be the cash bonus, the royalty rate, a share of the net profits, or an exploratory work commitment given in dollars (fig. 3.25). Some tracts are leased with a "sliding scale" royalty based on the volume of production. The royalty rate in all bids is never less than one-eighth, and the minimum share of net profits (when applicable) is never less than 30 percent. An annual rental is charged for each acre leased until production is established. The Secretary of the Interior has delegated bid acceptance or rejection authority to the appropriate Regional Director of the Minerals Management Service.

Federal leases currently supply about 15 percent of the oil produced in the United States. The federal estate includes over one-third of the estimated discoverable and recoverable reserves of oil

FINAL 5-YEAR OCS OIL & GAS LEASING SCHEDULE

U.S. Department of the Interior
Minerals Management Service

Proposed Sale Dates

July 1982

1982

RS-2	August
71 Diapir Field	September
52 North Atlantic	October
69 Gulf of Mexico	October
57 Norton Basin	November

1985

90 S. Atlantic	January
85 Barrow Arch	February
92 N. Aleutian Basin	April
98 C. Gulf of Mexico	May
111 Mid-Atlantic	June
102 W. Gulf of Mexico	August
91 C. & N. California	September
100 Norton Basin	October
94 E. Gulf of Mexico	November

1983

*70 St. George Basin	February
76 Mid-Atlantic	April
72 C. Gulf of Mexico	May
78 S. Atlantic	July
74 W. Gulf of Mexico	August
73 C. & N. California	September
79 E. Gulf of Mexico	November

1986

95 S. California	January
96 N. Atlantic	February
107 Navarin Basin	March
104 C. Gulf of Mexico	April
97 Diapir Field	June
105 W. Gulf of Mexico	July
99 Kodiak	October
101 St. George Basin	December

1984

80 S. California	January
82 N. Atlantic	February
83 Navarin Basin	March
81 C. Gulf of Mexico	April
87 Diapir Field	June
84 W. Gulf of Mexico	July
88 Gulf of Alaska/Cook Inlet	October
89 St. George Basin	December

1987

108 S. Atlantic	January
109 Barrow Arch	February
110 C. Gulf of Mexico	April
86 Shumagin	June

Secretary of the Interior

*The Department will consult with the Alaska Land Use Council following Issuance of the Proposed Notice of Sale.

Fig. 3.24. This five-year schedule for OCS leasing is issued by the Secretary of the Interior. This schedule may be changed as political and economic conditions change.

Fig. 3.25. Bidders come prepared to a federal oil and gas lease sale. (*Courtesy of American Petroleum Institute*)

and gas in the United States, underscoring the growing importance of federal leasing. Foreign companies may participate in federal lease sales where reciprocal agreements allow U.S. nationals to lease in that country. Securing the legal rights for resource development is becoming more sophisticated and international, reflecting global concerns and affecting everyone.

The United States, Canada, and other oil-producing nations have made a conscientious effort to allocate the exploration and exploitation of its mineral resources. Before a company can explore, drill, and produce oil and gas, it must obtain the legal rights to do so. In most oil-producing nations, mineral resources are owned by the national government, and companies must negotiate with government agencies in order to obtain contracts for petroleum development. In the United States, much of the mineral wealth is also publicly owned, although about one-third belongs to the private sector. The rights to explore and exploit these privately owned resources are negotiated by the lessee and lessor, although state and federal laws regulate leasing, as well as the drilling and production activites that follow.

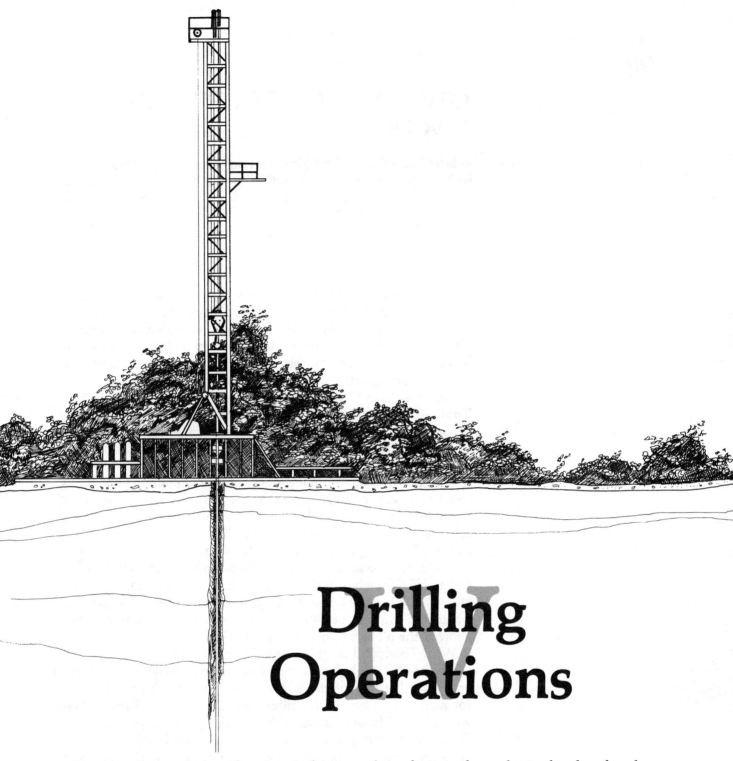

Drilling
Operations

IV

Once an oil company's exploration geologists and geophysicists have obtained and analyzed the data from a prospective petroleum site, the landman has secured a lease, and drilling permits and other preliminary papers are in order, the company turns its attention to the drilling phase of the petroleum industry. A drilling contract is negotiated, and the well is drilled. Drilling the well (known as *making hole* in the industry) according to the operator's specifications is the challenge of a drilling contractor and his crews. A drilling contractor is a company that owns drilling rigs and hires crews whose primary job is the drilling of wells.

DEVELOPMENT OF DRILLING FOR OIL

A brief history of drilling for oil reveals some of the obstacles that early U. S. oilmen faced in getting the oil out of the ground. As new areas were drilled and new problems arose, new methods were developed. Special tools and techniques are still being devised to solve the problems of drilling for today's oil and gas.

Drake's well   In 1857, James M. Townsend, a New Haven banker and the new president of the Pennsylvania Rock Oil Company of Connecticut, decided to send someone to Titusville, Pennsylvania—someone who could turn a headache into a money-making proposition. Townsend had been on the company's board of directors for several years. During this time, the company had leased the oil rights to an island that lay in Oil Creek, about a mile south of Titusville. An oil seep, or spring, bubbled to the surface on the property, but the oil was released in very small amounts.

At this time, the value of petroleum (rock oil) as a lubricant and illuminant was beginning to be recognized. Whale oil, the premium lubricant and main source of oil for illumination during this time, was in short supply and expensive. So it seemed to Townsend and others in his company that if ample quantities of rock oil could be recovered from the Oil Creek site, a profit could be made by selling it as a substitute for whale oil. The problem was that all efforts to accumulate the oil once it bubbled to the surface had failed. Trenches had been dug, dams constructed, and holes excavated by hand, but in every case either rainwater or ground-water had washed out the trenches, holes, and dams and had allowed the oil to run off into the creek.

What Townsend wanted was someone in Titusville to supervise a scheme that he believed would solve the problem. Townsend proposed that the company drill for oil just as others in the area for many years had been drilling for salt water, or brine. These wells allowed large volumes of brine to be pumped to the surface. Once on the surface, the water was evaporated to leave salt—a precious commodity in those days. By December, 1857, Townsend had found his man: an unemployed railroad conductor named Edwin L. Drake (fig. 4.1).

Drake did not have an easy time of it. Almost two years passed before he was able to carry out Townsend's idea. For one thing, Drake had a difficult time locating a driller who was willing to give up a profitable saltwater drilling business to drill for oil—a substance that as far as most drillers were concerned served only to contaminate salt. Then, when Drake finally hired a driller, technical problems began to crop up. The worst had to do with the way in which brine wells were drilled at that time. It was customary to use picks and shovels to dig out the topsoil at the site until bedrock was reached. Then the rig was erected over the hand-dug hole, or cellar, and the well was drilled. But at the Oil Creek site, groundwater persisted in filling up and caving in the cellar long before bedrock was reached, thus making it impossible to fully excavate the cellar.

To solve the problem, Drake got his drilling team—a blacksmith named Billy Smith and his son—to hammer steel pipe into the ground. The pipe, called casing, was driven down to bedrock and prevented the topsoil from caving in. Then the rig was built and drilling was started. By now it was April, 1859. To drill the well, Drake used a steam-powered cable-tool rig. The rig's steam engine turned large pulleys and belts, which in turn caused a large wooden beam—the walking beam—to move up and down, much as a child's rocking horse nods up and down when being ridden. A rope or cable was attached to the front of the walking beam, and a drill bit was connected to the cable. The bit was lowered into the

Fig. 4.1. The United States' first commercial oilwell was drilled near Titusville, Pennsylvania. Edwin L. Drake, in top hat and frock coat, oversaw the operation of this primitive rig in 1859.

Fig. 4.2. Using large pulleys and belts, the oaken walking beam (shown near the top of this photo of the Drake well reconstruction) alternately raised and dropped the bit to make hole.

hole with the cable and the walking beam actuated. As the walking beam rocked up and down, the bit was raised and dropped, raised and dropped, over and over. Each time the bit dropped to bottom, it pierced the rock and made hole (fig. 4.2).

In spite of Drake's being able to overcome the problems and get the well started, the citizens of Titusville remained thoroughly convinced that he was wasting his time and openly referred to the project as "Drake's Folly." Even Townsend had begun to lose faith.

Fig. 4.3. This oilfield on the Benninghoff Farm in Pennsylvania was one of many that sprang up during the 1865 oil boom. (*Courtesy of American Petroleum Institute*)

The company's stockholders had long since given up hope and refused to sink any more money into the project. Townsend had been the only source of capital for months, and in August, 1859, he too decided to call it quits. He sent a letter to Drake instructing him to abandon the well.

Meanwhile, in Titusville, Uncle Billy Smith and Drake continued to drill, unaware that Townsend's fateful letter was wending its way by stagecoach from New Haven. One Sunday afternoon before the letter arrived, Uncle Billy decided to visit the well to check on its progress. The hole was about 69 feet deep and had been drilled to that depth only with a great deal of difficulty. It is easy to imagine the anticipation he must have felt when he saw something glimmering in the pipe lining the well. Could it be oil? Sure enough, it was, and the boom began. The first commercial oilwell in the United States had been drilled. Within a few years, the area in and around Oil Creek was covered with derricks (fig. 4.3).

Since the days of Drake's discovery, drilling methods have changed drastically; indeed, the techniques used in Drake's time have been almost totally supplanted by newer methods. But the fact remains that Drake's well marked the beginning of a boom in oil that has not ceased. It is safe to say that Drake opened an era: the petroleum era.

Cable-tool drilling

One of the earliest methods of drilling, and the one that Drake used, is cable-tool drilling. Cable-tool drilling can be very effective, especially in hard-rock formations. In this drilling method, the drill, or bit, is suspended in the hole by a rope or cable. By means of a powered walking beam, the cable and attached bit are raised and then allowed to drop. This up-and-down motion is repeated over and over, and each time the bit drops, it hits the bottom of the hole with great force to pierce the rock. As it strikes the rock at the bottom of the hole, the chisel point of the bit usually penetrates quite deeply. Since the rate of penetration is high, the drilling rate is fast.

However, two features of cable-tool drilling can be disadvantageous. One is that drilling must be stopped frequently and the bit pulled from the hole so that pieces of rock, or cuttings, chipped away by the bit can be removed. If not removed, the cuttings will impede the bit's ability to drill ahead. The other disadvantage with cable-tool drilling is that the method cannot drill soft-rock formations. The splintered rock fragments tend to close back around the bit and wedge it in the hole.

Fig. 4.4. This cable-tool rig was used to make hole in Kansas in the 1950s.

Fig. 4.5. A wooden derrick was typical of the rigs used for cable-tool drilling from the late 1800s to the 1920s. (*Courtesy of Humanities Research Center, The University of Texas at Austin*)

In spite of its disadvantages, modern versions of cable-tool rigs are still used in a few instances, particularly where shallow wells are drilled or where, for other reasons, using another type of rig is not desirable (fig. 4.4). The rig still makes hole by the impact action of a bit suspended from steel drilling cable, and drilling still ceases when it becomes necessary to bail, or remove, cuttings from the hole.

Although cable-tool rigs are not used very much anymore, in their heyday they drilled a large number of wells. Stationary cable-tool drilling rigs with their pyramid-shaped wooden derricks (fig. 4.5) were a common sight in the oil patch from the 1860s to the 1920s. Then, the portable cable-tool rig, which was smaller and contained many steel components, became standard. But by the late 1950s, even portable cable-tool rigs had all but disappeared from the scene. In spite of the early popularity of the cable-tool method, it has been largely replaced by rotary drilling.

Rotary drilling

The first rotary drilling rig was developed in France in the 1860s, but it did not catch on at first. Because it was erroneously believed that most petroleum was associated with hard-rock formations, which could be very effevicly drilled with cable tools, cable-tool rigs dominated the scene. Then, in the 1880s, two brothers named Baker gained a reputation for drilling successful water wells in the soft formations of the Great Plains of the United States, an area where cable-tool rigs were not having much success. The rig the Bakers used was a rotary unit with a fluid-circulating system. The rotary technique proved equally successful in the unconsolidated soft rocks of Texas, where the Corsicana oilfield was discovered while the drillers were searching for water.

Finally, around 1900, several unsuccessful attempts were made with cable tools to drill the great Lucas well at Spindletop, which lay near Beaumont, Texas. Anthony Lucas, an Austrian-born mining engineer, was convinced that oil did indeed exist under the dome of Spindletop; the problem was how to drill for it. Rotary drilling provided the answer (fig. 4.6). With the advent of Spindletop, where some historians estimate that the well flowed over 80,000 barrels of oil per day (1 barrel equals 42 gallons), rotary drilling was launched in a big way (fig. 4.7).

In rotary drilling, the drilling action comes from pressing the teeth of a bit firmly against the ground and turning, or rotating, it. At the same time the bit is rotated, a fluid, usually a liquid concoction of clay and water called drilling mud, is forced out of special openings, or nozzles, in the bit. The mud jets out of the bit nozzles with great velocity. These jets of mud move cuttings made by the

Fig. 4.6. Rotary drilling at
Spindletop, near Beaumont,
Texas, revolutionized the
drilling industry.

Fig. 4.7. This 1903 oilfield is typical of early oil drilling and producing operations. (*Courtesy of American Petroleum Institute*)

bit teeth away from the teeth, and thereby continuously expose fresh, uncut rock to the teeth (fig. 4.8). Once the cuttings are lifted off bottom, the mud carries them up the hole and to the surface for disposal. Since the cuttings are continuously removed from the hole by the drilling fluid, drilling does not have to stop in order to remove cuttings. Further, because the cuttings, whether from soft or hard rock, are constantly removed, they do not impede the bit's ability to drill ahead. For these reasons, rotary drilling has virtually replaced cable-tool drilling.

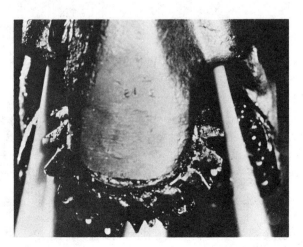

Fig. 4.8. Jets of drilling mud shoot out of the nozzles of a bit to move cuttings away from the bit teeth.

Drilling today   Rotary rigs are by far the most common rigs in the oil patch today (fig. 4.9). Most land rotary rigs are portable—that is, they are designed to be easily moved in and erected to drill the hole and then taken down and moved on to another drilling site. The discovery of oil offshore has led to the development of several different types of rotary rigs for use in marine environments. Some are mobile in that they are floated onto the drill site, the well is drilled, and then they are floated to the next site. Some are immobile in that they are placed on the site and remain there throughout the life of the field. Drilling in arctic regions of the world has led to the development of specially designed rotary rigs that are able to withstand the rigors of extreme cold and the effects of moving pack ice.

Fig. 4.9. A modern rotary drilling rig makes hole in North Texas.

DRILLING CONTRACTS

Regardless of where the well is to be located — on land, offshore, or in polar regions — it is usually drilled by a company known as a *drilling contractor*, which is hired by the oil company, called the *operator*. Because the operating company concentrates on finding and producing oil and gas, it usually does not own drilling rigs. Therefore, it hires a drilling company that has the personnel, rigs, and expertise to do the job. However, since the operator holds the rights to any oil and gas on a lease and is responsible for producing it, the operator owns the drilled well and thus sets the specifications for it.

Usually, the operating company has a person on the site at all times. Called the company representative, or *company man*, he works closely with the contractor's top man to assure that the well is drilled to specifications.

While it is true that a few operating companies own their own rigs and drill their own wells, most wells — about 98 percent — are drilled by drilling contractors.

The drilling contractor

Typically, a drilling contractor's crew consists of a toolpusher, driller, derrickman, and two or three rotary helpers, or rough-necks. Offshore, the contractor also hires several roustabouts.

The toolpusher is the contractor's top man on the drill site. He is responsible for the rig's overall operation and performance, and he must see to it that his crew drills the well to the operator's specifications. The driller is subordinate only to the toolpusher and is the one person who actually operates the rig. He also manages the day-to-day activities of the derrickman and rotary helpers.

The derrickman monitors and records the condition of the drilling mud except when drill pipe is being removed from or put into the hole. At such times, he manipulates the top of the pipe from a small platform high in the derrick or mast of the rig. Rotary helpers manipulate the bottom of the pipe on the rig floor when pipe is being removed from or put into the hole. At other times, they maintain and repair the tools and equipment on the rig. Roustabouts are employed offshore to assist in the loading and unloading of equipment and supplies brought to the rig by boat, and are responsible for keeping the entire rig painted, cleaned, and repaired.

Revised July, 1983

INTERNATIONAL ASSOCIATION OF DRILLING CONTRACTORS
DRILLING BID PROPOSAL
AND
FOOTAGE DRILLING CONTRACT — U.S.

TO: HARD ROCK DRILLING COMPANY
1234 Andrews Hwy.
Odessa, Texas 79799
_____
_____

Please submit bid on this drilling contract form for performing the work outlined below, upon the terms and for the consideration set forth, with the understanding that if the bid is accepted by
XYZ OIL AND GAS COMPANY, INC. _____
this instrument will constitute a contract between us. Your bid should be mailed or delivered not later than ___4:00___ P.M.
on ___July 15___, 19 _85_ to the following address:
4321 Wall Street
Midland, Texas 79701

        *   *   *   *   *   *

THIS AGREEMENT, made and entered into on the date hereinafter set forth by and between the parties herein designated as "Operator" and "Contractor".

OPERATOR: XYZ OIL AND GAS COMPANY, INC.
Address: 4321 Wall Street
Midland, Texas 79701
CONTRACTOR: HARD ROCK DRILLING COMPANY
Address: 1234 Andrews Hwy.
Odessa, Texas 79799

IN CONSIDERATION of the mutual promises, conditions and agreements herein contained and the specifications and special provisions set forth in Exhibit "A" and Exhibit "B" attached hereto and made a part hereof, Operator engages Contractor as an Independent Contractor to drill the hereinafter designated well in search of oil or gas on a footage basis.

For purposes hereof the term "footage basis" means Contractor shall furnish the equipment, labor, and perform services as herein provided to drill a well, as specified by Operator, to the contract footage depth. Subject to terms and conditions hereof, payment to Contractor at a stipulated price per foot of hole drilled is earned upon attaining such contract footage depth or other specified objective. While drilling on a footage basis Contractor shall direct, supervise and control drilling operations and assumes certain liabilities to the extent specifically provided for herein. Notwithstanding that this is a footage basis contract, Contractor and Operator recognize that certain portions of the operations as hereinafter designated, both above and below contract footage depth, will be performed on a daywork basis. For purposes hereof the term "daywork basis" means Contractor shall furnish equipment, labor, and perform services as herein provided, for a specified sum per day under the direction, supervision and control of Operator (which term is deemed to include any employee, agent, consultant, or subcontractor engaged by Operator to direct drilling operations). When operating on a daywork basis, Contractor shall be fully paid at the applicable rates of payment and assumes only the obligations and liabilities stated herein as being applicable during daywork operations. Except for such obligations and liabilities specifically assumed by Contractor, Operator shall be solely responsible and assumes liability for all consequences of operations by both parties while on a daywork basis, including results and all other risks or liabilities incurred in or incident to such operations.

1. LOCATION OF WELL:

Well Name and Number: JOHN DOE No. 1

Parish/County: Ector   State: Texas   Field Name: WILDCAT

Well location and land description: Sec. 19, Blk. 34, HID Survey

The above is for well and contract identification only and Contractor assumes no liability †whatsoever for a proper survey or location stake on Operator's lease.

2. COMMENCEMENT DATE:

Contractor agrees to use best efforts to commence operations for the drilling of well by the ___fifteenth___ day of ___August___, 19 _85_, or within 10 days of completion of site for drilling

3. DEPTH:

Subject to the right of Operator to direct the stoppage of work at any time (as provided in Par. 6), the well shall be drilled to the depth as specified below:

3.1 Contract Footage Depth: The well shall be drilled to ___13,000___ feet or 200' below top of Ellenberger
formation, or to the depth at which the ___5-1/2___ inch casing (oil string) is set, whichever depth is first reached, on a footage basis and Contractor is to be paid for such drilling at the footage rate specified below, which depth is hereinafter referred to as the contract footage depth.

3.2 Daywork Basis Drilling: All drilling below the above specified contract footage depth shall be on a daywork basis as defined herein and Contractor shall be paid for such drilling at the applicable daywork rate specified below.

3.3 Complete Daywork Basis Drilling: If all operations hereunder are performed at applicable daywork rates, provisions of this contract applicable to drilling on a "footage basis" shall not apply.

3.4 Maximum Depth: Contractor shall not be required to drill said well under the term. of this contract below a maximum depth of ___13,500___ feet.

4. FOOTAGE RATE, DAYWORK RATES, BASIS OF DETERMINING AMOUNTS PAYABLE TO CONTRACTOR:

Contractor shall be paid at the following rates for the work performed hereunder.

4.1 Footage Rate: For work performed on a footage basis the rate will be $ _21.50 + tax_ per linear foot of hole drilled determined by steel line measurement from the surface of the ground if Contractor provides cellar, or from the bottom of the cellar if Operator provides cellar, less footage made in regular size hole while working on daywork basis.

4.2 Operating Day Rate: For work performed on a daywork basis the daywork rate per twenty-four hour day with ___six___ man crew shall be:

Depth Intervals		Without Drill Pipe		With Drill Pipe	
From	To				
surface	13,500 ft.	$ 5,400 + tax per day		$ 5,600 + tax per day	
_____	_____	$ _____ per day		$ _____ per day	
_____	_____	$ _____ per day		$ _____ per day	

Using Operator's drill pipe $ _5,400 + tax_ per day.

(U.S. Footage Contract — Page 1)

Fig. 4.10. A footage drilling contract provides for payment by the number of feet drilled.

A drilling bid proposal and contract is the document usually employed to begin the process of getting a well drilled (fig. 4.10). The operator commonly sends the bid proposal and contract to several drilling companies who work in the area in which the well is to be drilled. Generally, the operator selects the contractor who responds with the lowest bid. However, a contractor's past performance and his proven capability to drill are also taken into consideration. Thus, the operator usually selects the contractor on the basis of his ability as well as his price.

Once the operator accepts a contractor's bid, the bid is signed by both parties and becomes a contract. The contract is an agreement between the operator and the contractor that spells out what each is expected to do and provide in order to get the well drilled to specifications. Clauses in the contract cover such items as the location of the well, the date on which the drilling is to commence, the well's depth, the basis of determining the amounts payable to the contractor, the time the amounts are payable, and so on.

One of the more important parts of the contract concerns the specifications of the well as determined by the operator. The contract lays out such specifications as the diameter and depth of each part of the hole, the drilling muds to be used, and the equipment and services to be furnished by the operator as well as the equipment and services to be furnished by the contractor. In short, by means of the contract, the operator states precisely what he expects from the contractor and, by using the same document, the contractor can then respond with a price figure that provides not only a well to meet specifications but also, if all goes well, a profit. As can be imagined, the business of bidding on wells requires skill, patience, experience, and no small measure of luck. Contract drilling is a highly competitive field.

Four basic types of contracts can be drawn up between an operator and a contractor: the *footage* contract, the *daywork* contract, the *turnkey* contract, and the *combination* agreement. Regardless of the type of contract, both the operator and the contractor are always concerned about the time required for completing the job; the safety of the personnel, equipment, and property throughout the operation; and the ability of both men and equipment to do acceptable work.

Bid proposals and specifications

The most commonly used contract is the footage contract. Under this contract, the operator agrees to pay the contractor a stipulated amount for each foot of hole drilled. Put another way, the operator pays the contractor so many dollars per foot of hole,

Footage contract

regardless of how long it takes the contractor to drill it. As an example, consider a hole drilled to a total depth of 10,022 feet. If the rate per foot of hole is $21.50, then the operator would pay the contractor a total of $215,473.00, regardless of how much time it took the contractor to drill to this depth. In footage contracts, the contractor assumes many of the risks of drilling a well. Should something happen that makes it impossible for the rig to drill, and it is not the operator's fault, the contractor stands to lose money, since he is being paid strictly on a footage-rate basis.

Daywork contract

A less commonly utilized contract is the daywork contract. In this case, the operator pays the contractor so much per day for the use of the rig, regardless of what work the rig is performing. Another way to look at a daywork contract is to think of the contractor's being paid by the hour instead of by the foot. Usually, daywork contracts stipulate rates of pay based on the operation the rig is performing. For instance, one rate may apply while the rig is actually drilling, and another may apply while the rig is capable of drilling but is shut down awaiting further orders from the operator.

In general, daywork contracts are drawn when the well to be drilled is in a high-risk area—one in which the formations are, for various reasons, more difficult than usual to drill. Since difficult-to-drill formations may present unusual delays and risks to the contractor and thus cost him more money, he is compensated for the extra costs by means of a daywork contract. Also, a daywork contract may be made when a wildcat well is to be drilled. A wildcat well is a well that is drilled in an area that has not been previously drilled. In such cases, the risks are pretty much unknown, and as a result the contractor may be compensated for the unknown risks with a daywork contract.

Turnkey contract

A turnkey contract requires the operator to pay an agreed-upon amount to the drilling contractor upon completion of the well. In this type of contract, the contractor furnishes all the equipment, material, and manpower needed to drill the well. Further, the contractor controls the entire drilling operation without any on-site supervision by the operator. The contractor assumes all the risks and adjusts the price he charges to reflect these risks. The operator benefits by not having to assume any risks and by receiving only one bill for the entire operation, thus relieving him of accounting expenses. Turnkey contracts are fairly rare but are sometimes awarded to contractors with whom the operator has worked closely in a particular area.

Often, the bases for payment are combined into the final agreement. For example, the operator may pay footage rates to a certain depth and then pay daywork rates for any drilling below that depth. Various clauses in a combination agreement can stipulate the daywork rate for any particular operation. For instance, a standby time rate can be established to compensate the contractor for days when his rig and crew are on the site and able to drill, but for reasons beyond his control no drilling is taking place. This situation can occur when the contractor is waiting for permission from the operator to start testing operations, for the arrival of equipment or material supplied by the operator, for muddy roads to become passable, and so forth. Since most footage contracts contain clauses concerning daywork rates, combination agreements are very popular.

Combination agreements

ROTARY DRILLING SYSTEMS

A rotary drilling rig can be thought of as a portable hole factory. Its sole purpose is to make holes in the ground, and since holes are drilled at many different locations, the rig must be movable. The primary job of a rig and its crew is to place a drill bit in the earth, put weight on it, and turn it. Further, since the cuttings made by the bit must be removed so that the bit teeth can continue to drill, drilling fluid must be circulated as the bit rotates.

As long as a sharp bit is on the bottom of the hole and rotating with weight applied to it and drilling fluid is available to move the cuttings away from the bit and up to the surface, a hole will be made. Making hole is the purpose of a rig, and many pieces of equipment and machinery are required for it to be able to do so. This equipment and machinery can be divided into four main systems: circulating, rotating, hoisting, and power (fig. 4.11).

The circulating system is needed to get drilling mud down the hole through special pipe called *drill pipe* and *drill collars*, out of the bit that is attached to the bottom of the bottommost drill collar, and to the surface (fig. 4.12).

Circulating system

112

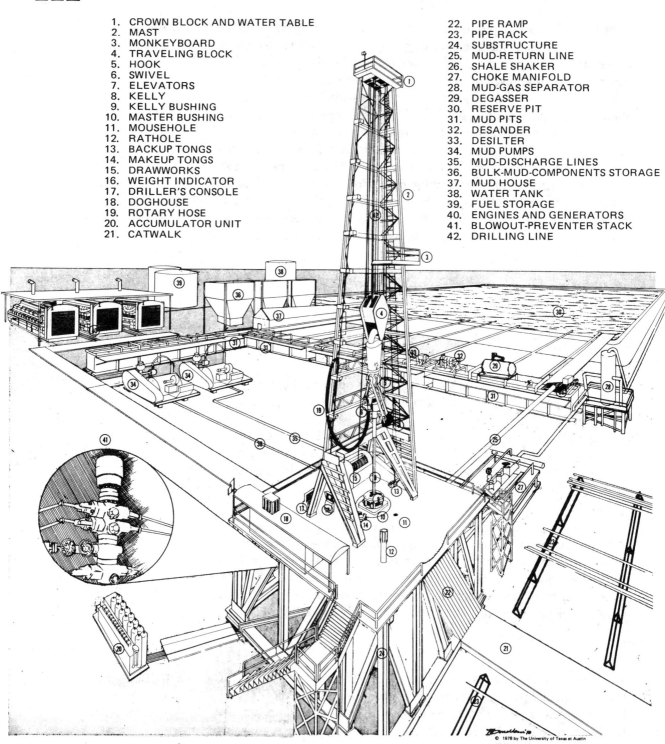

1. CROWN BLOCK AND WATER TABLE
2. MAST
3. MONKEYBOARD
4. TRAVELING BLOCK
5. HOOK
6. SWIVEL
7. ELEVATORS
8. KELLY
9. KELLY BUSHING
10. MASTER BUSHING
11. MOUSEHOLE
12. RATHOLE
13. BACKUP TONGS
14. MAKEUP TONGS
15. DRAWWORKS
16. WEIGHT INDICATOR
17. DRILLER'S CONSOLE
18. DOGHOUSE
19. ROTARY HOSE
20. ACCUMULATOR UNIT
21. CATWALK

22. PIPE RAMP
23. PIPE RACK
24. SUBSTRUCTURE
25. MUD-RETURN LINE
26. SHALE SHAKER
27. CHOKE MANIFOLD
28. MUD-GAS SEPARATOR
29. DEGASSER
30. RESERVE PIT
31. MUD PITS
32. DESANDER
33. DESILTER
34. MUD PUMPS
35. MUD-DISCHARGE LINES
36. BULK-MUD-COMPONENTS STORAGE
37. MUD HOUSE
38. WATER TANK
39. FUEL STORAGE
40. ENGINES AND GENERATORS
41. BLOWOUT-PREVENTER STACK
42. DRILLING LINE

Fig. 4.11. The major components of a rotary drilling rig work together to accomplish its function —making hole.

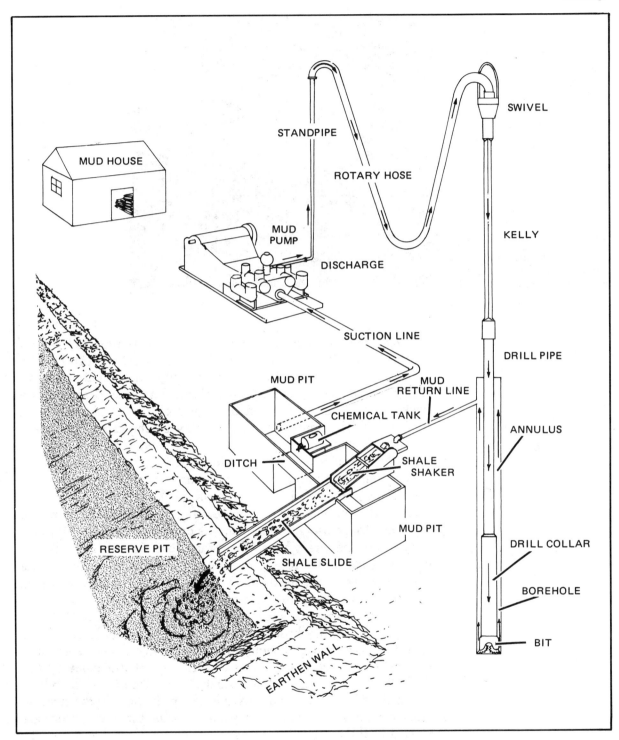

SWIVEL

STANDPIPE

ROTARY HOSE

KELLY

MUD HOUSE

MUD PUMP

DISCHARGE

DRILL PIPE

SUCTION LINE

MUD PIT

MUD RETURN LINE

ANNULUS

CHEMICAL TANK

DITCH

SHALE SHAKER

MUD PIT

RESERVE PIT

DRILL COLLAR

SHALE SLIDE

BOREHOLE

BIT

EARTHEN WALL

Fig. 4.12. The circulating system consists of a number of components, all of which serve to get mud down the hole and back to the surface.

Fig. 4.13. The mud pump pumps drilling fluid through the circulating system.

Circulating equipment. Large, heavy-duty pumps called *mud pumps* are the heart of the circulating system (fig. 4.13). Usually two pumps are installed even though only one at a time is normally in use during drilling operations. The second serves as a backup if the first requires repair. However, if the volume of mud required to be pumped is large, as may be the case with a very deep hole, then the two pumps can be used together, or compounded, to achieve the needed increase in pump capacity.

However, one pump or the other usually picks up mud from steel tanks, or pits, in which the mud is stored, and sends it through a *standpipe* and *rotary hose* (fig. 4.14). The standpipe is a rigid pipe that conducts mud from the pump, up one leg of the derrick, and to the rotary hose. The rotary hose, or kelly hose, is flexible and is connected to a device called the swivel. The rotary hose is flexible because it must move downward as the hole is drilled deeper and the swivel moves closer to the rig floor. It must also be able to move upward when drill pipe is added to the drill string.

Exiting the rotary hose, mud goes into the swivel to which the drill string is attached. The mud then goes down a special length of pipe called the kelly, and enters the drill pipe that is connected to the kelly. The mud goes down the drill pipe, into the drill collars, out of the bit nozzles, and moves back up the hole to the surface. Since the mud picks up cuttings made by the bit, the cuttings are carried in the mud as it returns to the surface. Mud and cuttings return to the surface in the *annulus*, or annular space, between the outside of the drill collars and drill pipe and the inside of the hole.

At the surface, the mud and cuttings leave the well through a pipe called the *mud return line*. The mud and cuttings flow out of

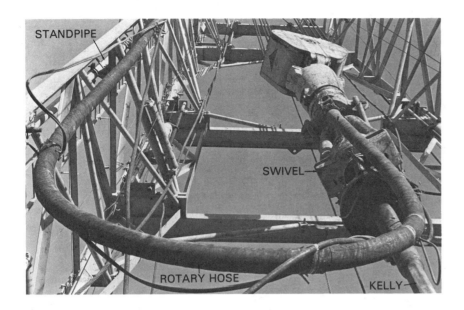

STANDPIPE

SWIVEL

ROTARY HOSE

KELLY

Fig. 4.14. A view from the rig floor shows the components that carry drilling mud to the hole: it travels up the standpipe, through the rotary hose, into the swivel, and down through the kelly.

the return line and onto a vibrating screen, or sieve, called the *shale shaker* (fig. 4.15). The cuttings fall onto the screen of the shaker and are disposed of, but mud falls through the screen and back into the pits. This clean mud is once again picked up by the pumps and sent back down the hole. Normally, mud circulation continues as long as the bit is on bottom and drilling.

Fig. 4.15. Cuttings carried by the mud are removed by two shale shakers on this location.

The cuttings that fall onto the shaker vibrate off and move down a slanted trough called the *shale slide.* They fall off the shale slide and into a pit that is dug out of the soil. This earthen-walled pit, which is often lined with plastic to protect the ground, is called the *reserve pit.* The reserve pit serves mainly as a disposal area. However, because the reserve pit is so much larger in area than the steel pits, sometimes drilling mud may actually be circulated through the reserve pit if the mud contains an unusually large amount of drilled solids. These solids, which are very fine particles of drilled formation, can be allowed to settle out in the reserve pit before the mud is recirculated back into the hole.

Maintaining desirable characteristics of the mud may require additional equipment. For example, a *desander* and *desilter* may be installed to remove very small solids that cannot settle out in the pits, and a *degasser* may be needed to remove entrained gas from the mud (fig. 4.16).

Drilling fluid. Drilling fluid, or mud, is very important to the rotary drilling process. Basically a mixture of water, clay, and special minerals and chemicals, mud performs many important jobs. In addition to removing cuttings from the hole, mud cools and lubricates the bit as it turns on bottom. Further, mud exerts pressure inside the hole. This pressure keeps fluids that may be in the formation from entering the hole and perhaps blowing out to the surface. In addition, pressure in the hole forces solid particles of clay in the mud to adhere to the sides of the hole as the mud circulates upward on its way to the surface. The solids form a thin, impermeable cake on the walls of the hole. This wall cake plasters the hole and, since it is impermeable, prevents the liquid part of the mud from going into the formations that the hole penetrates,

Fig. 4.16. Additional circulating equipment includes a degasser, desilter, and desander, which are located over the mud pits downstream from the shaker.

thereby eliminating the need to continually add water or other liquid to the mud. Wall cake also stabilizes the hole; that is, it prevents the hole from caving in (fig. 4.17).

Because it is so important, mud is carefully formulated to do the best job on the particular well being drilled. Once formulated, its properties are monitored and adjusted as necessary during the drilling of the well. Some of the mud properties that are closely monitored are its viscosity, or resistance to flow; its weight, or density; its filtration rate, or water-loss properties; and its solids content.

The mud viscosity is important because viscosity affects the ability of the mud to carry cuttings. Its weight affects its ability to prevent formation fluids from blowing out, and its filtration rate affects its ability to build an effective wall cake to prevent makeup water and other liquids from being lost to the formations. Its solids content is important because the amount of solids in the mud affects the rate of penetration of the bit: the more solids in the mud, the slower will be the rate of penetration. Excessive solids slow the penetration rate because they add unnecessary weight to the mud. Weight increases the mud pressure on bottom, and this pressure holds the cuttings made by the bit to the bottom, thus impeding the bit's ability to drill the formation.

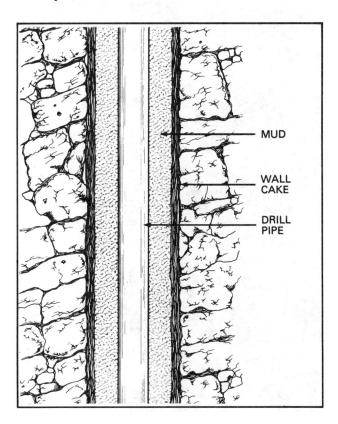

Fig. 4.17. Solid particles in the drilling mud plaster the wall of the hole and form an impermeable wall cake.

Rotating system

Since a large part of drilling involves rotating the bit, a rotating system is needed. The main part of this system is a powerful machine called the *rotary table.* Located on the floor of the rig, the rotary table is capable of creating a strong rotating force, or torque. Since the bit is suspended at the bottom of a hole that could be thousands of feet deep, and since the rotary table is on the surface, some method of transmitting the torque from the table to the bit is required. Additional equipment is therefore needed.

Rotating equipment from the surface to the bottom of the hole consists of the swivel, kelly, rotary table, drill pipe, drill collars, and bit. The kelly, drill pipe, and drill collars are collectively called the *drill stem.* The drill pipe by itself is termed the *drill string.* Frequently, however, drilling personnel refer to the entire drill stem as the drill string.

The swivel. The swivel is a piece of equipment to which the rotary hose and drill stem are connected. Even though the swivel does not rotate, it permits the drill stem to rotate and provides a passageway for mud to enter the drill stem. A large bail, or handle, on the swivel fits into a hook on a large set of pulleys, or sheaves, called the *traveling block.* The swivel and attached drill stem are thus suspended from the *hook* (fig. 4.18).

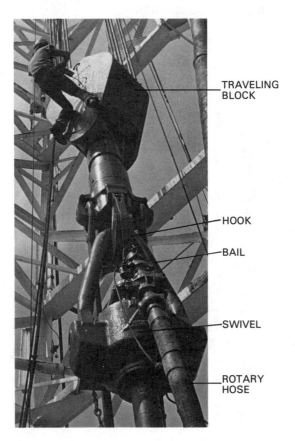

TRAVELING
BLOCK

HOOK

BAIL

SWIVEL

ROTARY
HOSE

Fig. 4.18. The drill stem is attached to the bottom of the swivel, which hangs from the hook on the traveling block. Drilling mud enters the swivel through the rotary hose.

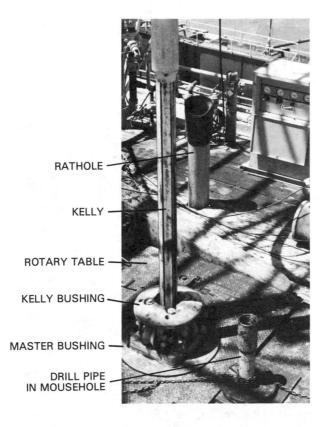

RATHOLE

KELLY

ROTARY TABLE

KELLY BUSHING

MASTER BUSHING

DRILL PIPE
IN MOUSEHOLE

Fig. 4.19. The kelly passes through the kelly bushing, which fits into the master bushing of the rotary table. Note the rathole where the kelly and kelly bushing can be stored when they are disconnected from the drill stem. Also, note the joint of drill pipe in the mousehole ready for a connection.

The kelly. Attached to a threaded connection on the bottom of the swivel is the kelly. The kelly is a special section of pipe that is available in lengths of 40, 46, and 54 feet. Unlike most pipe, the kelly is not round; rather, it has flattened sides so that in cross section it has a square or hexagonal shape.

The four- or six-sided kelly fits inside a corresponding square or hexagonal opening in a device called the *kelly bushing.* The kelly bushing in turn fits into a part of the rotary table called the *master bushing.* Powered gears in the rotary table rotate the master bushing. As the master bushing rotates, the kelly bushing also rotates. The square or hexagonal opening in the kelly bushing fits against the square or hexagonal kelly and causes the kelly to turn. The turning kelly rotates the drill stem and thus the bit (fig. 4.19). Since the kelly slides through the opening in the kelly bushing, the kelly can move down as the bit drills the hole deeper.

The rotary table. The rotary table, equipped with its master bushing and kelly bushing, supplies the necessary torque to turn the drill stem. In addition, when the kelly and kelly bushing are removed, the hole left in the master bushing accommodates a special set of gripping devices called *slips.* Slips have teethlike gripping elements called *dies* that are placed around the drill pipe to

Fig. 4.20. Crewmen grasp the slips by the handles as they set them in the master bushing.

keep it suspended in the hole when the kelly is disconnected (fig. 4.20). The rotary table can be locked to keep it from turning when desired, as when a new bit is installed on the drill collars (fig. 4.21).

Fig. 4.21. The rotary table locks in place, allowing a new bit to be installed.

Power swivel. A recent innovation in rotary drilling is a device called a power swivel, or top-drive system (fig. 4.22). This system eliminates the need for a kelly and a rotary table. A heavy-duty motor incorporated into the swivel turns a drive shaft that is connected directly to the drill stem. When the motor is actuated, the drill stem and bit are rotated by the motor. The main advantage of a power swivel over the conventional kelly and rotary table system lies in the area of pipe handling and drilling efficiency. With the conventional system, joints of pipe must be added to the drill stem one at a time as the hole deepens. With a power swivel, pipe can be added three joints at a time. Adding three-joint stands of pipe saves time in making connections.

Drill pipe and drill collars. Although aluminum drill pipe is available, most drill pipe and drill collars are steel tubes that rotate with the kelly or power swivel. Drilling mud is pumped to the bottom of the hole through them. Drill pipe and drill collars come in sections, or joints, about 30 feet long (fig. 4.23). Other dimensions of interest are the outside diameter and weight. Drill pipe is available in many different diameters, but the most commonly used diameters are 4, 4½, and 5 inches. Drill pipe weight varies with its diameter, but a commonly used 4½-inch drill pipe weighs 16.6 pounds per foot. Thus, a 30-foot joint of such drill pipe weighs 498 pounds. Since a drill collar joint is usually larger in

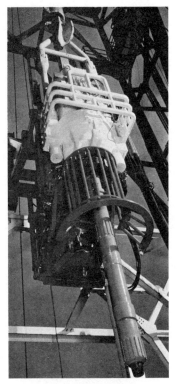

Fig. 4.22. A power swivel, or top-drive unit, eliminates the need for a kelly and rotary table. (*Courtesy of Hughes Drilling Equipment*)

Fig. 4.23. Available in 30-foot joints, this drill pipe is being set back on the rig floor.

diameter than a drill pipe joint and has thicker walls, a drill collar joint is heavier. For example, one available type of drill collar joint is 7 inches in diameter and weighs 125 pounds per foot. Thus, a 30-foot joint of a drill collar this size weighs 3,750 pounds, or over 1¾ tons. Drill collars are heavy because they are used to put weight on the bit.

Since several joints of drill pipe and drill collars must be joined together, or made up, as the hole gets deeper, threaded connections are provided. On drill pipe the threaded connections are called *tool joints.* Tool joints are steel rings, or collars, that are welded to each end of a joint of drill pipe. One tool joint is a pin (male) connection, and the other is a box (female) connection. The pin of one joint fits into the box of another joint, thus allowing long strings of pipe to be made up.

Because drill collars have thicker walls than drill pipe, adding tool joints to them is not necessary. Rather, the pin and box connections are machined directly into the steel of the drill collar wall. Just as tool joints on drill pipe allow the joints of pipe to be connected together, drill collar connections provide a way for the drill collar joints to be connected together.

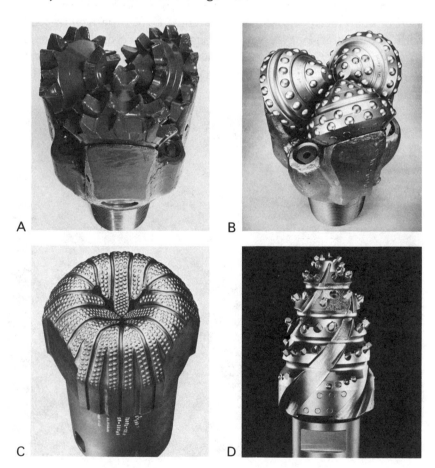

Fig. 4.24. Roller cone bits are available with milled teeth (*A*) or with carbide inserts (*B*). Several industrial diamonds are used as the cutting elements in a diamond bit (*C*). Polycrystalline diamond inserts are used in a PCD bit (*D*).

The bit. At the bottom of the drill stem is the bit, which drills the formation rock and dislodges it so that drilling fluid can circulate the fragmented material back up to the surface where it is filtered out of the fluid. In general, bits are chosen according to the hardness of the formation to be drilled. The two most common types of bits are roller cone bits and diamond bits (fig. 4.24). They range in size from 3¾ inches in diameter to 26 inches in diameter, but some of the most commonly used sizes are 17½, 12¼, 7⅞, and 6¼ inches. Bits are available in many different sizes because different sizes of holes must be drilled.

Roller cone bits usually have three (but sometimes two or four) cone-shaped steel devices that are free to turn as the bit rotates. Several rows of teeth on each cone scrape, gouge, or crush the formation as the teeth roll over it. The teeth are either machined out of the steel alloy comprising the cone, or very hard pellets of tungsten carbide are inserted into holes drilled into the cones.

The cones rotate on bearings. Some bits have ball and roller bearings (fig. 4.25); others have plain, or journal, bearings (fig. 4.26). Journal bearing bits generally do not wear out as fast as ball and roller bearing bits but are more expensive. Some ball and roller bearing bits and all journal bearing bits are sealed. The seals protect the bearings from being abraded by the drilling mud that surrounds the bit at the bottom of the hole. However, in unsealed bits, drilling mud provides the only lubrication to the bit bearings. Since drilling mud cannot lubricate sealed bearing bits, a small reservoir of grease is built into the body of sealed bits.

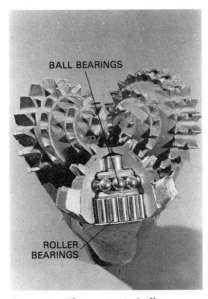

Fig. 4.25. The cones on ball and roller bearing bits rotate on the bearings.

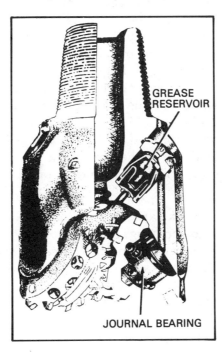

Fig. 4.26. The cones on this bit rotate on journal bearings. (*Courtesy of Hughes Tool Company*)

Most roller cone bits are jet bits in that drilling fluid exits from the bit through nozzles that are placed between the cones. Just as a nozzle on a garden hose creates a high-velocity stream of water, the nozzles on the bit create high-velocity jets of mud. The jets of mud strike the bottom of the hole and help lift cuttings away from the bit so that they will not impede drilling.

Diamond bits do not have cones or teeth. Rather, small industrial diamonds are embedded into the sides and bottom of the bit. When the bit is rotated on bottom with weight, the diamonds shear the formation rock. Diamond bits are available for drilling soft, medium, and hard formations. Their initial cost is higher than roller cone bits, but sometimes the higher cost can be offset because a diamond bit may not become dull as quickly as a roller cone bit.

A special type of diamond bit is the *polycrystalline diamond* (PCD) bit. A layer of synthethic polycrystalline diamonds is bonded to tungsten carbide inserts that are pressed into holes drilled in the body of the bit. PCD bits are relative newcomers to the oil patch but appear to perform well. In spite of their relatively high cost when compared to roller cone bits, in some cases they can stay on bottom and drill for longer periods of time, thus offsetting their high initial cost.

Hoisting system

In order for any type of bit to drill, it must have weight applied to it. The weight is applied by the drill collars, which are allowed to press down on the bit. Even though drill pipe is made up on top of the drill collars, drill pipe is not used to put weight on the bit. Because drill pipe has relatively thin walls, it would bend and flex to the point of failure if it were used to put weight on the bit. Therefore, drill pipe must be kept in tension; an upward pull must be maintained on it. The hoisting system is responsible for keeping tension on the drill pipe while weight is applied to the bit by the drill collars. The hoisting system is also required for tripping the drill stem and bit in and out of the hole. The hoisting system is made up of the traveling and crown blocks, the drilling line, the drawworks, and the mast or derrick (fig. 4.27).

Blocks and drilling line. The swivel to which the drill stem is made up hangs from a hook on the bottom of the traveling block. The traveling block is a set of sheaves through which is reeved (threaded) a large wire rope called the *drilling line*. The drilling line is reeved through the sheaves of the *traveling block* and the sheaves of the *crown block* (fig. 4.28). The crown block is a stationary set of sheaves that is located at or near the top of the mast or derrick. The line is reeved several times between the traveling

125

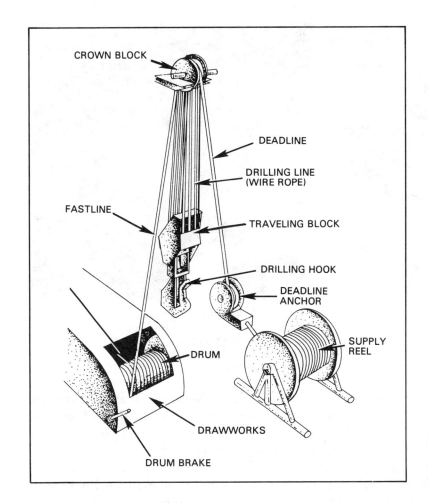

CROWN BLOCK

DEADLINE

DRILLING LINE
(WIRE ROPE)

FASTLINE

TRAVELING BLOCK

DRILLING HOOK

DEADLINE
ANCHOR

SUPPLY
REEL

DRUM

DRAWWORKS

DRUM BRAKE

Fig. 4.27. A rotary rig hoisting system is shown with derrick or mast removed for clarity.

A

B

Fig. 4.28. Drilling line passes several times through the traveling block (A) and the crown block at the top of the mast (B) through grooved sheaves in each.

Fig. 4.29. Drilling line is spooled around the drum, or spool, in the drawworks.

Fig. 4.30. A deadline anchor on the rig substructure holds the deadline firmly in place. The loop at the bottom goes to the supply reel.

block sheaves and the crown block sheaves so that even though the line is one continuous piece, the effect is that of several lines—8, 10, or 12. The hoisting capacity of the line is multiplied several times by running it between the blocks.

One end of the line is wound, or wrapped, several times around a *drum*, or *spool*, in the drawworks that rests on the rig floor, or in some cases on the ground (fig. 4.29). The end of the line that is wrapped onto the drawworks drum is the *fastline*, so called because it rapidly moves on and off the drawworks drum as the traveling block is raised or lowered.

The other end of the drilling line runs downward from the crown block and is firmly secured to a *deadline anchor* that is usually located in the substructure of the rig (fig. 4.30). The part of the line that runs from the crown block to the anchor is called the *deadline* because it does not move during hoisting or drilling operations. The drilling line comes off the deadline anchor and onto a *supply reel* that is usually located on the ground near the rig (fig. 4.31).

With perhaps 2,200 feet of wire rope in use as drilling line, the remainder on the supply reel is used in scheduled slip-and-cut programs that change locations of wear and stress to lengthen wire rope life. When the line has moved a ton of load over the distance of 1 mile, it is said to have given a ton-mile of service. Ton-mile records are carefully kept in order to employ a satisfactory cutoff program for good service life.

Fig. 4.31. This skid-mounted reel supplies wire rope for the rig. The electric motor returns the drilling line to the reel when the rig is moved to a new location.

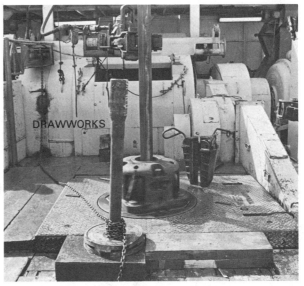

Fig. 4.32. The drawworks is a large hoist that rests on the floor of this rig. In the foreground can be seen a joint of pipe in the mousehole; behind it is the kelly and kelly bushing.

Drawworks. The drawworks is a large hoist with a drum around which drilling line is wrapped (fig. 4.32). A brake on each end of the drum holds it stationary and sustains the great weight of the traveling block and drill stem. By releasing the brake and applying power to the drum, the drill stem can be raised and lowered. If line is taken onto the drum, the traveling block is raised. If line is let off the drum, the traveling block is lowered.

Extending out of each end of the drawworks is a powered shaft called a *catshaft*. On each end of the catshaft is attached a *cathead*. On one side of the drawworks is the makeup cathead (fig. 4.33), and on the other side is the breakout cathead. The makeup cathead is used to apply tightening force to large wrenches called *tongs* that are used to make up and tighten joints of drill pipe and drill collars. A chain attached to the end of the tongs runs to the makeup cathead, and when the cathead is actuated, the chain is pulled tight. Pulling in the chain pulls on the tongs and causes them to tighten the pipe on which the tongs are latched.

The breakout cathead is used to apply loosening force to break out joints of pipe. A wire rope attached to the end of the tongs runs to the breakout cathead and applies loosening torque to the tongs and pipe. The makeup and breakout catheads are often called automatic, or mechanical, catheads.

Fig. 4.33. The makeup cathead is attached to a catshaft coming out of the drawworks. Note the chain entering the cathead; it is used to apply tightening (makeup) force to drill pipe.

A small spool is also attached to the catshaft next to each automatic cathead. The spool forms a friction cathead around which large cloth-fiber rope, or soft line, can be wrapped. With a friction cathead and soft line, a crew member can move fairly heavy items of equipment on the rig floor. However, many contractors install small, pneumatically operated hoists called *air hoists*, or air tuggers, to perform many of the light hoisting tasks on the rig floor that formerly were done with friction catheads (fig. 4.34).

Derrick or mast. A derrick or mast is the steel towerlike structure that makes a drilling rig so distinctive (fig. 4.35). The purpose

Fig. 4.34. A crewman uses an air hoist to move light equipment on the rig floor.

Fig. 4.35. This mast supports the hoisting system of the rig.

of the derrick or mast is to support the traveling and crown blocks and the enormous weight of the drill stem. A derrick or mast also supports the drill pipe and drill collars when they are pulled out of the hole and set back.

A derrick differs from a mast in that the legs of the derrick rest on each corner of the rig floor, and once assembled it must be disassembled in order for it to be moved. On the other hand, a mast fits into an A-frame that rests on the rig floor or on the ground. Further, a mast is portable; that is, once it is assembled, it may be folded or telescoped down and transported as a single unit. (However, the mast is sometimes taken apart in sections to facilitate moving it.) Although a mast and derrick appear to be much the same, and both serve the same purpose, a close examination of the two reveals many differences (fig. 4.36).

Most land rigs utilize masts because masts are so much easier to move than derricks. On offshore rigs, however, derricks are widely used, because the derrick moves as the entire rig moves, and assembling and disassembling the derrick is unnecessary.

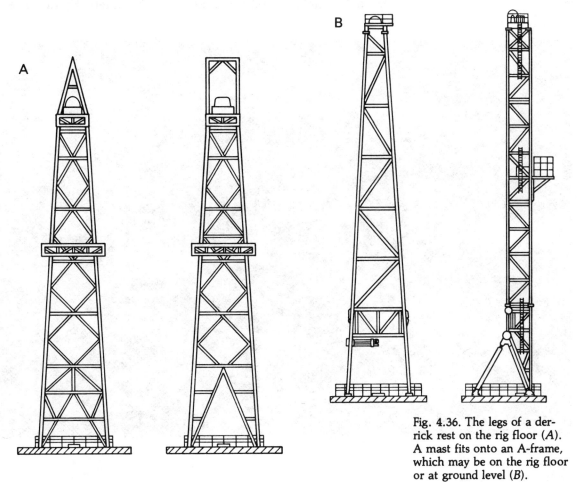

A

B

Fig. 4.36. The legs of a derrick rest on the rig floor (A). A mast fits onto an A-frame, which may be on the rig floor or at ground level (B).

Power system

A drilling rig needs power to run the circulating, rotating, and hoisting systems. Usually, internal-combustion engines are used as *prime movers* (prime power sources). Also, some method of transferring the power made by the engines to a particular component, such as the mud pumps, drawworks, or rotary table, is required.

Engines. Since the power requirements of a rig are large, several engines are usually employed because one engine cannot provide all the needed power. Most rigs require from 1,000 to 3,000 horsepower (hp), which is provided by two or more engines, depending on well depth and rig design (fig. 4.37). Shallow or moderate-depth drilling rigs will need from 500 to 1,000 hp for hoisting and circulation. Heavy-duty rigs for 20,000-foot holes are usually in the 3,000-hp class. Auxiliary power for lighting and the like may be from 100 to 500 hp.

Today, most prime movers are diesel engines. Diesel engines are compression-ignition engines in that they do not use spark plugs to ignite the fuel-air mixture in the engine cylinders. Rather, heat that

Fig. 4.37. Four engines located in the substructure provide the power for this rig.

is generated by compressing the fuel-air mixture is used as a source of ignition. Natural gas and liquefied petroleum gas (LPG) engines are spark-ignition engines in that spark plugs are required to ignite the fuel-air mixture. Diesel engines have become more popular than gas or LPG engines because diesel fuel is more readily available and is more easily transported than gas or LPG.

Power transmission. In general, two methods are used to transmit power from the engines to the components that require the power. Both methods have advantages and disadvantages, and the choice of which type to use is mainly a matter of preference on the contractor's part. One is the *diesel-electric system* in which diesel engines drive electric generators that are mounted directly on the engine (fig. 4.38). As the engine runs, the generator generates electric power. This electric power is transmitted by cables to electric motors that are mounted on or near the component requiring the power (fig. 4.39). The motors drive the equipment. Electric power from each engine-generator set to the motors

Fig. 4.38. A high-voltage electric generator is driven by the engine to which it is coupled.

Fig. 4.39. A heavy-duty electric motor is mounted on or near the component requiring power. Note the electric cables at left that receive electric power from the generators.

132

is controlled by the driller from his position on the rig floor (fig. 4.40). If more power is needed for a particular component than can be provided by one engine, the driller can engage controls at his work station to compound the engines—that is, the power from two or more engines can be assigned to the motors on the component.

Fig. 4.40. Electric power is assigned to each component by the driller from his position on the rig floor.

The other way in which engine power is transmitted to components is by means of a *mechanical compound.* Large pulleys are mounted to each diesel or gas engine, and these pulleys are connected by chains and belts to each rig component requiring power. The pulleys, chains, and belts are called the compound because they allow the power from each engine to be combined, or compounded (fig. 4.41). As is the case with diesel-electric drive, mechanical-drive rigs are also controlled by the driller from his position on the rig floor.

Fig. 4.41. Three engines are compounded by pulleys, belts, and chains. This package is being assembled in a shop prior to being mounted on a rig.

ROUTINE DRILLING OPERATIONS

Once the legal aspects of drilling a well have been taken care of—leases acquired, contracts signed, and so on—the operating company determines an exact spot on the surface for the well to be drilled. At this time it becomes necessary to prepare the drill site, rig up, and begin making hole.

To accommodate the rig and equipment, the drill site must be prepared. If necessary, the land is cleared and leveled, access roads are built, and the reserve pits are dug. In environmentally sensitive areas, such as Alaska and California, a large effort is made not to alter the surface area comprising the drill site more than is necessary. For example, reserve pits may not be dug. Instead, large steel bins are placed on the site to receive the cuttings and other materials that are normally dumped into the reserve pits. These bins can then be trucked away from the site and the material inside them disposed of properly. Also, even in areas where reserve pits are excavated, they are often lined with thick plastic sheeting to prevent any contaminated water or other materials from seeping into the ground.

Preparing the drill site

Since a source of fresh water is required for the drilling mud and for other purposes, a water well is sometimes drilled prior to moving the rig onto the location. If other sources are available, the water may be piped or trucked to the site.

At the exact spot on the surface where the hole is to be drilled, a rectangular pit called a *cellar* is dug, or culvert-like pipe is driven into the ground. If the cellar is dug, it may be lined with boards, or forms may be built and concrete poured to make walls for the cellar. The cellar is needed to accommodate drilling accessories that will be installed under the rig later.

In the middle of the cellar, the top of the well is started, sometimes with a small truck-mounted rig. This hole—the conductor hole—is large in diameter, perhaps as large as 36 inches or more, about 20 to 100 feet deep, and is lined with conductor casing, which is also called conductor pipe. If the topsoil is soft, the conductor pipe may be driven into the ground with a pile driver. In either case, the conductor casing keeps the ground near the surface from caving in. Also, it conducts drilling mud back to the surface from the bottom when drilling begins; thus the name *conductor pipe* (fig. 4.42).

Usually, another hole considerably smaller in diameter than the conductor hole is dug beside the cellar and also lined with pipe. Called the *rathole*, it is used as a place to store the kelly when it is

Fig. 4.42. This cellar is lined with boards with the conductor pipe in place.

134

temporarily out of the borehole during certain operations. Sometimes on small rigs, a third hole, called the *mousehole*, is dug. On large rigs, it is not necessary to dig a mousehole because of the rig floor's height above the ground. In either case, the mousehole is lined with pipe and extends upward through the rig floor and is used to hold a joint of pipe ready for makeup.

Rigging up With the site prepared, the contractor moves in the rig and related equipment. The process, known as rigging up, begins by centering the base of the rig—the substructure—over the conductor pipe in the cellar. The substructure supports the derrick or mast, pipe, drawworks, and sometimes the engines. If a mast is used, it is placed into the substructure in a horizontal position and hoisted upright (fig. 4.43). A standard derrick is assembled piece by piece on the substructure. Meanwhile, other drilling equipment such as the mud pumps are moved into place and readied for drilling.

Other rigging-up operations include erecting stairways, handrails, and guardrails; installing auxiliary equipment to supply electricity, compressed air, and water; and setting up storage facilities and living quarters for the toolpusher and company man. Further, drill pipe, drill collars, bits, mud supplies, and many other pieces of equipment and supplies must be brought to the site before the rig can make hole.

Fig. 4.43. A mast is raised to an upright position, using the drawworks.

With all the preparations complete, drilling is ready to begin; in oilfield parlance, the hole is ready for *spudding in*. To spud in, a large bit, say 17½ inches in diameter as an example, is attached to the first drill collar and is lowered into the conductor pipe by adding drill collars and drill pipe one joint at a time until the bit reaches the bottom. With the kelly attached to the top joint of pipe, the pump is started to circulate mud, the rotary table is engaged to rotate the drill stem and bit, and weight is set down on the bit to begin making hole. Normally, when rotating the bit, the rotary table turns to the right. Thus, when the bit is on bottom and making hole, the expression "on bottom and turning to the right " indicates that drilling is proceeding.

As the bit drills ahead, the kelly moves downward through the kelly bushing, and since the ground near the surface is normally fairly soft, the kelly is soon *drilled down*; that is, its entire length reaches a point just above the bushing.

To drill the hole deeper, a *connection* must be made; another joint of drill pipe must be added to the string to make it longer. To make a connection, the kelly and attached drill string are picked up off bottom by means of the hoisting system. When the tool joint of the topmost joint of pipe clears the rotary — passes through the opening in the rotary table to a point just above the rig floor — the slips are placed around the pipe and into the opening in the master bushing. The slips grip the pipe and keep it from falling back into the hole when the kelly is broken out, or removed, from the drill string.

To break out the kelly, two sets of tongs are used. One set of tongs is latched around the bottom of the kelly where it joins the drill pipe. These tongs are the *breakout* tongs. The other set is latched around the tool joint of the drill pipe and are the *backup* tongs. A line is attached to the end of the breakout tongs and goes to the breakout cathead on the drawworks. When the cathead is engaged, it pulls on the line to cause the tongs to apply loosening torque on the kelly. A line is also attached to the backup tongs. It goes to a secure mounting point on the rig and keeps the backup tongs and the pipe on which they are latched from turning as torque is applied by the breakout tongs.

Once the kelly is broken out, the tongs are removed and the drill pipe is spun out of the kelly by turning the rotary table. When the kelly spins out of the drill pipe and releases, it is moved over to a 30-foot joint of drill pipe resting in the mousehole. The pin of the kelly is then stabbed into the box of the new joint, and the two are screwed together, or *made up*. In many cases, a *kelly spinner*, which is an air-actuated motor mounted near the top of the kelly, is used to make up the kelly on the joint of pipe. Backup tongs are latched onto the joint of pipe in the mousehole to keep it from turning as the kelly spinner turns the kelly into the joint.

Once the kelly is made up tightly to the joint, the two are picked up and moved from the mousehole to the rotary table. The bottom of the new joint of pipe is stabbed into the top of the joint of pipe coming out of the borehole, and the kelly spinner is again actuated to make up the joints. With the new joint made up, the slips are pulled, and the pipe is lowered until the bit nears the bottom. Then the pumps are started, rotation is begun, weight is applied to the bit, and another 40 feet or so of hole is drilled, depending on the length of the kelly. Each time the kelly is drilled down, a connection has to be made.

Eventually, at a depth that could range from hundreds of feet to a few thousands of feet, drilling comes to a temporary halt, and the drill stem is pulled from the hole. This first part of the hole is known as the surface hole. Even though the formation that contains the hydrocarbons may lie many thousands of feet below this point, drilling ceases temporarily because steps must now be taken to protect and seal off the formations that occur close to the surface. For example, freshwater zones must be protected from contamination by drilling mud. To protect them, special pipe called *casing* is run into the hole and cemented.

Tripping out

The first step in running casing is to pull the drill stem and bit out of the hole. Pulling the drill stem and bit out of the hole in order to run casing, change bits, or perform some other operation in the borehole is called *tripping out.* To trip out, the driller stops rotation and circulation. Then, by manipulating controls on the drawworks, he raises the drill stem off the bottom of the hole until the top joint of drill pipe clears the rotary table. As the pipe is held there, the rotary helpers, or floormen, set the slips around the drill pipe to suspend it in the hole. Next, by using the tongs, the floormen break the kelly out of the drill string. Finally, the kelly is put into the rathole. Since the kelly bushing, swivel, and rotary hose are left on the kelly when it is placed in the rathole, the area above the rotary where the top of the drill string protrudes from the hole is left clear (fig. 4.44). Only the traveling block hangs above the drill pipe suspended in the hole.

Attached to the traveling block are a set of drill pipe lifting devices called *elevators.* The elevators usually remain attached to the traveling block at all times and swing downward into position when the swivel is removed from the hook of the traveling block. Elevators are gripping devices that can be latched and unlatched around the tool joints of the drill pipe. The crew latches the elevators around the drill pipe, and the driller raises the traveling

Fig. 4.44. While tripping out, the kelly and related equipment rest in the rathole (*center*), and the stand of pipe just removed from the hole is set back (*at left*).

Fig. 4.45. Tripping out nears completion as the bit comes out of the hole and floormen attach tongs to the drill collar to break it loose from the bit.

block to pull the pipe upward. When the third joint of pipe clears the rotary table, the rotary helpers set the slips and use the tongs to break out the pipe. The pipe is usually removed in stands of three joints. Removing pipe in three-joint stands, rather than in single joints, speeds the tripping out process. With the stand of pipe broken out, the crew guides it into position on the rig floor to the side of the mast or derrick.

Once the bottom of the stand of pipe is set down on the rig floor, the derrickman goes into action. Standing on a small platform called the *monkeyboard* that is about 90-feet high in the mast or derrick, he pulls the pipe toward him, unlatches the elevators from the top of the stand, and guides it back into the fingerboard. Safety belts prevent the derrickman from falling as he leans out from the monkeyboard to manipulate the pipe. Working as a close-knit team, the driller, rotary helpers, and derrickman continue tripping out until all the drill pipe, the drill collars, and the bit are out of the hole (fig. 4.45). At this point the only thing in the hole is drilling mud because mud was pumped into the hole while pipe was tripped out.

138

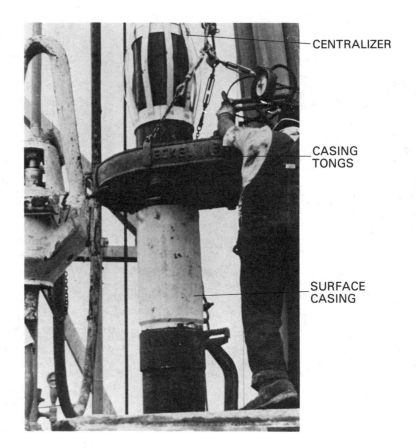

CENTRALIZER

CASING
TONGS

SURFACE
CASING

Fig. 4.46. Surface casing is run into the hole one joint at a time. Note the centralizer above the special casing tongs. (*Courtesy of Bakerline*)

Running surface casing

Once the drill stem is out, often a special casing crew moves in to run the surface casing. Casing is large-diameter steel pipe and is run into the hole with the use of special heavy-duty casing slips, tongs, and elevators. Casing accessories include centralizers, scratchers, a guide shoe, a float collar, and plugs. Centralizers keep the casing in the center of the hole so that when the casing is cemented, the cement can be evenly distributed around the outside of the casing. Scratchers help remove mud cake from the side of the hole so that the cement can form a better bond. The guide shoe guides the casing past debris in the hole and has an opening in its center out of which cement can exit the casing. The float collar serves as a receptacle for special cementing plugs and allows drilling mud to enter the casing at a controlled rate. The plugs begin and end the cementing job and serve to keep cement separated from the mud so that the mud cannot contaminate the cement. The casing crew, with the drilling crew available to help as needed, runs the surface casing into the hole one joint at a time (fig. 4.46). Like drill pipe, casing is available in joints of about 30 feet. Once the hole is lined from bottom to top with casing, the casing is cemented in place.

An oilwell cementing service company usually performs the job of cementing the casing in place. As in running casing, the men and equipment of the drilling contractor are usually available for the cementing operation. The cement used to cement oilwells is not too different from the cement used as a component in ordinary concrete. Basically, oilwell cement is portland cement that usually contains special materials to give it the necessary characteristics to make it suitable for cementing a particular well. High temperatures, which are sometimes found in oilwells, make cement set (harden) faster than normal. Thus, adding a retarder—a material that slows down the setting time of cement—makes it possible to successfully cement high-temperature wells.

Cementing service companies stock various types of cement and use special trucks to transport the cement in bulk to the well site. Bulk cement storage and handling at the rig location make it possible to mix the large quantities needed in a short time. The cementing crew mixes the dry cement with water, often using a recirculating mixer (RCM) (fig. 4.47). This device thoroughly mixes the water and cement by recirculating part of the already

Fig. 4.47. Cement slurry is mixed on site and pumped through the topmost casing into the wellbore by powerful truck-mounted pumps.

140

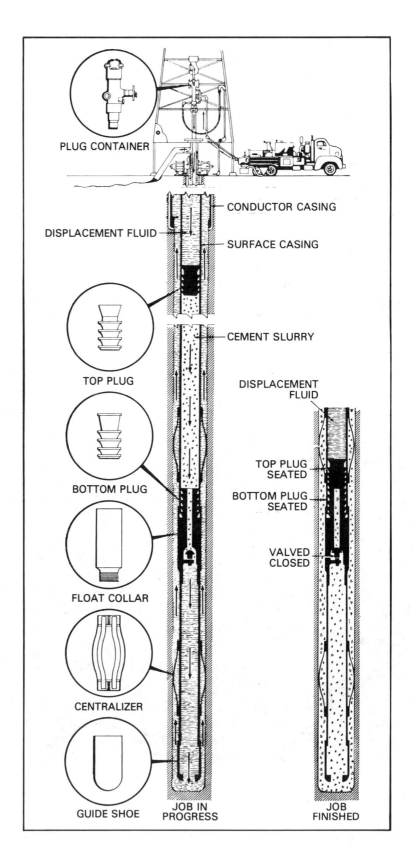

Fig. 4.48. This diagram of a casing cementing job shows the route of the slurry as well as the role played by the various cementing accessories. (*Courtesy of Halliburton Services*)

PLUG CONTAINER

DISPLACEMENT FLUID

CONDUCTOR CASING

SURFACE CASING

TOP PLUG

CEMENT SLURRY

BOTTOM PLUG

DISPLACEMENT FLUID

FLOAT COLLAR

TOP PLUG SEATED

BOTTOM PLUG SEATED

CENTRALIZER

VALVED CLOSED

GUIDE SHOE

JOB IN PROGRESS

JOB FINISHED

mixed components through a mixing compartment. Powerful cementing pumps move the liquid cement (slurry) through a pipe to a special valve made up on the topmost joint of casing. This valve is called a cementing head, or plug container. As the cement slurry arrives, a plug—the bottom plug—is released from the cementing head and precedes the slurry down the inside of the casing. The bottom plug keeps any mud that is inside the casing from contaminating the cement slurry where the two liquids interface. Also, the plug wipes off mud that adheres to the inside wall of the casing and prevents it from contaminating the cement.

The plug travels ahead of the cement until it reaches the float collar. At the collar the plug stops, but continued pump pressure breaks a seal in the top of the plug and allows the slurry to pass through a passageway in it. The slurry flows out through the guide shoe and starts up the annulus between the outside of the casing and the wall of the hole until the annulus is filled.

A top plug is released from the cementing head and follows the slurry down the casing. A liquid, usually salt water and called displacement fluid, is pumped behind the top plug (fig. 4.48). The top plug keeps the displacement fluid from contaminating the cement slurry. When the top plug comes to rest on the bottom plug in the float collar, the pumps are shut down and the slurry is allowed to harden. Allowing time for the cement to set is known as waiting on cement (WOC) and varies in length. In some cases, it may be only a matter of a few hours; in other cases, it may be 24 hours or even more, depending on well conditions. Adequate WOC time must be given to allow the cement to set properly and bond the casing firmly to the wall of the hole. After the cement hardens and tests indicate that the job is good—that is, that the cement has made a good bond and no voids exist between the casing and the hole—drilling can be resumed.

Tripping in

To resume drilling, the drill stem and a new, smaller bit that fits inside the surface casing must be tripped back into the hole. The bit is made up on the bottommost drill collar. Then, working together, the driller, floormen, and derrickman make up the stands of drill collars and drill pipe and trip them back into the hole. To make up a stand of pipe, the floormen set the slips around the pipe in the hole. The derrickman then latches the elevators to a stand set back in the mast or derrick, and as the driller picks up the stand, the floormen swing it over to the pipe in the hole and stab it. Next, the backup tongs are latched around the pipe in the hole, and the new stand is spun up. To spin up the stand, a spinning chain is used (fig. 4.49).

142

Fig. 4.49. Using a spinning chain, floormen spin a new stand of pipe into the stand suspended in the hole.

The spinning chain is a Y-shaped chain. The tail of the Y is attached to the makeup cathead, and one of the arms is attached to the end of the makeup tongs. The final arm is carefully wrapped around the tool joint of the pipe in the rotary just above the backup tongs. With a deft flick of the wrist, one of the floormen flips the chain upward causing it to unwrap from the joint in the rotary and wrap around the joint of the new stand. The driller then actuates the makeup cathead, which pulls the chain from the stand. As the chain unwraps from the stand, it spins up the stand into the joint in the rotary. The makeup tongs are then latched around the spun-up stand. Continued pull on the chain by the makeup cathead pulls on the end of the makeup tongs, causing them to apply final makeup torque to the stand. The backup tongs prevent the suspended pipe from turning as makeup torque is applied. The driller then picks up the made-up stand, the floormen remove the slips, and the pipe is lowered into the hole. This making-up process continues until the bit reaches bottom.

When the drill bit reaches bottom, circulation and rotation are begun and the bit drills through the small amount of cement left in the casing, the plugs, the guide shoe, and into the new formation below the cemented casing. As drilling progresses and hole depth increases, formations tend to get harder; as a result, several *round trips* (trips in and out of the hole) are necessary to replace worn bits.

Controlling formation pressure

As the next section of hole is drilled and indeed during all phases of drilling, an important consideration is well control. Well control is preventing the well from blowing out by using proper procedures and equipment. A blowout is the uncontrolled flow of fluids—oil, gas, water, or all three—from a formation that the hole has penetrated.

Blowouts are relatively rare events, but when they happen, they can be spectacular. Lives and property are often threatened, and pollution of the environment is possible. The great Lucas well drilled at Spindletop in 1901 is one of the more famous blowouts of all time in that it revealed the presence of theretofore unheard-of amounts of oil and gas. Some historians estimate that as many as 100,000 barrels of oil per day were spewing from the gusher when it first blew in (fig. 4.50). Since Spindletop, large strides

Fig. 4.50. The Lucas gusher at Spindletop, near Beaumont, Texas, sprayed perhaps 100,000 barrels of oil a day for nine days in 1901 before the well could be controlled.

have been made in understanding what causes blowouts and how to prevent them from happening. Today, rig crews receive extensive training in how to recognize impending blowouts and what to do should one threaten.

The key to well control is understanding pressure and its effects. Pressure exists in the borehole because it contains drilling mud and in some formations because they contain fluids. All fluids— drilling mud, water, oil, gas, and so forth—exert pressure. The denser the fluid (the more the fluid weighs), the more pressure the fluid exerts. A heavy mud exerts more pressure than a light mud. For effective control of the well, the pressure exerted by the mud in the hole should be higher than the pressure exerted by the fluids in the formation.

Pressure exerted by mud in the hole is called *hydrostatic pressure*. Pressure exerted by fluids in a formation is called *formation pressure*. The amount of hydrostatic pressure and formation pressure depends on the depth at which these pressures are measured and the density, or weight, of each fluid. Regardless of the depth, hydrostatic pressure must be equal to or slightly greater than formation pressure, or the well *kicks*. The well kicks—formation fluids enter the hole—if hydrostatic pressure falls below formation pressure. Thus, one of the crew's main concerns during all phases of the drilling operation is to keep the hole full of mud whose weight is sufficiently high to overcome formation pressure.

However, unexpectedly high formation pressures can be encountered, formation fluids can be swabbed, or pulled, into the hole by the pistonlike action of the bit as pipe is tripped out of the hole, or the mud level in the hole can fall so that the hole is no longer full of mud. Whatever the reason, when hydrostatic pressure falls below formation pressure, crew members have a kick on their hands, and they must take quick and proper action to prevent the kick from becoming a blowout.

Helping the crew keep an eye on the rig's operation are various control instruments located on the driller's console (fig. 4.51). Some rigs have data processing systems that utilize slave computer display terminals, or CRTs (short for cathode ray tubes), on the rig floor, in the mud logging trailer, in the toolpusher's trailer and in the company man's trailer (fig. 4.52). When limits that have been programmed into the system are exceeded, the system goes into an alarm condition.

Whether the kick warning signs come from electronic monitors, a computer printout, or from the behavior of the mud returning from the hole, an alert drilling crew detects the signs and takes proper action to shut the well in. To shut a well in, large valves called *blowout preventers*, which are installed on top of the cemented casing, are closed to prevent further entry of formation

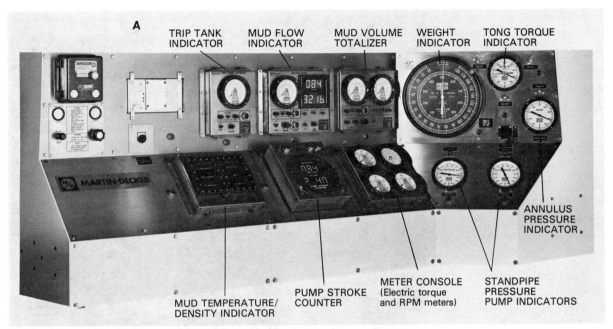

A

TRIP TANK INDICATOR
MUD FLOW INDICATOR
MUD VOLUME TOTALIZER
WEIGHT INDICATOR
TONG TORQUE INDICATOR

MARTIN-DECKER

ANNULUS PRESSURE INDICATOR.

MUD TEMPERATURE/ DENSITY INDICATOR
PUMP STROKE COUNTER
METER CONSOLE (Electric torque and RPM meters)
STANDPIPE PRESSURE PUMP INDICATORS

B

MARTIN-DECKER   DRILLING CONSOLE

Fig. 4.51. Instruments linked to various parts of the rig let the driller keep up with downhole pressure, mud volume, weight on bit, and other vital signs by reading recordings on his console. The instrument panel readouts may be analog (A) or digital (B). (*Courtesy of Martin-Decker*)

Fig. 4.52. This CRT and printer keep the toolpusher in touch with what's going on downhole. (*Courtesy of Martin-Decker*)

Fig. 4.53. A blowout pre-venter stack is used to shut the well in if it kicks.

fluids into the hole (fig. 4.53). Once the well is shut in, procedures are begun to circulate the intruded kick fluids out of the hole. Also, weighting material is added to the mud to increase its density to the proper amount to prevent further kicks, and the weighted-up mud is circulated into the hole. If the mud has been weighted the proper amount, then normal operations can be resumed.

Running and cementing intermediate casing

At a predetermined depth, drilling stops again in order to run another string of casing. Depending on the depth of the hydrocar-bon reservoir, this string of casing may be the final one, or it may be an intermediate one. In general, where the depth of the reser-voir is relatively shallow — say 10,000 feet or less — only one more casing string is required. On the other hand, in wells where the reservoir is deep — perhaps up to 20,000 feet or more — at least one intermediate casing string is usually needed. Intermediate casing is smaller than surface casing because it must be run inside the surface string and to the bottom of the intermediate hole. In general, it is run and cemented in much the same way as surface casing.

In deep wells, it is quite likely that the borehole will encounter so-called troublesome formations. Troublesome formations are those that could cause a blowout or contain shale that sloughs off and fills the hole. Such formations can be successfully drilled by carefully controlling the properties of the drilling mud. However, troublesome zones must be cased off and cemented so that they do not cause problems later, when the well is drilled to final depth.

Using a still smaller bit that fits inside the intermediate casing, the next part of the hole is drilled. Often, the next part of the hole is the final part of the hole unless more than one intermediate string is required. In any case, the bit and drill stem are tripped in, the intermediate casing shoe is drilled out, and drilling resumes, usually with the pay zone in mind—that is, a formation capable of producing enough gas or oil to make it economically feasible for the operating company to complete the well.

Drilling to final depth

In some cases, the company pretty well knows that it is drilling a well to be a producer because it is being drilled in an already existing field. In other cases, however, if the well is a wildcat (a well being drilled in an area where no oil or gas is known to exist), then it is necessary for the company to determine whether the well has indeed struck oil or gas. To determine whether hydrocarbons have been encountered, the well must be evaluated.

To help the operator decide whether to abandon the well or to set a final, or production, string of casing, several techniques can be used. A thorough examination of the cuttings made by the bit helps the operator determine whether the formation contains sufficient hydrocarbons. A geologist catches cuttings at the shale shaker and analyzes them in a portable laboratory at the well site. He often works closely with a mud logger—a technician who monitors and records information brought to the surface by the drilling mud as the hole penetrates formations of interest (fig. 4.54).

Evaluating formations

Wireline well logging is another valuable method of analyzing downhole formations. Using a mobile laboratory, well loggers lower sensitive tools to the bottom of the well on wireline and then pull them back up the hole. As they pass back up the hole, the tools measure and record certain properties of the formations and the fluids (oil, gas, and water) that may reside in the formations.

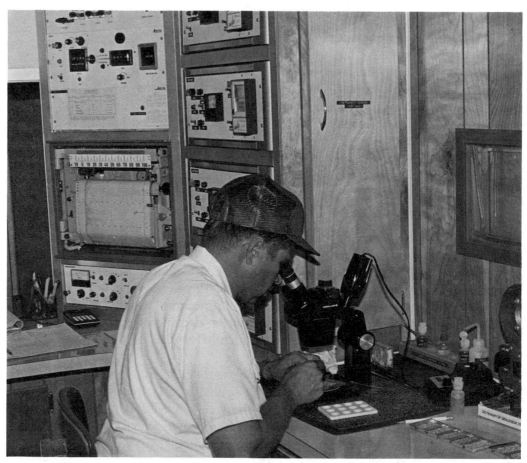

Fig. 4.54. The mud logger
examines cuttings in his portable laboratory.

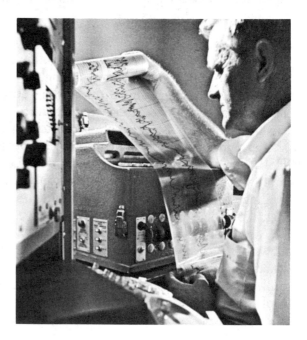

Fig. 4.55. An electric log helps
determine the presence of oil
and gas.

Many types of logs, as detailed in the exploration chapter, are generated for study and interpretation by experienced geologists and engineers to determine the presence of oil or gas (fig. 4.55).

Another evaluation technique is the drill stem test (DST). A DST tool, which is made up on the bottom of the drill stem, is set down on the bottom of the hole where downhole formation pressure and fluids enter the tool and activate a recorder. The tool is retrieved, and the recovered fluids and recorded pressure graph can be analyzed for indications of the presence and volume of hydrocarbons.

In addition to well logging and drill stem testing, formation core samples can be taken from the hole and examined in a laboratory.

After the drilling contractor has drilled the hole to final depth and the operating company has evaluated the formations, the company decides whether to set production casing or plug and abandon the well. The company may doubt that a well will produce enough oil or gas to pay for casing and completing the well, which can be a costly proposition. If the well is judged to be a dry hole — that is, not capable of producing oil or gas in commercial quantities — the well will be plugged and abandoned. Several cement plugs will be put in the well to seal it permanently. Plugging and abandoning a well is considerably less expensive than completing it.

On the other hand, if evaluation reveals that commercial amounts of hydrocarbons exist, the company may decide to set casing and complete the well. Casing will be hauled to the job; the drilling crew will pull the drill stem from the hole and lay it down one joint at a time so that it can easily be transported to the rig's next drilling location; the services of a cementing company will once more be arranged for; and the production casing will be run and cemented in the well.

The drilling contractor nears the completion of his job when the hole has been drilled to total depth and production casing has been set and cemented. In some cases, the rig and crew remain on the location and run *tubing* (a string of relatively small-diameter pipe through which the hydrocarbons are produced), set the wellhead, and bring in the well, following the instructions of the operating company. In other cases, the drilling contractor moves his rig and equipment to the next location after the production casing is cemented; in such cases, the operator may hire a special completion rig and crew to finish up the job.

Setting production casing

DEVELOPMENT OF
OFFSHORE DRILLING

Equipment and techniques used in drilling wells from offshore locations are very similar to those used in drilling wells on land. The biggest differences are in the appearance of the rigs used to drill offshore wells and in the specialized methods of operation that have been developed to deal with the problems presented by marine environments.

Offshore operations in the petroleum industry began as extensions of land operations. The first offshore well in the United States was drilled in 1897 off the coast of southern California. A wooden pier that extended about 300 feet into the Pacific Ocean was built from the shore. Near the end of the pier a drilling rig was erected, and a well was drilled to tap oil and gas that lay in a subsurface reservoir below the water. Because several successful wells had already been drilled and produced on the beach, it was only natural for oilmen in the area to extend their drilling seaward.

Early barges and platforms

By the late 1930s seismic surveys had been made in the coastal marshlands, bayous, and shallow bays adjacent to the Gulf of Mexico, and many of the surveys showed subsurface structures that were favorable to the accumulation of hydrocarbons. To drill exploratory wells into these potential reservoirs, a channel 4 to 8 feet deep was dredged in the marshes and bays, and a barge was towed into the channel. The barge was submerged and secured in place by wooden pilings, and a rig was erected on the deck of the barge, which remained above the waterline.

Another method of drilling wildcats involved building a wooden platform on timber piles and erecting a rig on the platform. Supplies were brought to the platform on a barge, or they were trucked to the rig on trestles that were built from the shore outward. As offshore exploration techniques began to reveal the presence of possible reservoirs in deeper waters, the existing barges and platforms began to show limitations. For example, a barge could not be submerged in waters deeper than about 10 feet without its deck being covered by water. Moreover, while platforms could be erected in deeper waters than barges, they were not mobile; once built, platforms could not be moved without being totally disassembled. Mobility was important to the drilling of wildcat wells because usually several wells had to be drilled in many different locations in an area before an oil or gas strike was made. Since most wildcat wells turned out to be dry holes, building a new platform every time a well was drilled was not economical.

Because some type of mobile, or movable, offshore drilling rig was required for drilling in deeper waters, engineers and naval architects put their minds to the problem, and in 1948 the first mobile offshore rig was designed. Several steel beams, or posts, were attached to the deck of a barge, an upper deck was laid on top of the posts, and drilling equipment was placed on the upper deck. The rig was floated out to the drill site near the mouth of the Mississippi River, and water was allowed to enter the barge hull at a controlled rate so that the unit slowly submerged and came to rest on the seafloor. The posts extended the drilling deck well above the waterline, and a stable platform was provided for the drilling operation. After the first well was successfully drilled in about 18 feet of water, the water in the barge hull was pumped out, and once again the entire unit floated on the water's surface. It was then towed to a second location, submerged, and another well drilled. So successful was this mobile design, called a posted barge rig, that a few units very similar to the original are still being built today for drilling in the relatively shallow waters of bays, inlets, lakes, and marshes.

First mobile drilling rig

As offshore drilling moves into greater water depths and more hostile environments, all costs increase rapidly. One of the most striking effects is the increase in size of petroleum reserves required to justify the development of a field. In some cases a reserve estimated at 100 million barrels of oil must be considered marginal; that is, if only 100 million barrels are estimated to be recoverable from a reservoir, then that reservoir cannot be profitably developed due to the high costs of drilling, producing, and transporting the oil to shore. Commercial reserves begin at 300 million barrels of anticipated production for some fields. However, exploring for offshore oil and gas is growing as improvements in drilling and production techniques reduce costs and increase the chances of success.

Offshore drilling today

The distinction between exploratory drilling and development drilling is important to the understanding of offshore operations. *Exploratory* drilling is almost always done from mobile rigs. *Development* wells—wells drilled to produce a reservoir that has been discovered by wildcat drilling—are often drilled from fixed platforms with production and well maintenance facilities. However, more and more development wells are being drilled with mobile rigs utilizing subsea production methods that eliminate the need for and the high cost of fixed platforms.

MOBILE OFFSHORE DRILLING UNITS

To drill an offshore exploratory well, the operating company most often employs a drilling contractor who owns mobile offshore rigs, or as the rigs are more formally termed, *mobile offshore drilling units* (MODUs). Several types of MODUs are available, and they can be classified in many different ways. Perhaps the most convenient is to divide them into *bottom-supported units* and *floating units*. Bottom-supported units include submersibles and jackups. Floating units include inland barges, drill ships and ship-shaped barges, and semisubmersibles.

Many factors are involved in choosing an offshore rig, but two of the most important are the water's depth and the weather at the drill site. For example, floating units must be used to drill wells in

Fig. 4.56. The barge hull of this posted barge submersible rests on the seafloor. The drilling platform and equipment rest on steel posts that extend upward from the barge hull.

Fig. 4.57. Bottles at each of the four corners of this bottle-type submersible are flooded so that the rig rests on the seafloor.

very deep water—water over 350 feet in depth. In general, the different types of MODUs were developed to meet the requirements imposed by the environment in which they work.

When a bottom-supported unit is drilling a well, part of its structure is in contact with the seafloor, while the remainder of it is supported above the water's surface. Only when bottom-supported units are moved do they float on the water's surface. In contrast, floating units are not supported by the seafloor and share the common feature of floating on or slightly below the water's surface when they are on site and drilling. With the exception of inland barges, floating units are usually employed to drill wells in waters that exceed the maximum operating depth of jackups. Floaters can also be used to drill in areas where the sea is so rough that jackups cannot withstand the wave and wind forces.

Submersibles are bottom-supported MODUs and can be further divided into posted barges, bottle-type submersibles, and arctic submersibles.

Submersibles

Posted barges. The earliest submersible design is the posted barge. It consists of a barge hull to which are attached several steel posts. A deck is laid across the top of the posts, and the drilling equipment is installed on the deck (fig. 4.56). Posted barges are confined to drilling in relatively shallow waters that do not exceed about 30 feet.

Bottle-type submersibles. The bottle-type submersible is an early design that has several steel cylinders, or bottles, on top of which a deck is laid for the drilling equipment (fig. 4.57). When the bottles are flooded (when water is allowed to enter the bottles at a controlled rate), the rig submerges and comes to rest on the ocean bottom. When it is time to move the rig to its next drilling location, water is pumped out of the bottles until the unit floats on the surface. Then, towboats move the rig to the new site. Modern bottle-type submersibles are designed to drill in maximum water depths of about 100 feet, although one was built in 1962 that can work in water as deep as 175 feet.

Arctic submersibles. Arctic submersibles are unique in that they are specifically designed for drilling in waters where moving pack ice could damage or destroy conventional submersibles. Pack ice occurs in arctic areas where winter temperatures become so low that seawater freezes. Subjected to strong currents, the ice tends to break apart and move with the currents. Moving ice can be very destructive to any stationary vessel unless the vessel is constructed to withstand its forces.

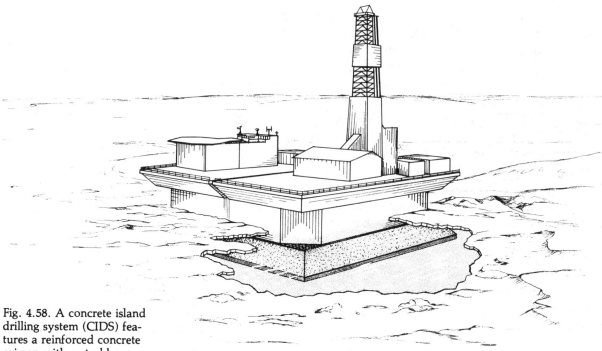

Fig. 4.58. A concrete island drilling system (CIDS) features a reinforced concrete caisson with a steel base.

Several different types of arctic submersibles are available, but all share a common feature—a steel or concrete caisson that rests on the seafloor when the rig is in the drilling mode (fig. 4.58). The caisson is a rectangular- or conical-shaped, watertight chamber on top of which the drilling equipment rests. The walls of the caisson protect the drilled wells from damage by deflecting moving ice. Also, the deck, or platform, upon which the drilling rig is mounted is reinforced to withstand the ravages of moving ice. Arctic submersibles can drill in water depths of up to about 150 feet.

Jackups   Jackups, also bottom-supported rigs, are more widely used than submersible units; indeed, they are the most widely used of all mobile offshore drilling units. Jackups usually cost less than other types of MODUs, yet they provide a very stable platform from which offshore wells can be drilled. Further, jackups can be moved relatively easy from one location to the other, and they are generally used to drill in water depths up to 350 feet. A variation of the jackup is designed to drill as deep as 600 feet and withstand 130-mph winds, giving jackups widespread offshore drilling capability.

The legs of a jackup can be either columns or trusses. Columnar legs are steel cylinders (fig. 4.59). Open-truss legs resemble a derrick or mast because they have several steel members, or trusses,

Fig. 4.59. Three cylindrical legs give a columnar-leg jackup unit its bottom support.

Fig. 4.60. This jackup rig uses open-truss legs that are lowered to rest on the sea-floor.

Fig. 4.61. An inland barge rig works in the marshy waters of the Gulf Coast.

Fig. 4.62. A drill ship steams down a channel on its way to a remote drill site.

that crisscross between structural corners (fig. 4.60). Columnar legs are less expensive to fabricate than truss legs but are more susceptible to twisting stresses. Because of this weakness, jackups with columnar legs are not able to operate in waters quite as deep as jackups with open-truss legs.

Both types of jackups have watertight barge hulls that can float on the surface of the water while the unit is being moved between drill sites. Most jackups are designed to be towed from one location to another, but at least one jackup is self-propelled; that is, it has engines and propellers (called *screws*) that can power the unit as it floats. Also, large flat-decked ships are available that are capable of carrying one or two jackup rigs. The rig is secured on the ship's deck, and the ship transports it to its destination. Ship transport is primarily used when the rig must be moved a long distance—say, several thousands of miles. Regardless of how a jackup is moved, its legs are jacked up to maximum height above the barge hull. Jacking the legs upward brings them out of the water (or above the deck of a ship) to facilitate the move.

When the jackup is positioned at the drill site, the legs are jacked down until they rest on the seabed. The hull is then jacked up above the water's surface until a sufficient air gap exists to permit operations to be carried out unhampered by tides and waves.

Inland barges

An inland barge rig, generally considered a floater, is shaped very much like a traditional barge in that it is flat-bottomed and flat-sided. The drilling rig is erected on the deck of the barge, and the entire unit is towed to the drill site. At the site, the unit is moored (anchored), and drilling is commenced (fig. 4.61). Or sometimes, in very shallow water, the unit is submerged and is thus utilized as a submersible unit. Inland barges, which are also called swamp barges, are most often employed to drill wells in the relatively shallow waters of swamps and bays or in large inland bodies of water such as Lake Maracaibo in Venezuela, where vast amounts of hydrocarbons have been discovered.

Ship-shaped barges and drill ships

The primary difference between a ship-shaped barge and a drill ship is that a drill ship is self-propelled (capable of moving under its own power), and a ship-shaped barge must be towed from site to site. Both have the appearance of a ship with a drilling rig mounted on deck (fig. 4.62). Even though drill ships have the advantage of being highly mobile, they have the disadvantage of requiring a ship's captain and crew in addition to the usual drilling crew; therefore, drill ships are more expensive to operate than ship-shaped barges. However, because drill ships are so mobile, they are often used to drill wells in remote areas, quite distant from land.

Fig. 4.63. A semisubmersible
drilling unit floats on hulls
that are flooded and sub-
merged just below the water's
surface.

Both drill ships and ship-shaped barges are often termed surface units in that they float on the water's surface at all times. Because they float on the surface, both types are very susceptible to wave and wind motion, and even though such motion can be compensated for, it cannot be entirely eliminated. As a result, drill ships and ship-shaped barges are most often employed in offshore areas where extreme weather is confined to short periods.

Drill ships and ship-shaped barges can also drill in very deep waters. They can be anchored on location, much as any ocean-going vessel is anchored, or a system called *dynamic positioning* can be used to keep the vessel on station (on the drill site). Dynamic positioning does not involve the use of anchors; instead, special propellers, or *thrusters*, are mounted on the vessel's hull below the waterline. Controlled by an onboard computer that receives data from sensors attached to the well being drilled and from sensors on the vessel itself, the thrusters are activated to maintain the unit precisely on station. In short, wind, wave, and current data are transmitted to the computer, and the computer automatically controls the thrusters.

Semisubmersibles

Because drill ships and ship-shaped barges are so susceptible to wave motion, and because oil and gas exist in formations that lie below rough seas, semisubmersible MODUs were developed. Semisubmersibles, classified as floating units, often resemble bottom-supported submersibles; indeed, the semisubmersible evolved from the bottle-type submersible. It was found that if the bottles of a submersible were only partially flooded, some buoyancy remained, and the rig floated below the water's surface but well above the seafloor. In this semisubmerged state, the unit was not affected by wave motion to the extent that surface units were. When anchored in position on the drill site, a semisubmersible provided a quite stable floating platform from which wells could be drilled (fig. 4.63).

Modern semisubmersibles consist of two or more pontoon-shaped hulls to which are attached several vertical columns. Because of the vertical columns, such semisubmersibles are termed *column-stabilized units*. A deck, the Texas deck, is laid across the top of the columns, and a derrick or mast and other drilling equipment are placed on the deck. When the unit is being moved, water is removed from the hulls and columns so that it floats on the sea surface. Usually, the unit is towed to the drill site, although some semisubmersibles are self-propelled. At the site, the hulls and columns are flooded with sufficient water to cause the unit to submerge to the required depth below the surface. Anchors or dynamic positioning is utilized to keep the vessel on station.

Semisubmersibles are among the largest and most expensive of all MODUs and can be found working in all parts of the world. However, they are most often employed in areas where the seas are rough because they are more stable than surface units. Semisubmersibles can also be used to drill in deep water; in fact, a semisubmersible holds the current record for having drilled a well in the deepest water. The well was drilled off the eastern coast of the United States in water almost 7,000 feet deep. Little doubt exists that in the near future this record will be surpassed as the oil industry extends the offshore search for oil and gas.

OFFSHORE DRILLING PLATFORMS

Although important exceptions exist, as a general rule MODUs are used to drill exploratory wells—wells drilled to confirm the suspected presence of hydrocarbons in formations below the water. Once hydrocarbons are known to be present, several additional wells must be drilled in order to develop, or exploit, the hydrocarbons. Offshore reservoirs must be large and contain large amounts of oil or gas in order to be economical to develop. Therefore, a large number of wells are needed to effectively produce the oil and gas from the reservoir. Even though MODUs are sometimes used to drill development wells, most are drilled from immobile, permanent structures called drilling platforms. A drilling platform may be rigid or compliant and may be built of steel or concrete. Types of *rigid platforms* include the steel-jacket platform, the concrete gravity platform, and the caisson-type platform. Types of *compliant platforms* include the guyed-tower platform and the tension-leg platform.

Rigid platforms are the oldest design in offshore platforms and are most often utilized in development drilling in water depths that do not exceed about 1,000 feet. Rigid steel platforms are extensively used in the Gulf of Mexico, in the Pacific Ocean off the coast of California, in the North Sea, and in other areas of the world where development drilling is occurring.

The use of rigid platforms in waters much over 1,000 feet deep is not practical; construction costs become prohibitively high and fabrication and installation become very difficult or impossible. Therefore, compliant platforms, which contain fewer steel parts and are lighter than rigid steel-jacket platforms, are being utilized. Because they are relatively lightweight and actually yield to wind and water movements, compliant platforms are used for deep-water drilling and production.

The most common type of platform is a rigid platform called a steel-jacket platform (fig. 4.64). Basically, it consists of the jacket, which is a tall vertical section fabricated from tubular steel members, and additional sections that are placed on top of the jacket. The steel jacket serves as the foundation of the platform, and it is pinned to the seabed by means of piles that are driven deeply into the soil composing the seafloor. The jacket extends upward so that the top lies well above the waterline. Placed on top of the jacket are the additional sections that provide space for crew quarters, a drilling rig, and additional equipment as required. The height of the jacket depends on the water's depth, but at least one unit stands in over 1,000 feet of water. With the additional sections on top of the jacket, the entire structure reaches a height of over 1,200 feet.

Platform jackets are usually constructed on land, placed onto a barge, and towed out to the site. Several of the tubular members of the jacket are sealed airtight so that when launched into the water,

Steel-jacket platforms

Fig. 4.64. A steel-jacket platform is a rigid structure designed to drill development wells.

the jacket floats on its side. Then, large barge-mounted cranes are rigged to the jacket to stabilize it and raise it to an upright position as the tubular legs are flooded with water. When the legs come to rest on the seafloor, piles are driven into several of the legs to pin the jacket firmly to bottom. Finally, the remaining elements of the platform are placed on the jacket, and the unit is readied for drilling operations.

Concrete gravity platforms

Another type of rigid platform is the concrete gravity platform, which has found wide use in the North Sea (fig. 4.65). Steel-reinforced concrete is used in its construction. Tall, smokestacklike columns, or caissons, are the dominant feature of concrete gravity platforms. The platform is built in a sheltered location and floated out to sea on its air-filled caissons. Once the platform is on site, the caissons are flooded like the hull of a submersible MODU until they rest on the seafloor. Because they are extremely heavy, the force of gravity alone is sufficient to keep them in place, eliminating the need for pilings. Crew quarters, drilling equipment, and other equipment are installed on a deck on top of the caissons. Frequently, special concrete cylinders are arranged around the base of the caissons on the seafloor. The cylinders can store up to a million barrels of oil. Oil storage capacity is advantageous when no pipeline exists for transporting oil to shore for refining.

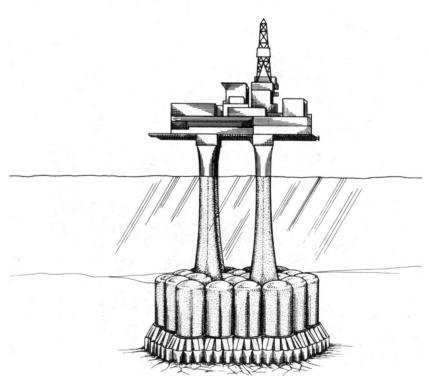

Fig. 4.65. A concrete gravity platform is so heavy that the force of gravity alone is sufficient to keep it in place.

Fig. 4.66. This steel-caisson platform rests in the Cook In-let of Alaska.

Steel-caisson platforms

A third, specialized type of rigid platform is the steel-caisson plat-form (fig. 4.66). Designed specifically for use in the Cook Inlet of Alaska, where fast-moving tidal currents carry ice floes that can destroy conventional steel-jacket platforms, steel-caisson plat-forms have proved to be successful. One design has a caisson at each of the four corners of the rectangular platform, firmly affixed to the bottom of the inlet. The drilling and production decks are laid on top of the caissons. The caissons are made of two layers of thick steel to prevent ice floe damage.

164

Guyed-tower platforms

A guyed-tower platform is compliant but similar to a rigid steel-jacket platform in that it has a jacket that rests on the seafloor. However, the guyed jacket is much slimmer and does not contain as much steel as a rigid jacket. Further, it weighs less and is less expensive to build. Several guy wires are attached to the jacket relatively close to the waterline and are spread out evenly around it. The guy wires are anchored to the seafloor in a unique way by means of *clump weights*. Clump weights are a series of weights that are put together in such a way as to behave much like a bicycle chain. The weights lie flat on the ocean floor, but as the platform moves, they are lifted off the floor. Then as the platform moves back, the clump weights once again lie on bottom. In short, clump weights allow the guy wires to move with the movement of the platform, yet still provide firm anchoring for the wires (fig. 4.67).

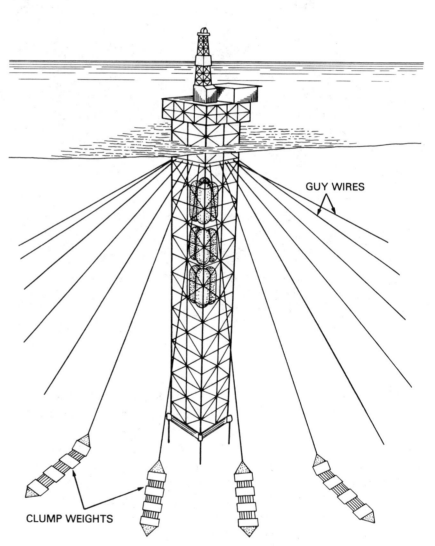

Fig. 4.67. On a guyed-tower platform, several clump weights anchor the guy wires, yet allow them to move as the platform jacket moves.

GUY WIRES

CLUMP WEIGHTS

A second type of compliant platform is the tension-leg platform. Like guyed towers, tension-leg platforms move with wind, waves, and currents. Resembling a semisubmersible drilling unit, a tension-leg platform has several steel tubes that are firmly attached to the ocean floor. Since the platform itself is buoyant, this buoyancy applies tension to the steel tubes (fig. 4.68).

Tension-leg platforms

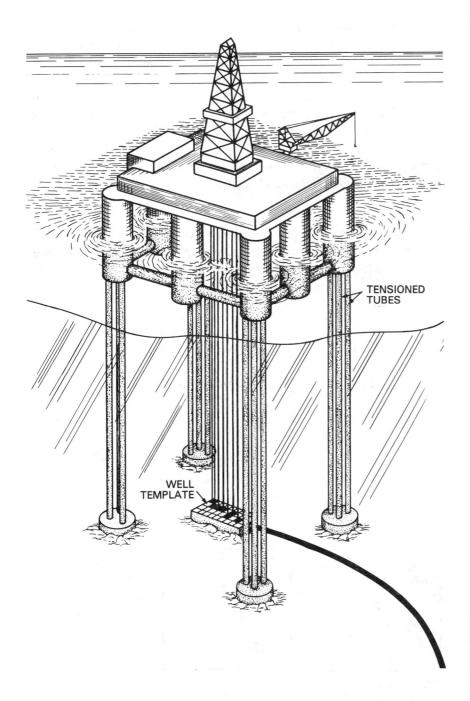

TENSIONED TUBES

WELL TEMPLATE

Fig. 4.68. Several steel tubes are tensioned by the buoyancy of the platform on a tension-leg platform.

166

DIRECTIONAL DRILLING

Although wellbores are normally planned to be drilled vertically, many occasions arise that make it necessary or advantageous to drill at an angle, especially in offshore operations. This deviation from drilling a straight hole, known as directional drilling, makes it possible to accomplish many things that could not be done with straight holes.

Directional wells are drilled straight to a predetermined depth, and then are gradually curved to penetrate the reservoir at several different points. Since the curvature of each well is gradual—only about 2 or 3 degrees per 100 feet of well depth—the drill stem and casing are able to follow the curve of the well without difficulty. Even though the curve is gradual, directional wells can be deflected off vertical to a very high degree; indeed, sometimes they achieve a horizontal direction in the subsurface. It is important to remember, however, that such high angles are built up in small increments of 2 or 3 degrees at a time.

Uses

When development wells are drilled from rigid and compliant platforms, directional drilling is usually employed. In this technique, several wells (as many as forty or more) are drilled from the platform without the platform's having to be moved (fig. 4.69).

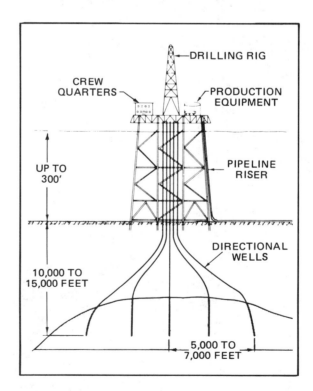

Fig. 4.69. Several directional wells are drilled from an offshore platform to penetrate the reservoir at different points.

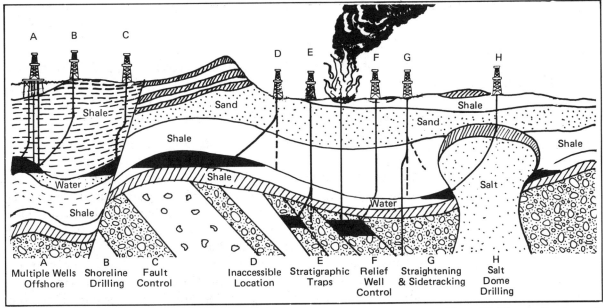

Fig. 4.70. The applications of controlled directional drilling are many.

In addition to offshore development drilling, directional drilling has several other uses (fig. 4.70). A directional well may be drilled from a shoreline and deflected to reach a pay zone that lies a short distance offshore. Directional drilling is used when it is necessary to avoid drilling along a fault line, which can cause hole problems. When it is impossible to locate the drilling rig over the desired spot because of a river, hill, or some other obstruction, the rig can be erected at one side of the obstruction and the hole deviated to the pay zone.

Directional holes can also be used in exploratory drilling in cases where the straight hole misses the reservoir. By deflecting the hole from the original dry well, it may be possible to intersect a pay zone.

Another use is in killing a *wild well*, a well that has blown out, caught on fire, and cratered. A relief or offset well is drilled and deviated so that it bottoms out near the borehole of the blown-out well. Then mud can be pumped down the relief well to kill the wild well.

A directional well can be used to straighten a crooked hole or to sidetrack around a *fish* (an object lodged in the borehole) that cannot be removed. Finally, directional drilling is very useful when drilling into traps associated with salt domes. Typically, the traps around salt domes are arranged on the edges of the dome and are difficult to pinpoint accurately. A directional well makes it possible to drill into the trap if the initial straight hole misses.

Tools and techniques

Special tools and techniques are required for directional drilling. Several ways exist to deviate the hole from vertical, but the most common involves the use of a *bent sub* and *downhole motor.*

A bent sub is a short piece of pipe that is threaded on both ends and bent in the middle. Typically, the bend ranges from 1 to 3 degrees off vertical. It is installed in the drill stem between the bottommost drill collar and a downhole motor.

A downhole motor, driven by drilling fluid, imparts rotary motion only to the drilling bit connected to the tool, thus eliminating the need to rotate the entire drill stem in order to make hole. Shaped like a piece of pipe, a downhole motor can have turbine blades, or it can have a spiral steel shaft that turns inside an elliptically shaped opening in the housing. In the case of a turbine tool, the force of the circulating drilling fluid inside the tool strikes the turbine blades to make a shaft turn. Since the bit is attached to the shaft, it turns when the shaft turns. With the other tool, circulating drilling mud flows through the elliptical opening and forces the spiral shaft to rotate. The rotating shaft turns the bit.

In order to curve the hole toward the target, the deflected drill stem must be turned in a predetermined direction while drilling proceeds. Orientation can be done in several ways. One method uses a directional instrument containing a magnetic or gyroscopic compass and an inclinometer (an instrument that measures the angle of the hole). Drilling must be halted while the instrument is run down to the bit and retrieved. Most of the downtime of this method can be avoided by using instead a continuous readout instrument known as a *steering tool* to send directional information uphole via wireline to a rig-floor monitor. The newest device, coming into common use in some areas, is the mud pulse generator, a wireless, self-contained instrument that transmits sonic signals uphole, via the drilling fluid in the drill stem, to a readout device at the surface. This system, commonly called *measurement while drilling* (MWD), is useful for maintaining straight hole as well as for directional drilling.

FISHING

Fishing is the drilling term for retrieving any object — or *fish* — from the wellbore. The fish can be anything from part or all of the drill stem stuck or lost in the hole to smaller pieces of equipment, called *junk*, such as bit cones, hand tools, pieces of steel, or any other nondrillable item in the hole.

Drill pipe or drill collars can get stuck in the hole for several reasons: (1) the hole can collapse around the pipe; (2) the pipe can get stuck in a *keyseat* — a small-diameter (undergauge) portion of the hole worn into a bend, or *dogleg;* or (3) pressure differential can hold the drill collars so securely to the wall of the hole that no amount of pulling can free the pipe.

The most common reason for the wall of the hole collapsing around the pipe is that interstitial salt water (water contained in the interstices, or pores, of the formation rock) can attract, under certain conditions, the water in the drilling mud. If the formation happens to be made of shale and the water in the mud is in contact with the water in the shale, the water in the mud has a tendency to transfer to the shale. The transferred water causes the shale to expand; then small sheets of shale slough off into the hole, the hole eventually fills with debris, and the pipe sticks.

Pipe can also get stuck in a keyseat. A keyseat is caused by a dogleg, which is a severely crooked section of hole. ("It's as crooked as a dog's hind leg" is the expression that gives rise to the term.) The drill pipe tends to lean against the side of the dogleg, and as the pipe rotates, it digs out a new, smaller hole in the side of the main borehole. Then, when the drill stem is pulled from the hole, the tool joint of the drill pipe or the wider drill collars can be jammed into the keyseat so hard that they cannot be freed by normal means (fig. 4.71).

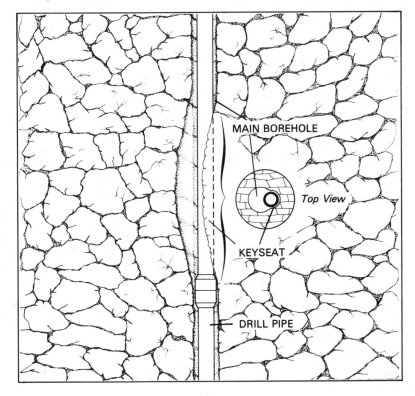

Fig. 4.71. Drill pipe is stuck in a keyseat.

Pipe can get stuck in the hole when the formation pressure and the pressure within the wellbore differ too much—that is, when the pressure in the wellbore is quite a bit higher than the pressure in the formation. In drilling through permeable layers without adjusting the properties of the mud to compensate for the permeable layers, the higher pressure in the wellbore can cause a thick cake of mud solids to build up on the inside wall of the hole. Also, when the drill stem is stationary, as when a connection is being made or during a trip, the higher pressure in the wellbore can force the drill collars into the thick cake and stick them very firmly to the wall of the hole.

To free wall-stuck pipe, the crew will probably *spot oil* around the stuck portion and *jar* on the drill stem. To spot oil means to circulate oil or other lubricant down the drill stem and into the annulus to the stuck portion. To jar on the drill stem means to install a special device—a *drilling jar* or a *bumper sub*—on top of the drill stem, allowing the driller to strike very heavy upward or downward blows on the stuck pipe. Spotting oil and jarring usually free wall-stuck pipe.

To free pipe that is stuck by sloughing shale or to free wall-stuck pipe that remains stuck because spotting oil and jarring are not successful, the first step is to use a device called a *free-point indicator*. A free-point indicator, in spite of its name, is used to determine the point at which the pipe is stuck. It is lowered inside the drill stem, and the drill stem is stretched by being picked up. The indicator induces a magnetic field in the pipe and sends a signal to a meter on the surface. Where the pipe is free, the pipe stretches a large amount, and strong signals are sent to the meter; therefore, the meter needle moves a great deal. But where the pipe is stuck, the pipe stretches very little, and weak signals are sent to the meter; therefore, the needle moves very little. By noting the depth at which the needle movement is slight, the stuck point can be determined (fig. 4.72).

Once the stuck point is determined, a *string shot* is positioned opposite a tool joint several joints above the stuck point. A string shot is a long, stringlike explosive charge that is usually run below the free-point indicator. When detonated, the string shot helps loosen the threads of the pipe, just as banging on the lid of a jar that is difficult to open helps loosen it. As the string shot is set off, the driller turns the rotary to the left to back off (unscrew) the pipe. Then the free pipe above the stuck pipe is tripped out of the hole. The pipe is usually backed off several joints above the stuck point because additional tools will be run outside the pipe and this clear pipe serves as a guide for the additional tools.

The stuck pipe is left dangling in the hole, and a special kind of pipe called *washover pipe*, or *washpipe*, is lowered into the hole.

The washpipe is lowered over the stuck drill stem (the fish), and circulation and rotation begin. A rotary shoe on the bottom of the washpipe drills the shale or wall cake that is causing the pipe to stick (fig. 4.73). However, when all the material holding the fish is removed, the fish could fall to bottom. To prevent this, a *backoff connector* is latched inside the washpipe. The connector is made up on the top of the fish before the washover begins; as washover proceeds, the connector is designed to remain stationary on top of the fish. When the fish is washed free and starts to fall, the connector is activated to firmly grip the inside of the washpipe and prevent the fish from falling. After all the shale or wall cake is drilled up, the washpipe and the drill stem (secured inside the washpipe by the backoff connector tool) are pulled out of the hole simultaneously.

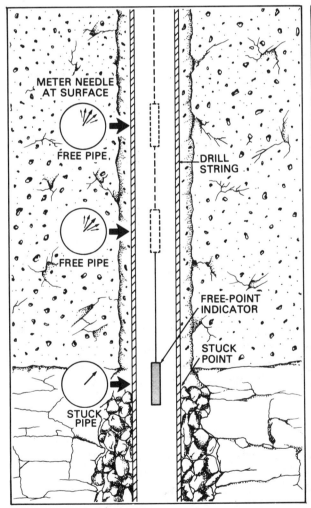

Fig. 4.72. A free-point indicator is used to locate stuck pipe.

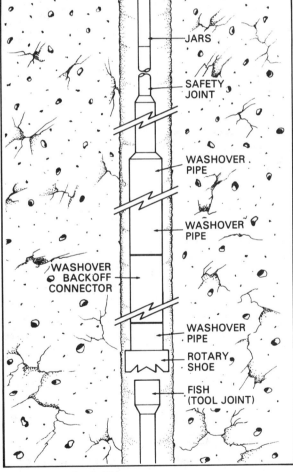

Fig. 4.73. A backoff connector is used on the stuck pipe inside the washover pipe to keep the drill pipe from falling when freed.

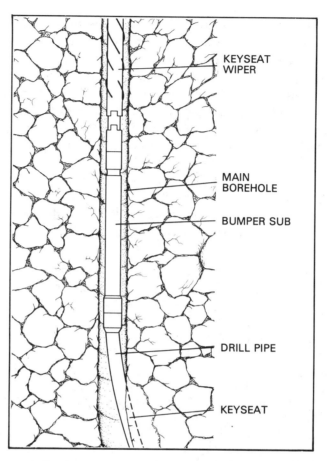

KEYSEAT WIPER

MAIN BOREHOLE

BUMPER SUB

DRILL PIPE

KEYSEAT

Fig. 4.74. A keyseat wiper and bumper jar can be used to free pipe stuck in a keyseat.

If the drill stem is stuck in a keyseat, the fishing process is slightly different; however, it starts with sending the free-point indicator and string shot down inside the pipe as before. When the stuck point is located, all but the last five or six joints above the stuck drill pipe or drill collar are removed, so the top of the fish is in the main borehole and not in the keyseat. Then a bumper jar is lowered down the hole, is made up on the fish, and is actuated to jar the fish loose from the keyseat. A reaming device called a *keyseat wiper* is run on top of the jar so that the keyseat can be reamed out to normal size. In this way the tool joints and drill collars will pass through the reamed-out keyseat (fig. 4.74).

Retrieving twisted-off pipe

Another fishing problem that can happen with drill pipe and drill collars is that the pipe twists off. Because of fatigue or damage, the pipe literally breaks in two; such breaks are called twistoffs. Fishing for twisted-off pipe can be a relatively simple operation, or it can be difficult or impossible, depending on the situation. In most instances, however, retrieving a twisted-off fish is not complicated.

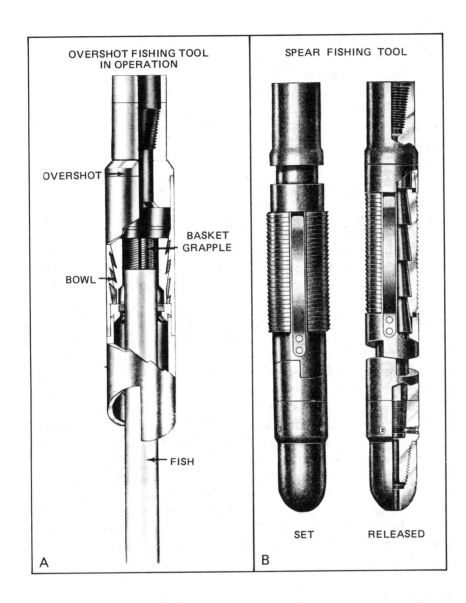

OVERSHOT FISHING TOOL
IN OPERATION

OVERSHOT

BASKET
GRAPPLE

BOWL

FISH

A

SPEAR FISHING TOOL

SET   RELEASED

B

Fig. 4.75. An overshot (A) and a spear (B) are two commonly used tools for retrieving fish.

Two commonly used tools for retrieving twistoffs are the *overshot* and the *spear* (fig. 4.75). An overshot or spear is run into the hole on drill pipe until it contacts the top of the fish. An overshot goes over the outside of the fish and grips it firmly. A spear goes inside the fish and strongly grips the inside of the pipe. A spear is generally used only when it is difficult or impossible to go outside the fish with an overshot, such as when a drill collar, which is very close to being the same size as the hole, is involved. In either case, when the spear or overshot grips the fish, both the tool and the fish are pulled out of the hole together. Special mills are sometimes used to smooth the top of the fish prior to running the overshot or spear. Milling the top of the fish makes attaching the fishing tool easier.

174

Fishing for junk Small nondrillable pieces that find their way to the bottom of the hole and cause drilling to cease are known as *junk*. Many fishing tools, including powerful magnets and special baskets through which mud can be circulated, are used to retrieve junk. In fact, the only limit to fishing tools and methods seems to be the imagination and ingenuity of the people faced with the problem of removing fish from the hole.

AIR OR GAS DRILLING

Circulating with air or gas instead of mud as a drilling fluid is an alternate method of rotary drilling that has spectacular results but limited application. Penetration rates are higher, footage per bit is greater, and bit cost is lower with air or gas. Air or gas cleans the bottom of the hole more effectively than mud, and for that primary reason the rate of penetration is faster. Mud is denser than air or gas and tends to hold the cuttings on the bottom of the hole. As a result, the bit cannot make hole as efficiently because the bit redrills some of the old cuttings instead of being constantly exposed to fresh, undrilled formation.

Air or gas drilling has several negative aspects that are considered more significant than the positive ones. The hazard of fire or explosion is always present. Because air or gas does not impose significant bottomhole pressure, high-pressure formations cannot be safely encountered in drilling. It is impossible to prevent the entry of formation fluids into the well, and most deep wells encounter water-bearing formations sooner or later. If the wall of the hole tends to slough, or cave, into the hole, air or gas circulation is impossible because the drill stem may stick. Also, corrosion has been a problem with air or gas drilling, although chemicals have been developed to take care of this problem.

To drill with air or gas, large compressors and related equipment are moved to the site. Usually, only part of the hole will be drilled with air or gas; then the rig will be changed over to drilling mud. Actually, air or gas is not circulated in the sense that it is used over and over again; rather, it makes one trip from the compressors, down the drill stem, out the bit, and up the annulus back to the surface, where it is blown out a *blooey line*, or vent pipe (fig. 4.76).

Aerated mud—that is, mud to which air or gas has been purposely added—has been used successfully to prevent lost circulation. Lost circulation occurs when drilling mud leaks out of the borehole and into a subsurface formation. Thus, mud does not return to the surface but is lost downhole. Air or gas in the mud reduces the amount of pressure exerted by the mud on downhole formations and thus relieves one of the causes of lost circulation.

BLOOEY LINE

Fig. 4.76. Skid-mounted compressors furnish the high-pressure air used on this regular rotary rig for drilling. The air can be seen blowing a cloud of dust out the blooey line.

Drilling in offshore, arctic, and remote regions of the world has led to the development of specially designed rigs that are able to withstand the rigors of extreme cold, the hazards of marine environments, and the difficulties encountered in underdeveloped lands. New tools and techniques will continue to be developed as new areas are drilled and new problems arise. Electronically operated instruments and computer applications have made an impact on drilling operations just as they have on other phases of the industry. As oil and gas reservoirs of the future become more difficult to reach, computer technology will play an even more important role in making hole.

Production

In the petroleum industry, production is defined as the phase of operation that deals with bringing the well fluids to the surface and preparing them for their trip to the refinery or processing plant. Since this phase begins after the well is drilled, the first step is to complete the well — that is, to perform whatever operations are necessary to start the well fluids flowing to the surface. Routine maintenance operations, such as replacing worn or malfunctioning equipment — known as servicing — are standard during the well's producing life. More extensive repairs — known as workovers — may also be necessary to maintain the flow of oil and gas. Well fluids are usually a mixture of oil, gas, and water and must be separated into these components because each is handled differently. The water must be disposed of, while oil and gas must be treated, measured, and tested before they are transported. Thus, it can be seen that production is a combination of operations: bringing the fluids to the surface; doing whatever is necessary to keep the well producing; and taking the fluids through a series of steps to purify, measure, and test them.

EARLY PRODUCTION METHODS

The earliest methods of well completion were probably used by the Chinese. They are thought to have completed gas, water, and saltwater wells as early as 1000 B.C. Their technology passed down through the centuries with few changes, and it was not until the 1800s that the innovation and inventiveness that mark the petroleum industry today began.

It all started in 1808 with a saltwater well that two brothers, David and Joseph Ruffner, were working on to get salt. They had a problem with the West Virginia well: the less-concentrated saltwater sands higher up in the well were diluting the concentrated salt water down below. In order to extract more of the concentrated salt water from the bottom of the well, the brothers decided that running a pipe or tube from the well bottom to the surface would prevent entry of the water from the upper zones. They also realized that a packing implement of some sort, placed just above the bottommost sand, would keep the shallower water sands that migrated to the bottom from coming up through the pipe. Since no metal tubing was available, they rigged up two strips of wood, two properly sized long half-tubes, fitted them closely together, and wrapped small twine all around the device. After fitting a watertight bag snugly near the lower end, the brothers placed the homemade tube carefully in the well. With this invention, brine flowed freely through the tube, and the undesired waters were excluded. This West Virginia experiment was the first step toward the tubing, casing, and packers that are now so vital to the oil industry.

Another important operational procedure borrowed from the water-well industry was the use of the reciprocating walking beam to operate a bottomhole pump after well completion. In the decade after Drake's discovery well in 1859, a cable-tool drilling rig remained on a well in order for the walking beam to be used for pump operation. Its bull wheel and sand reel were employed for well servicing. These two large spools had cable wrapped around them, and tools attached to the cables could be lowered and raised from the well. This equipment was known as the "standard rig front" and was mostly composed of wood and driven by steam.

Central pumping power, developed by the oil industry around 1880, was the first major innovation in production equipment. Central pumping consisted of using a single prime mover to pump several wells. More developments followed at the turn of the century. Rotary drilling evolved, internal-combustion engines began to replace steam engines, and wooden sucker rods were being pushed out by the new iron sucker rods. All of these inventions laid the groundwork for continuing improvements in equipment and procedures. Adoption of new devices was slow, however, with operators utilizing old equipment as long as possible. By the early 1920s, though, the demand for a better method to replace the standard rig front led the oil industry to focus on the beam pumper, a self-contained unit, mounted on the well at the surface, that operated a downhole pump. After a slow start, the transition was quite rapid. The demand for better methods of operation also fostered the search for alternate ways to lift oil artificially because of problems with sucker rod pumping. The electric submersible pump was developed in 1930 as an acceptable alternative, followed by marketable gas lift devices a few years later. But even today, about 80 percent or more of all wells on artificial lift are using beam pumping units, sucker rods, and sucker rod pumps.

The industry also had a problem with what to do with the oil after it was lifted from the well. Often, the only lease facilities available were wooden barrels, and these were used for the collection, storage, shipment, and even the measurement of petroleum (fig. 5.1). Sometimes, mere earthen pits were dug into the ground and the well fluids routed into them. Lease operators were left to their own devices for the most part, having to find their own methods for handling and separating oil, gas, sediments, and water. Handling and separation became even more difficult when considerable amounts of water and sand came mixed with the produced oil. Wooden barrels were slowly replaced by tanks, first those of wood, then riveted iron, and finally bolted or welded steel.

Fig. 5.1. Wooden barrels were used for early lease facilities.

After 1920, large advances were made in the operation of lease facilities. Petroleum engineers improved methods for oil, gas, and water separation, chemical and electrical emulsion treatment, and water handling. Although producing oil and gas has utilized many technological advances, modern petroleum-producing facilities still retain some similarity to old-time designs. The emphasis now, however, is placed on using efficient techniques and methods to improve lease operation.

WELL COMPLETION

After a well has been drilled to the projected depth and the productive formations have been evaluated as economical to produce, the work of setting the casing, preparing the well for production, and bringing in the oil or gas begins. Completion equipment and the methods employed are quite varied, and the decisions for an individual well are usually based on the type of oil or gas accumulations involved, the requirements that may develop during the life of the well, and the economic circumstances at the time the work is done. Low-pressure pipe, sometimes secondhand, will be employed if the oil accumulation has a marginal payout, and other expenditures will be scaled down accordingly. If high pressure is anticipated and well life is expected to be long, however, the best grade of pipe will be needed.

Downhole strings   Many oil and gas wells require four concentric strings of large pipe: conductor pipe, surface casing, intermediate casing, and production casing (fig. 5.2). The latter is often called the *oil string* or the *long string* in the oil patch. The conductor pipe prevents the hole from caving in at the surface and endangering the drilling rig foundation. If the ground permits, the conductor pipe will be driven into place by a pile driver. If not, a small drilling rig will be used to drill a hole into which the conductor pipe is cemented. When the surface casing is set and cemented in place, it provides protection for freshwater formations. It also prevents loose shale and sand or gravel from falling into the hole and affords a means of controlling the flow of fluid from the well. Setting depths may vary from 500 feet to 5,000 feet. Intermediate casing may be needed if troublesome zones are encountered below the surface casing and above the final depth of the well. The final casing for most wells is the production casing. The producing formation is usually completely cased off, but in rare instances the production casing is set near or just on top of the potential pay zone.

Another type of pipe that is not uncommon in wells over 10,000 feet is called a *liner.* Liners are really just like casing—that is, they serve the same purpose—but they do not extend all the way to the surface. Instead, a liner is suspended from the larger casing above it by means of a slip device called a *liner hanger.* A liner can function as production casing, in which case it is termed a *production liner.* Since it does not go to the surface, which is sometimes a considerable distance, the operator has a lower pipe cost.

The final string of pipe usually run in a producing well is the *tubing.* Tubing is nearly always freely suspended in the well from the tubing head. In a flowing well its small diameter produces more efficient results than casing. Also, tubing is considerably easier to remove than casing when it becomes plugged or damaged. Tubing, when used in conjunction with a *packer*, keeps well fluids away from the casing because the packer seals the space between the tubing and the casing (fig. 5.3). Well fluids corrode the casing and thus cause costly repair jobs later.

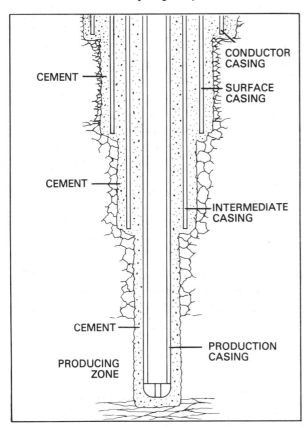

Fig. 5.2. Conductor, surface, intermediate, and production casing are cemented in the well. Note that the production casing is set through the producing zone.

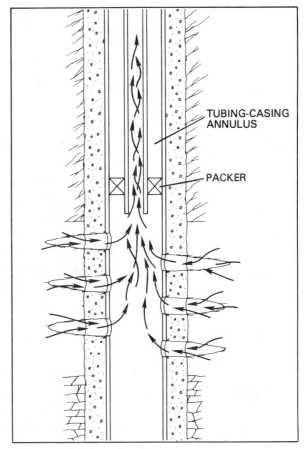

Fig. 5.3. A packer placed in the tubing string keeps well fluids out of the tubing-casing annulus.

The packer consists of a pipelike device through which well fluids can flow, a rubber sealing element that forms a fluid-tight seal, and gripping elements (called *slips*) that hold the packer in the tubing-casing annulus just above the producing zone. Since the packer seals off the space between the tubing and the casing, formation fluids flowing into the well are forced into and up the tubing.

Another device frequently installed in the tubing string near the surface is a *subsurface safety valve*. The valve remains open as long as fluid flow is normal. When the valve senses something amiss with the surface equipment of the well, it closes, preventing the flow of fluids.

Secondary cementing

Cementing is one of the most critical operations performed during the drilling and completion of a well. Secondary cementing is often a remedial step taken to solve problems caused by faulty primary cementing. Primary cementing is the cementing that takes place immediately after casing has been run into the hole during the drilling operation. A good primary cementing job is one in which the cement completely fills the annulus around a centralized string of casing so that no voids, or spaces, occur in the cement. If a deficiency occurs in this first cementing job, then secondary cementing — a costly and time-consuming operation — may be needed before a well can be completed. Its purpose, like that of primary cementing, is to mechanically anchor the casing in place and to seal the borehole in order to prevent vertical movement of fluids from zone to zone.

Two main types of secondary cementing are *squeeze* cementing and *plug-back* cementing. Cement plugs are used when a well is to be abandoned, when a zone needs to be isolated, or when directional drilling, lost circulation control, or formation testing is required. The principal purpose of squeeze cementing, so named because the process uses pressure to force cement slurry into specific intervals in a well, is to provide a seal. This seal is vital in preventing fluid migration between zones after the well is perforated.

Completion methods

A well servicing contractor may move in a smaller rig to perform the operations necessary to put the well into production, or the work may be done before the drilling rig moves off. The type of completion method used is determined by the characteristics of the reservoir and its economic potential. Among the various methods are open-hole, perforated, wire-wrapped screen, tubingless, and multiple completions.

183

Open-hole completion. An open-hole, or barefoot, completion has no production casing or liner set opposite the producing formation (fig. 5.4). Instead, reservoir fluids flow unrestricted into the open wellbore. This type of completion, which is rarely used and is generally restricted to limestone reservoirs, is useful where only one productive zone and low-pressure formations exist. In an open-hole completion, casing is set just above the pay zone, and drilling proceeds into the productive zone as far as necessary to complete the well.

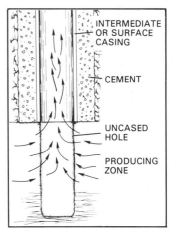

Fig. 5.4. An open-hole completion allows well fluids to flow into the uncased hole.

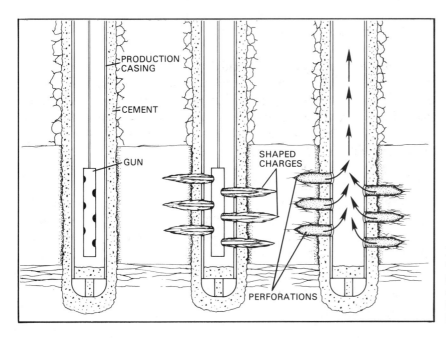

Fig. 5.5. A perforating gun is used to create holes or perforations in the casing, cement, and producing formation, allowing fluids to enter the well.

Perforated completion. The perforated completion — by far the most popular method of completing a well — requires a good cementing job and the proper perforating method. Perforating is the process of piercing the casing wall and the cement to provide openings through which formation fluids may enter the wellbore. Perforating is accomplished by lowering a perforating gun down the production casing or the tubing until it is opposite the zone to be produced. The gun is fired to shoot *bullets* or, more often, to set off special explosive charges known as *shaped charges* (fig. 5.5). A shaped charge is designed so that an intense, directional explosion is formed. Because the explosion is actually a jet of high-energy gases and particles, it is called *jet perforating*. If a production liner is used rather than production casing, the liner is perforated to complete the well (fig. 5.6).

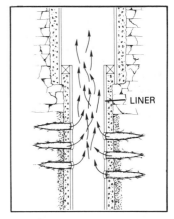

Fig. 5.6. In a perforated liner completion, a liner is perforated opposite the producing zone.

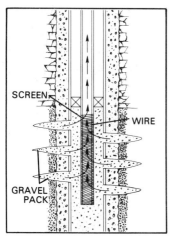

Fig. 5.7. A wire-wrapped screen completion is often run with a gravel pack inside perforated casing.

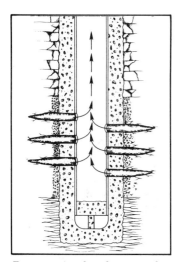

Fig. 5.8. A tubingless completion has no inner tubing string. Well fluids are produced through small-diameter casing.

Fig. 5.9. A multiple completion, such as this triple, usually has a separate tubing string and packer for each producing zone.

Wire-wrapped screen completion. Another type of completion involves a *wire-wrapped screen*, which is a short length of pipe that has openings in its sides and is wrapped with a specially shaped wire. One method involves attaching the screen to the bottom of the tubing string and lowering it into a well that has been perforated. Usually, wire-wrapped screens are run in conjunction with a *gravel pack* — that is, gravel is placed in the hole outside the screen. Well fluids flow through the gravel pack, through the wire-wrapped screen, and into the tubing (fig. 5.7). Wire-wrapped screen completions are most often utilized in cases where the producing zone or zones are likely to produce sand as well as oil and gas. The screen and gravel pack minimize the entry of sand into the wellbore, thus preventing problems caused by sand.

Tubingless completion. Although most wells are completed with tubing, a small-diameter well that uses small-diameter casing may be completed without the tubing. The small casing is cemented and perforated opposite the producing zones (fig. 5.8). Tubingless completions are used mostly in small gas reservoirs that produce few, if any, liquids and are low in pressure.

Multiple completions. Multiple completions can be used when one wellbore penetrates two or more producing zones. Usually, a separate tubing string with packers is run inside the production casing for each producing zone. For example, in a triple completion, three tubing strings and three packers can be utilized in a single production string (fig. 5.9).

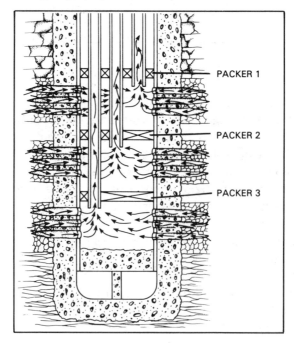

PACKER 1

PACKER 2

PACKER 3

The wellhead is the equipment used to confine and control the flow of fluids from the well. It forms a seal to prevent well fluids from blowing out or leaking at the surface. The conditions expected to be encountered in the individual well determine the type of wellhead that is needed. Sometimes, all that is required is a simple assembly to support the weight of the tubing in the well. In other cases, the control of formation pressure is necessary, and a high-pressure wellhead is required. Pressures greater than 20,000 pounds per square inch have been found in some fields. Basically, the wellhead is made up of a combination of parts called casinghead(s), tubing head, and Christmas tree (fig. 5.10).

The wellhead

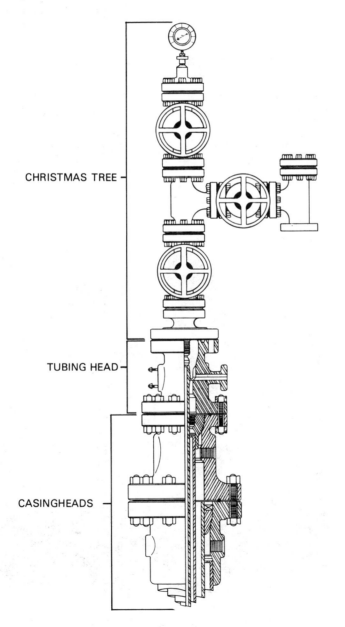

CHRISTMAS TREE

TUBING HEAD

CASINGHEADS

Fig. 5.10. A wellhead has one or more casingheads, a tubing head, and a Christmas tree.

186

Casinghead. The casinghead is a heavy steel fitting at the surface to which the casing is attached. During drilling and workover operations, the casinghead is used as an anchor for the pressure control equipment that may be necessary. If several strings of casing are used in the well, more than one casinghead may be used on a wellhead assembly.

Tubing head. Similar in design and use to the casinghead, the tubing head supports the tubing string, seals off pressure between the casing and the inside of tubing, and provides connections at the surface with which the flowing liquid or gas can be controlled. The tubing head is supported by the casinghead if a casinghead is used on the well.

Christmas tree. The assembly of control valves, pressure gauges, and chokes located at the top of a well is known as the Christmas tree, so named because of its treelike shape, with a large number of fittings branching out above the wellhead (fig. 5.11). The valves are opened and closed to control the flow of oil and gas from the well after it has been completed. The master valve can shut off the flow entirely. Also in the Christmas tree is a *choke*, or

Fig. 5.11. Valves on the Christmas tree control the flow of fluids from the well.

restriction in the line, that controls the amount of fluids flowing from the well. The pressure gauges reveal casing and tubing pressures. By knowing these pressures under various operating conditions, better well control is possible.

An operation that temporarily lowers the fluid level in the well so that it can begin to produce is sometimes necessary if the well does not flow after perforating. This is known as *swabbing* (fig. 5.12). The purpose of this operation is to swab, or lift, enough fluid out of the tubing so that the hydrostatic pressure is reduced to a value below that of the formation pressure. If it is successful, formation fluids will start flowing immediately. If the well does not begin to produce, it may need to be stimulated, or a pump may have to be installed as a permanent lifting device to bring the oil to the surface.

Swabbing

Fig. 5.12. Swabbing is done with a swab cup, which picks up fluid in the tubing and raises it to the surface by means of a continuous pull on a cable attached to a hoisting winch.

WELL TESTING

A variety of well tests are conducted to determine production rates for oil and gas wells. Each test reveals certain information about a particular well and the reservoir in which it is completed. Accuracy is, of course, very important, and these test data form the case history of a well.

Potential test

The potential test is a measurement of the largest amount of oil and gas that a well will produce in a 24-hour period under certain fixed conditions. This test is made on each newly completed well and at other times during the well's producing life. Potential test information is generally required by the state regulatory group, which uses it to establish the producing allowable for the well.

Bottomhole pressure test

A bottomhole pressure test measures the reservoir pressure of the well at a point opposite the producing formation. The test is usually conducted after the well has been shut in for 24 to 48 hours. When this test is conducted at scheduled intervals, valuable information about the decline or depletion of the zone in which the well is producing will be gathered.

Productivity test

A productivity test is conducted to determine the effects of different flow rates on the pressure within the producing zone of the well. This reveals certain physical characteristics of the reservoir, and the maximum potential rate of flow can be calculated without risking the damage that might occur if the well were produced at its maximum possible flow rate. The procedure for the test is first to measure the closed-in bottomhole pressure of the well and then to measure the flowing bottomhole pressure at several stabilized rates of flow. This type of testing is done on both oil and gas wells and is the most widely accepted method of determining the capacity of gas wells. In many states, the regulation of gas production is based upon a test of this type.

RESERVOIR STIMULATION

The term *reservoir stimulation* encompasses several processes used to enlarge old channels or to create new ones in the producing formation. Since oil usually exists in the pores of sandstone or the cracks of limestone formations, enlarging or creating new channels causes the oil or gas to move more readily to a well. An early method of stimulating wells used nitroglycerine. Using high explosives to improve a well's productive capacity began in the late 1800s and continued until *acidizing* and *hydraulic fracturing* were developed in the 1940s. Although nuclear explosives were used in recent experiments, the two most commonly used stimulation methods today are acidizing and hydraulic fracturing.

Acidizing is treating oil-bearing formations with acid to enlarge the pore spaces and passages by dissolving rock, thus enlarging existing channels and opening new ones to the wellbore (fig. 5.13). Oilfield acids must create reaction products that are soluble; otherwise, solid materials would be precipitated and plug the pore space in the rocks. The acid must also be relatively safe to handle, and, since large volumes are used, it must be fairly inexpensive. Reservoir rocks most commonly acidized are limestone (calcium carbonate) and dolomite (a mixture of calcium and magnesium carbonates).

Acidizing

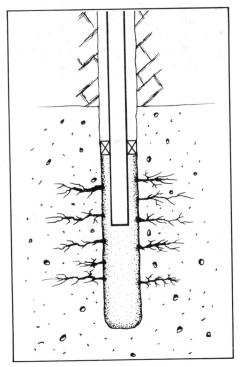

Fig. 5.13. Acid enlarges existing channels or makes new ones.

Fig. 5.14. Many truck-mounted pumps are used to fracture a formation by applying hydraulic pressure. (*Courtesy of Halliburton Services*)

Fig. 5.15. Sand is one proppant used to hold fractures open.

Additives are used with oilfield acids for many reasons, but one of the most important is to prevent or delay corrosion—that is, to inhibit the acid from attacking the steel tubing or casing in the well. A *surfactant*, or surface active agent, is another type of additive. It is mixed in small amounts with an acid to make it easier to pump the mixture into the rock formation and to prevent spent acid and oil from forming emulsions. Other common additives are *sequestering agents*, which prevent the precipitation of ferric iron during acidizing, and *antisludge agents*, which prevent an acid from reacting with certain types of crude and forming an insoluble sludge that blocks channels, or reduces permeability.

Most limestone and dolomite formations have low permeabilities. Acid injection into these low-permeability formations, even at moderate rates, usually results in a fracture type of acid treatment, meaning that the pressure is high enough to cause the formation to crack. Another type of treatment—interstitial, or matrix, acidizing—also results in production increases. Interstitial acidizing consists of treating at a rate and pressure low enough to avoid fracturing the formation. This technique is generally used when formation damage is present or when a water zone or gas cap is nearby and fracturing might result in high water or excessive gas production.

Hydraulic fracturing

Formation fracturing by hydraulic pressure has gained wide acceptance. Since hydraulic fracturing does for sandstone reservoirs what acid treatment does for limestone or dolomite reservoirs, this type of stimulation may result in commercial production in an area where it was not feasible before.

Essentially, the process consists of applying hydraulic pressure against the formation by pumping fluid into the well (fig. 5.14). This pressure actually splits the rocks. Hydraulic fracturing is used to accomplish four basic jobs: (1) create penetrating reservoir fractures to improve the productivity of a well, (2) improve the ultimate recovery from a well by extending the flow channels farther into the formations, (3) aid in improved recovery operations, and (4) increase the rate of injection of brine and industrial waste material into disposal wells. When used, fracturing is usually done on initial completion. Refracturing to restore productivity of a well is a regular procedure. Extension of the existing fractures will usually improve well productivity.

During early experimental work, it was discovered that a hydraulically formed fracture tends to heal, or lose its fluid-carrying capacity, after the parting pressure is released unless the fracture is propped open in some manner. *Proppants*, or propping agents, are used to hold the fractures open. Sand, nutshells, and beads of aluminum, glass, and plastic may be used as proppants (fig. 5.15).

Spacer materials are used between the particles of the proppant to ensure its optimum distribution.

The fracturing fluid must not only break down the formation but also extend and transport the propping agent into the fracture. The oil-base, water-base, or acid-base fracturing fluid and the additives used to modify its properties, along with the proppant and spacers, make up a very complex substance. Choice of the most suitable base depends on the chemical nature of the rock, its physical characteristics, and the nature of the reservoir fluid.

RESERVOIR DRIVE MECHANISMS

After the well has been completed, gas and oil begin their journey from the reservoir to the surface. This first period in the producing life of a reservoir is called primary recovery, or primary production. During this stage, natural reservoir energy, either by itself or in combination with an artificial assist, displaces the hydrocarbons from the pores of a formation and drives it toward production wells and up to the surface. Reservoir drive mechanisms, as the reservoir energies are called, include dissolved-gas drive, gas-cap drive, water drive, combination drive, and gravity drainage.

Dissolved-gas drive

In a dissolved-gas drive (known also as solution-gas drive), the lighter hydrocarbon components that exist as a liquid in the reservoir before it is produced come out in the form of gas as the reservoir is produced. The dissolved gas coming out of the oil expands to force the oil into the wellbore (fig. 5.16). In dissolved-gas drive reservoirs, pressure declines rapidly and continuously, and wells generally require pumping or some other artificial lift at an early stage. Because of this characteristic, both dissolved-gas and gas-cap drives are termed *depletion drives*. The gas-oil ratio is low initially, then rises to a maximum and drops. Except from edge wells that may have penetrated the oil-water contact, little or no water is produced. Recovery efficiency (how much of the oil in place is ultimately produced) varies from as little as 5 percent to as much as 30 percent.

Gas-cap drive

A gas-cap drive is a depletion drive in a reservoir that has a gas cap. As pressure is reduced in the oil zone by withdrawal, the gas cap expands and pushes oil out ahead of it (fig. 5.17). Performance in this type of reservoir is similar to dissolved-gas drive, but pressure may decline more slowly because the gas cap provides a lot of drive energy. Gas-oil ratios rise continuously in upstructure wells with this

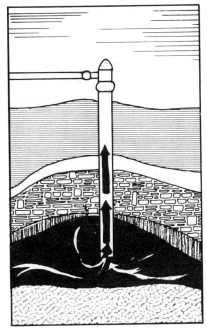

Fig. 5.16. In a dissolved-gas drive reservoir, gas comes out of the oil, expands, and lifts oil to the surface.

Fig. 5.17. In a gas-cap drive reservoir, the free gas in the cap and the gas dissolved in the oil expand to move oil up to the surface.

Fig. 5.18. In a water drive reservoir, salt water under the oil pushes oil to the surface.

drive, with little or no water production except in those wells that penetrate the reservoir near its edge. Because the whole system has more energy, such a reservoir may have a long flowing life, depending on the size of the gas cap. Oil recovery may be from 20 to 40 percent of the original oil in place.

Water drive

Water drive occurs when there is enough energy available from water in the reservoir to move the hydrocarbons out of the reservoir, into the wellbore, and up to the surface (fig. 5.18). The water in most water-bearing formations is under fluid pressure proportional to the depth beneath the surface; in other words, the deeper the water, the higher the pressure.

Water is quite efficient at displacing oil from reservoir rock. As the oil is driven out of the reservoir, the water moves in to replace it. The process is about the same as emptying a tank of oil by displacing the oil with water injected at the bottom. The pressure remains high as long as the volume of oil withdrawn is replaced by an approximately equal volume of water. If the reservoir pressure remains high, the surface gas-oil ratio remains low because little or no free gas is evolved in the reservoir. Because high reservoir pressure is maintained, wells usually flow on their own until water production becomes excessive and kills the well. Water production may start early and increase to an appreciable amount as water encroaches into the oil and into the producing wells. Expected oil recovery with water drive is generally higher—sometimes 50 percent or more of the oil originally in place—because of the greater displacement efficiency of water over gas.

Water drive reservoirs can have bottom-water drive or edgewater drive. In a *bottom-water drive* reservoir, the oil accumulation is totally underlain by water. A well drilled anywhere through a reservoir of this type penetrates oil first and then water. In an *edgewater drive* reservoir, the oil accumulation almost completely fills the reservoir. Water occurs only on the edges of the reservoir, so only wells drilled along the edges penetrate water. Wells drilled near the top of the structure penetrate oil only.

Combination drive

Depletion and water drives can be characterized as pure drive mechanisms; however, another drive is one that can best be described as a combination drive. One such drive has a gas cap above the oil and water below it. Both the gas cap and the water drive the oil into and up the wellbore to the surface. Another type of combination drive has gas dissolved in the oil with the water below it. Both the water and the gas coming out of solution drive the oil to the surface.

Gravity drainage

A less common type of drive is gravity drainage. The force of gravity is, of course, always at work in a reservoir. Usually, gravity causes oil to migrate upward by pulling the heavier water down beneath it. However, in shallow, highly permeable, steeply dipping reservoirs and in some deeper, nearly depleted reservoirs, the oil may flow downhill to the wellbore.

ARTIFICIAL LIFT

When a well is first completed, the fluid is expected to flow to the surface by natural reservoir energy for some period of time. At some time during their economic life, however, most oilwells will require some form of artificial lift to help raise the fluid to the surface and obtain the maximum recovery of oil for maximum profit to the producer. The most common methods of artificial lift are those that use gas and those that use pumps. Types of pumps used are sucker rod pumps, subsurface hydraulic pumps, and electric submersible pumps.

If a supply of gas is economically available and the amount of fluid will justify the expense, gas lift is commonly used (fig. 5.19). Gas lift is especially suitable for offshore use because platform space is limited and gas lift equipment is largely downhole. In the gas lift process, gas is injected into the fluid column of a well to lighten and raise the fluid by expansion of the gas. Injected gas aerates the fluid to make it exert less pressure than the formation does; consequently, the higher formation pressure forces the fluid out of the wellbore. Gas may be injected continuously or intermittently, depending on the producing characteristics of the well and the arrangement of the gas lift equipment.

Gas lift

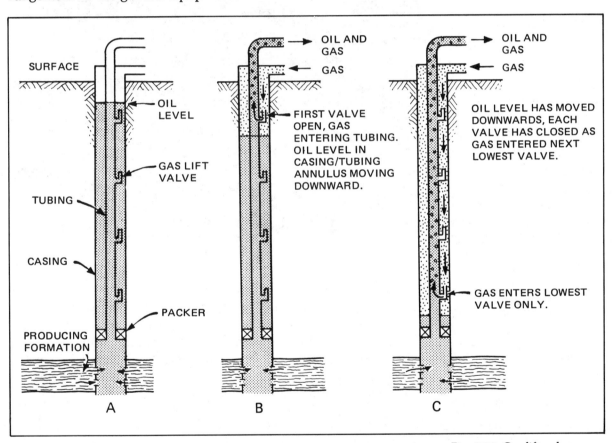

Fig. 5.19. Gas lift valves are installed on the tubing string at various depths. Gas enters the annulus and progresses down through the gas lift valves.

Fig. 5.20. A beam pumping unit with wellhead, sucker rods, and sucker rod pump is one method of artificial lift.

Sucker rod pumps

The artificial lift method that involves sucker rod pumps is commonly known as rod pumping, or *beam pumping* (fig. 5.20). Surface equipment used in this method imparts an up-and-down motion to a sucker rod string that is attached to a piston, or plunger, pump submerged in the fluid of a well. Most beam pumping units have the same general operating principles (fig. 5.21).

Subsurface hydraulic pumps

A hydraulic pump is so called because it is operated by a hydraulic, or fluid-driven, motor in the unit at the bottom of the well. The fluid used to drive the motor is the oil from the well itself. The motor, in turn, drives a pump that pumps the oil to the surface.

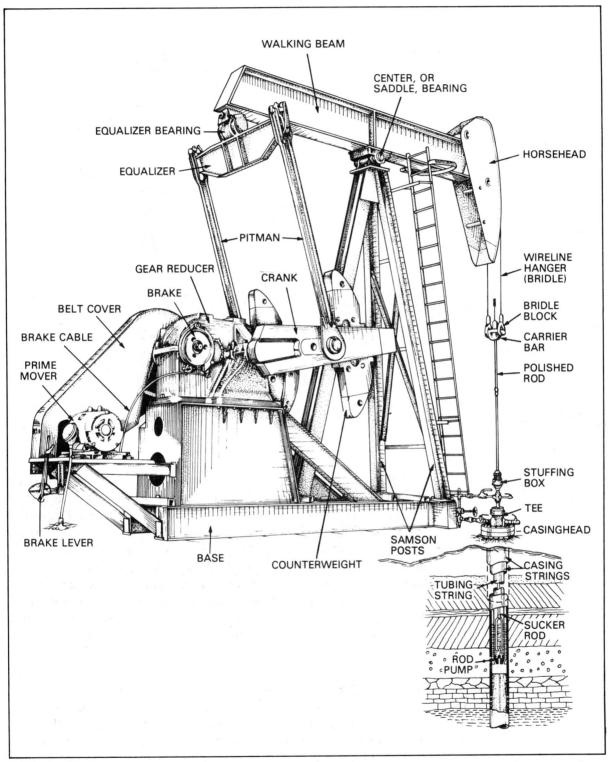

WALKING BEAM

CENTER, OR
SADDLE, BEARING

EQUALIZER BEARING

EQUALIZER

HORSEHEAD

PITMAN

GEAR REDUCER

CRANK

BRAKE

WIRELINE
HANGER
(BRIDLE)

BELT COVER

BRIDLE
BLOCK

BRAKE CABLE

CARRIER
BAR

PRIME
MOVER

POLISHED
ROD

STUFFING
BOX

TEE

CASINGHEAD

BRAKE LEVER

SAMSON
POSTS

BASE

COUNTERWEIGHT

CASING
STRINGS

TUBING
STRING

SUCKER
ROD

ROD
PUMP

Fig. 5.21. A beam, or rod,
pumping unit has many components.

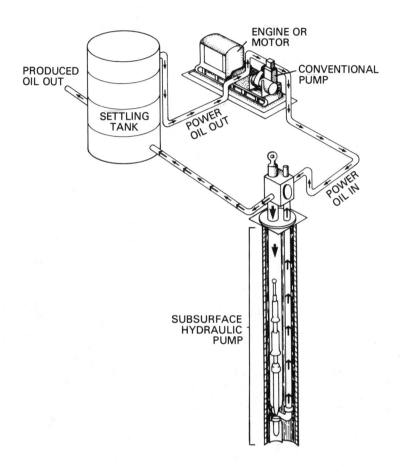

ENGINE OR MOTOR

PRODUCED OIL OUT

CONVENTIONAL PUMP

SETTLING TANK

POWER OIL OUT

POWER OIL IN

SUBSURFACE HYDRAULIC PUMP

Fig. 5.22. A hydraulic pumping system has surface and subsurface components.

One type of hydraulic pump is the free pump. This pump is installed in the bottom of the tubing and is operated by oil taken from a *settling tank* at the surface and pumped downward through the tubing (fig. 5.22). The power oil is returned to the surface through the small tubing, along with new oil taken from the formation. Other types of hydraulic pumps are available, and all of them use the same basic fluid motor and pump in the bottom of the hole. The motor and pumping unit are lowered into the well on a string of tubing, and the pump and power-oil tubing can be run either inside the regular well tubing or inside the casing.

Electric submersible pumps

A submersible pump is essentially a centrifugal pump that has blades, or impellers, attached to a long shaft. The shaft is connected to a long electric motor that is submerged in the well. The pump is usually installed in the tubing just below the fluid level. Since both the pump and the pump motor are submerged in the well fluid, electric current is supplied through a special heavy-duty armored cable (fig. 5.23).

A submersible pump can lift as much as 25,000 barrels of fluid per day, making it especially suitable for wells that have large volumes of fluid to be removed, such as older wells that produce a large volume of water in relation to the volume of oil.

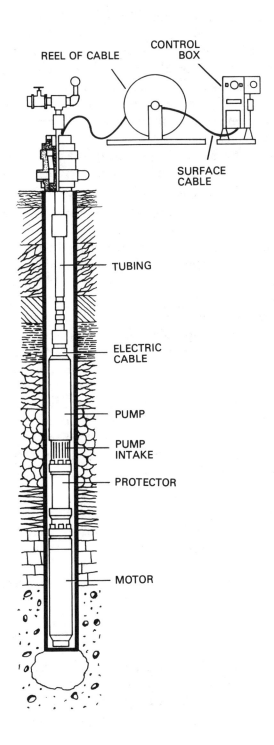

REEL OF CABLE

CONTROL BOX

SURFACE CABLE

TUBING

ELECTRIC CABLE

PUMP

PUMP INTAKE

PROTECTOR

MOTOR

Fig. 5.23. In an electric submersible pump, an electric motor turns impellers in the pump. A protector insulates the motor from the pump.

IMPROVED RECOVERY TECHNIQUES

A reservoir may approach the end of its primary life having produced only a small fraction of the oil in place for many reasons. Production of the reservoir may have been started before good development and production practices were known. Unknown problems such as a casing leak could have resulted in wasted reservoir energy, or the owner might not have been willing to invest more money for maintenance while wells were producing profitably. Even with the best primary production methods, however, as much as 75 percent of the oil may be left behind. To recover this oil, additional oil recovery methods are employed.

When operations are employed to rejuvenate a depleted or nearly depleted reservoir, the reservoir is said to be in the enhanced oil recovery (EOR) stage of production. Probably because of the dynamic growth in the field of improved oil recovery, terminology has been used inconsistently in referring to various categories of recovery. In the past, the term *secondary recovery* was nearly synonymous with pressure maintenance methods such as waterflooding, and the term *tertiary recovery* with advanced EOR methods that improved oil displacement. As the field has developed, however, this distinction has not held true. Most EOR methods can be used during more than one stage of recovery. To further complicate matters, in current programs of government-assisted EOR projects, enhanced oil recovery is defined to include only advanced EOR methods (which do *not* include waterflooding and immiscible gas injection) for which the operator may be reimbursed. Therefore, it might be helpful to discuss improved oil recovery methods as types rather than as stages of recovery.

The major methods of improved oil recovery are waterflooding, gas injection, chemical flooding, and thermal recovery. Gas injection may be immiscible or miscible — that is, not capable or capable of mixing — with the oil in the reservoir. Waterflooding and immiscible gas injection are pressure maintenance methods and generally used in secondary recovery operations. Table 5.1 gives a brief description of each method's process and its main use.

Waterflooding

Waterflooding is the least expensive and most widely used improved recovery method, but it is not considered an advanced recovery method because it is used typically in secondary recovery. In this method, water is introduced into the formation through injection

TABLE 5.1

IMPROVED RECOVERY METHODS

Method of Recovery		Process	Use
WATER-FLOODING	Water	Water is pumped into the reservoir through injection wells to force oil toward production wells.	Method most widely used in secondary recovery.
IMMISCIBLE GAS INJECTION	Natural gas, flue gas, nitrogen	Gas is injected to maintain formation pressure, to slow the rate of decline of natural reservoir drives, and sometimes to enhance gravity drainage.	Secondary recovery.
MISCIBLE GAS INJECTION	Carbon dioxide	Under pressure, carbon dioxide becomes miscible with oil, vaporizes hydrocarbons, and enables oil to flow more freely. Often followed by injection of water.	Secondary recovery or tertiary recovery following waterflooding. Considered especially applicable to West Texas reserves because of carbon dioxide supplies located within a feasible distance.
	Hydrocarbons (propane, high-pressure methane, enriched methane)	Either naturally or under pressure, hydrocarbons are miscible with oil. May be followed by injection of gas or water.	Secondary or tertiary recovery. Supply is limited and price is high because of market demand.
	Nitrogen	Under high pressure, nitrogen can be used to displace oil miscibly.	Secondary or tertiary recovery.
CHEMICAL FLOODING	Polymer	Water thickened with polymers is used to aid waterflooding by improving fluid-flow patterns.	Used during secondary recovery to aid other processes during tertiary recovery.
	Micellar-polymer (surfactant-polymer)	A solution of detergentlike chemicals miscible with oil is injected into the reservoir. Water thickened with polymers may be used to move the solution through the reservoir.	Almost always used during tertiary recovery after secondary recovery by waterflooding.
	Alkaline (caustic)	Less expensive alkaline chemicals are injected and react with certain types of crude oil to form a chemical miscible with oil.	May be used with polymer. Has been used for tertiary recovery after secondary recovery by waterflooding or polymer flooding.
THERMAL RECOVERY	Steam drive	Steam is injected continuously into heavy-oil reservoirs to drive the oil toward production wells.	Primary recovery. Secondary recovery when oil is too viscous for waterflooding. Tertiary recovery after secondary recovery by waterflooding or steam soak.
	Steam soak	Steam is injected into the production well and allowed to spread during a shut-in soak period. The steam heats heavy oil in the surrounding formation and allows it to flow into the well.	Used during primary or secondary production.
	In situ combustion	Part of the oil in the reservoir is set on fire, and compressed air is injected to keep it burning. Gases and heat advance through the formation, moving the oil toward the production wells.	Used with heavy-oil reservoirs during primary recovery when oil is too viscous to flow under normal reservoir conditions. Used with thinner oils during tertiary recovery.

wells in order to move oil to the production wells. Although water for injection may be supplied from water wells drilled specifically for this purpose, the water that is produced along with the oil may also be used.

Regardless of whether the water is injected into formations for disposal, pressure maintenance, or waterflood, the fluid must meet certain requirements. The injected water must be clear, stable, and similar to the water in the formation where it is being injected. It also must not be severely corrosive and must be free of materials that may plug the formation. If the water is severely corrosive, it may be treated with inhibitors or other chemicals, or corrosion-resistant materials may be used in the disposal equipment and wells. De-aerating, softening, filtering, chemical treating, stabilizing, and testing are common water-treatment processes.

The use of waterflooding as an improved recovery method in the frozen Arctic has posed many technological challenges to the industry. The harsh environment, unfriendly to man and machine, has necessitated the development of new methods. Until the mid-1980s, the oil- and gas-rich Prudhoe Bay area on Alaska's North Slope flowed on its own natural reservoir drives. Except for gas reinjection, no pressure maintenance or enhanced recovery programs were used until treated seawater was injected into the Sadlerochit formation at rates of up to 2 million barrels per day. The seawater treatment program is still being used in that region. Other enhanced recovery measures such as infill drilling, gas lift programs, and low-pressure separation are also being put into operation.

The waterflood system designed for this region incorporates several special features, including the building of the largest water injection pumps in existence and a seawater intake structure specifically adapted to the shallow Beaufort Sea to minimize the loss of marine life. Such a system must be reliable under harsh conditions, since an unplanned flow stoppage can shut down the system for as long as a year. Equipment breakdown of even a minor sort is expensive in this part of the world because of high labor costs. Also, the presence of permafrost means that water cannot be distributed through underground pipes, the usual practice in most waterfloods operating today. Instead, the Prudhoe Bay system employs specially insulated and protected aboveground pipes for water movement. Fluids are kept flowing with special operations dealing with emergency circulating and heating modes, disposal, evacuation, filling, and cleaning. Many other aspects of the system have been investigated with great care, especially a plan to automate sequential steps in case an emergency requiring the evacuation of the system arises.

Immiscible gases—those that will *not* mix with oil—include natural gas, flue gas, and nitrogen. The natural gas produced with the oil can be reinjected into the well to maintain formation pressure. Immiscible gas injected into the well behaves in a manner similar to that in a gas-cap drive: the gas expands to force additional quantities of oil to the surface. Gas injection requires the use of compressors to raise the pressure of the gas so that it will enter the formation.

Immiscible gas injection

The petroleum industry first began to use miscible gas injection projects in the 1950s while searching for a miscibility process that would effectively recover oil during secondary and tertiary production. The injection fluids used in this method include liquefied petroleum gases (LPGs) such as propane, methane under high pressure, methane enriched with light hydrocarbons, nitrogen under high pressure, and carbon dioxide used alone or followed by water. All of these are effective in displacing most of the trapped reservoir oil, but their usefulness is somewhat diminished by practical field application problems and certain economic considerations.

Miscible gas injection

LPGs are appropriate for use in many reservoirs because of their miscibility with crude oil on first contact. However, liquefied petroleum gases are in such demand as a marketable commodity in their own right that their employment in improved recovery appears limited.

Methane also has a good market value, even though enriched methane and methane under high pressure can be used economically as injection fluids in areas deprived of ready natural gas markets.

In the 1970s carbon dioxide began to be used frequently as an injection gas. Carbon dioxide has a greater viscosity under pressure than many other gases and achieves miscible displacement at low pressures. When used as an injection gas in areas well supplied with carbon dioxide, it is a very attractive economic choice because it is less costly than the LPGs and methane.

Natural drive fluids such as water and gas leave a large supply of oil in the reservoir even under the best of conditions. Because these fluids are immiscible with oil, the oil resists being displaced from the rock pores. Also, these drive fluids have densities and mobilities that are incompatible with oil. For this reason they sweep only certain portions of the reservoir and finger past the oil. The improved recovery method known as chemical flooding adds chemicals to the water in order to overcome these problems.

Chemical flooding

Polymer flooding, a type of chemical flooding, is one way to control drive-water mobility and fluid flow patterns in reservoirs. Polymers—long, chainlike, high-weight molecules—have three important oil recovery properties. They increase water viscosity, decrease effective rock permeability, and are able to change their viscosity with the flow rate. Small amounts of water-dissolved polymers increase the viscosity of water. This higher viscosity slows the progress of the water through a reservoir and makes it less likely to bypass the oil in low-permeability rock.

Another chemical method, called *micellar-polymer flooding*, can be broken down into two parts. Detergentlike chemicals are used to free residual oil trapped in reservoir rock. Synthetic detergents and soaps have an ingredient known as a surfactant, or surface-active agent. Surfactant molecules are dually attracted to both water and oil. They thus have the ability to reduce surface tension and to break up oil into tiny droplets that can be drawn from rock pores by water. A surfactant-water solution is injected into a reservoir to release the oil, and then polymer-thickened water is injected to push the oil toward producing wells.

Micellar flooding and *alkaline flooding*, two other chemical flooding methods, are used to form a drive fluid that can be mixed with oil. These methods can be employed after waterflooding during secondary or tertiary recovery.

Thermal recovery

Thermal recovery methods lower the viscosity of oil and increase its flow by introducing heat into a reservoir. High-viscosity and high-density crude oils respond better to heat application than to other recovery methods. Heat is generated on the surface and is transmitted to the formation in the form of steam or hot water. Hot fluid is pumped into the injection wells and moved toward the producing wells.

Steam injection, also known as steam drive or continuous steam injection, is better at heating reservoir oil and produces less water than hot waterflooding.

Another approach, *fire flooding*, or *in situ combustion*, generates heat in a reservoir by injecting air into the well and starting a fire in the formation close to an injection well. The fire and the air flow move simultaneously toward the production wells. Known as forward combustion, this method allows for the use of air as the injected heat carrier (dry combustion) or a combination of air and water injection (wet combustion).

A seldom-used variant is *reverse combustion*, in which the fire moves from the production well to the injection well. The direction of the fire flow in this method is counter to the flow of the injected air.

SURFACE HANDLING
OF WELL FLUIDS

Oil and gas are not usually merchantable as they come from the wellhead. Typically, a well stream is a high-velocity, turbulent, constantly expanding mixture of hydrocarbon liquids and gases mixed with water and water vapor, solids such as sand and shale sediments, and sometimes contaminants such as carbon dioxide and hydrogen sulfide. Several steps are necessary to get oil or gas ready to transport to its next stop (fig. 5.24).

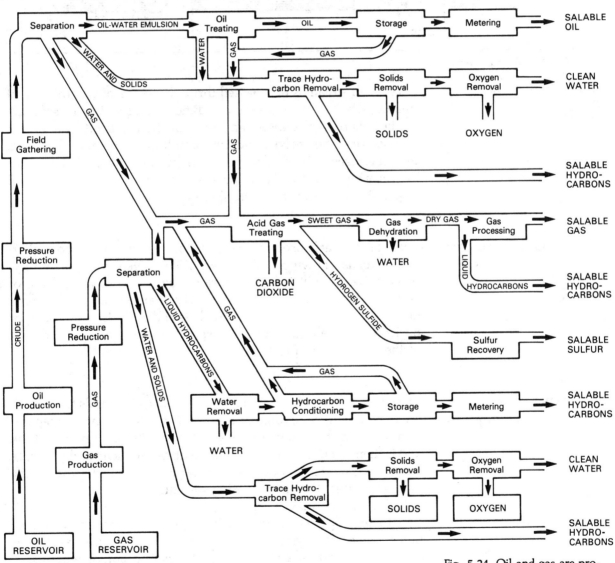

Fig. 5.24. Oil and gas are processed through many stages to become merchantable.

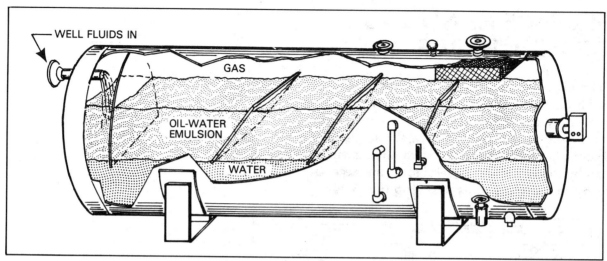

WELL FLUIDS IN

GAS

OIL-WATER
EMULSION

WATER

Fig. 5.25. A horizontal, three-phase, free-water knockout provides space for the free water to settle out and for the gas to rise.

The well stream is first passed through a series of separating and treating devices to remove the sediments and water, to separate the liquids from the gases, and to treat the emulsions for further removal of water, solids, and undesirable contaminants. The oil is then stabilized, stored, and tested for purity. The gas is tested for hydrocarbon content and impurities, and gas pressure is adjusted to pipeline or other transport specifications.

Removing free water

Some of the water produced with the oil may not be mixed with it; this water is known as *free water*. When given the opportunity, this free water will readily separate from the oil by the force of gravity alone, since the water is heavier than the oil. A free-water knockout (sometimes abbreviated as FWKO) is a vertical or horizontal vessel that provides a space for free water to settle out of the well stream (fig. 5.25).

Separating liquids from gases

The simplest type of equipment used to separate liquids from gases is a tank in which the force of gravity is used to achieve the separation. Oil, which is heavier than gas, falls to the bottom of the tank and is then removed for additional treatment or sent to the storage tanks. The lighter element, gas, is removed from the top of the tank and enters the gas gathering system.

Vertical and horizontal separators use centrifugal force in addition to gravity to achieve the best possible separation of well fluids. Separators may be two-phase or three-phase devices. A two-phase vessel separates the well fluids into liquids and gas. A three-phase vessel separates the fluids into oil, gas, and water; however, the oil and gas may still contain some water at this stage of treatment. A three-phase separator may be used for well fluids that contain relatively small amounts of free water.

Hydrocarbons from a well troubled with paraffin (a white, waxy hydrocarbon sometimes found in petroleum) can also mean problems for the surface equipment used to treat the well fluids. Paraffin may not only reduce the efficiency of oil and gas separators but also make the equipment inoperable by partially filling the vessel or blocking fluid passages. While the best place to treat paraffin problems is downhole, paraffin blockage can also be handled in separators. Steaming and using solvents are effective, for example, just as they are downhole. Another method that can be used involves coating all internal surfaces of the separator with a plastic to which the paraffin will not stick and thus not build up to a harmful thickness.

Vertical separators. Well fluids enter the vertical separator at about midpoint of the tank (fig. 5.26). Here, while they are swirled

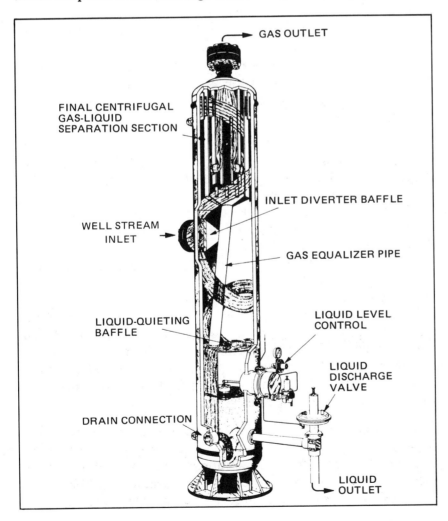

Fig. 5.26. Both gravity and centrifugal force are used to separate well fluids in a vertical separator.

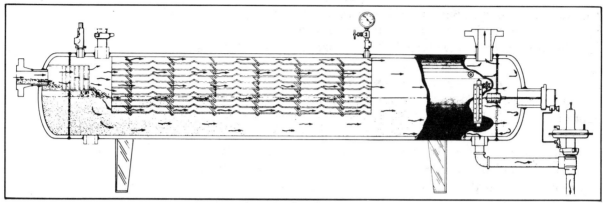

Fig. 5.27. This horizontal separator has a single tube, though some separators have more.

around, two forces work to separate the oil and gas: *gravity* causes the heavier oil to drop to the lower part of the separator, and *centrifugal force* of the whirling action causes the heavy oil particles to collect on the walls of the separator. The gas, which still contains some liquid particles, rises through the chamber where it is again swirled, and the same forces operate to remove more liquid. The gas then rises again and passes through the scrubber dome where the last liquid particles are removed. After this final process, the gas is removed through the gas outlet near the top of the separator, and the heavier oil is removed through an outlet at the bottom of the separator.

Horizontal separators. A horizontal separator operates in much the same way as a vertical separator, though, as the name implies, the unit is in a horizontal position instead of upright (fig. 5.27). The same gravity and centrifugal forces are used, and the oil is removed from the bottom of the unit and the gas from the top.

Horizontal separators may be of either the single-tube or the double-tube design. The double-tube horizontal separator has two horizontal units mounted one above the other and joined by flow channels near the ends of the units. The well fluids enter at one end of the upper unit and are swirled, and the liquids fall through the flow pipe into the liquid reservoir in the lower portion of the bottom unit. The separating process continues in both the upper and the lower units. Gas removed from the liquid in the lower unit rises through the flow channel and joins the gas stream leaving the upper separator at the gas outlet. Oil is discharged through a connection at the lower part of the bottom tube.

Multistage separation. In order to obtain more complete recovery of liquids, more than one stage of separation is often desirable. When two-stage separation is used, the well fluids pass through either a vertical or a horizontal separator for the first stage of separation. The liquid resulting from this stage is then sent into a second separator that operates at a lower pressure, and more gas is removed from the liquid. Any number of separators may be used in stage separation as long as each stage operates at successively lower pressures. However, separators commonly remove free water and gas, leaving an emulsion as the liquid.

An *emulsion* is a mixture in which one liquid is uniformly distributed (usually as minute globules) in another liquid. In an oil-water emulsion, the oil is dispersed in the water; in a water-oil emulsion, the reverse is true. The two liquids that form an emulsion, oil and water in this case, are immiscible liquids; that is, they will not mix together under normal conditions (fig. 5.28). These two liquids will form an emulsion only if there is sufficient agitation to disperse one liquid as droplets in the other and if there is an emulsifying agent, or emulsifier, present. Emulsifying agents commonly found in petroleum emulsions include asphalt, resinous substances, and oil-soluble organic acids.

To break down a stable emulsion into its components, some form of treating is necessary. An emulsion may be referred to as

Treating oilfield emulsions

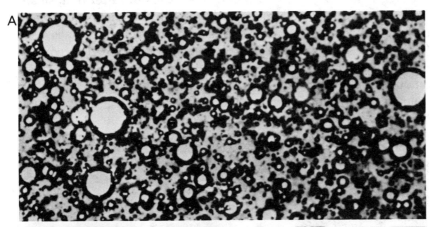

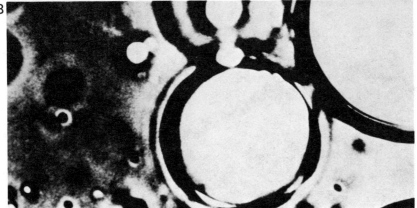

Fig. 5.28. This photomicrograph of a water-oil emulsion shows water droplets (clear circles) dispersed in oil (*A*). A closeup shows that water droplets touch but are unable to merge because of the film around them (*B*).

210

tight (difficult to break) or *loose* (easy to break). Whether an emulsion is tight or loose depends on the properties of the oil and water, the percentage of each found in the emulsion, and the type and amount of emulsifier present.

Treating facilities may use a single process or a combination of processes to break down an emulsion, depending upon the emulsion being treated. To break down a water-in-oil emulsion (the most common type), the properties of the emulsifying agent must be neutralized or destroyed so that the droplets of water may unite. Among the treatments developed to accomplish this process are the application of gravity, chemicals, heat, and electricity. To apply any or a combination of these treatments, the emulsion is sent through a treating tank or vessel. In addition to a special holding tank, heater-treaters and electrostatic treaters are used.

Use of chemicals. After separation from the gas and free water, the remaining emulsion is often piped into special tanks, or vessels, for treatment. Certain chemicals called emulsion breakers, or demulsifiers, may be added to the emulsion in order to make the droplets of water merge, or coalesce. When droplets merge, they get bigger, and big, heavy water drops settle out faster than small, light ones. To help determine which chemical is the most efficient demulsifier for the emulsion being treated, a bottle test is used (fig. 5.29). Results from a bottle test also indicate the required ratio of treating compound to emulsion—that is, the smallest amount of the proper chemical needed to satisfactorily break the volume of emulsion being produced. Chemicals may be applied to the emulsion through a chemical-injection pump.

Fig. 5.29. A bottle test helps determine the best chemical to use as a demulsifier for treating an emulsion.

Fig. 5.30. Well fluids are processed through various vessels on the lease site for separation and treatment.

Use of heat. The emulsion may also be heated. When heated, the viscosity of the emulsion is reduced, and the water and oil molecules move about rapidly, causing the water droplets to strike each other. When the force and frequency of the collision are great enough, the surrounding film of the emulsifying agent is ruptured, and the water drops merge and separate from the oil. Heat alone may not cause an emulsion to break down. Usually the application of heat is an auxiliary process to speed up separation. If at all possible, the use of heat is minimized or eliminated entirely from the treating process.

Use of electric current. Electricity is also used to treat emulsions, usually in conjunction with heat and chemicals. The film around the water droplets formed by the emulsifier is composed of polar molecules; that is, they have a positive and a negative end, very much like a bar magnet. When an electric current disturbs this film of polar molecules, the molecules rearrange themselves. The film is no longer stable, and adjacent water droplets coalesce freely until large drops form and settle out by gravity.

Emulsion treaters. Commonly used emulsion treaters are heater-treaters and electrostatic treaters. These devices combine various pieces of equipment used to treat an emulsion in one vessel. Thus, an emulsion treater is the vessel in which the effects of chemicals, heat, settling, and often electricity are applied to an emulsion (fig. 5.30). A treater may be a combination oil-gas separator, free-water knockout, heater, and filter. Any one of these functions may be emphasized more than the other, depending on the service for which the treater is designed. In addition, treaters are available in a number of sizes and shapes so that different volumes of well fluids may be handled. Some treaters are designed for use in extremely cold climates; other models are especially designed to treat foaming oil.

Treaters can be operated at atmospheric pressure, but they often operate under low working pressure, and it is often advantageous to use the treater as a low-pressure, second-stage separator as well as a treating unit. Where flow-line pressures are low, it can be used

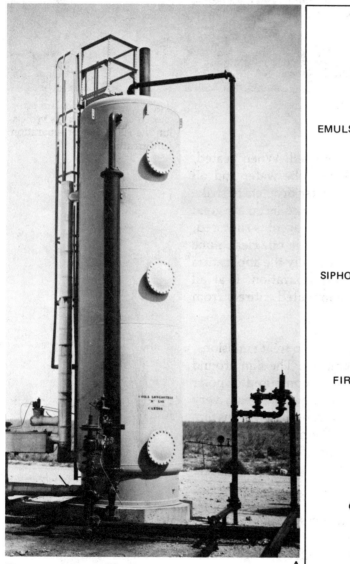

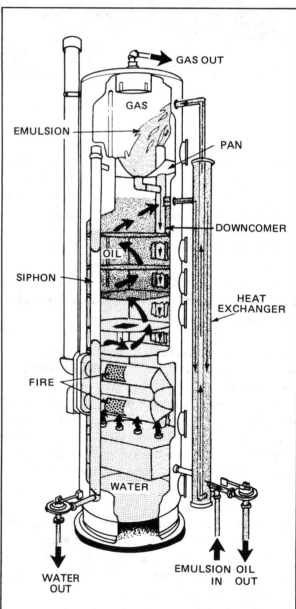

Fig. 5.31. A vertical heater-treater (A) separates an oil-water emulsion as shown in the cutaway view (B).

as a primary separator, thus eliminating the need for a regular separator. Flow-line treaters are either vertical or horizontal vessels constructed so that the emulsion enters the vessel with treating chemical already added.

In a *vertical heater-treater* (fig. 5.31), the chemically treated emulsion flows through a heat exchanger. Cold, incoming emulsion is preheated by warm, already treated crude oil leaving the vessel through the heat exchanger. The emulsion then flows to the top of the treater, splashes over a pan, and falls downward through a downcomer. Gas breaks out of the emulsion as it splashes over the pan. When the emulsion reaches the lower section of the treater, free water and sediment fall out. Emulsion rises through the water, which is heated by a fire tube, and settles midway in the vessel. There, the oil-water emulsion separates. The water falls to the bottom, while the crude oil rises upward, passes through an oil outlet to the heat exchanger, and warms incoming emulsion. As the clean crude exits the heater-treater, it is piped to a stock tank.

A *horizontal heater-treater* (fig. 5.32) is similar in operation to a vertical one but has the advantage of a larger treating section and

Fig. 5.32. The treating action of a horizontal heater-treater is like that of a vertical treater.

can handle larger volumes of fluid. However, a horizontal treater requires more space and cannot handle sediment as well as a vertical treater.

An *electrostatic treater*, similar in design and operation to a horizontal heater-treater, features a high-voltage electric grid. The emulsion rises to the grid, and the water droplets receive a charge that causes them to collide, merge, and settle out.

Handling natural gas

Once the gas has been separated from free liquids such as crude oil, hydrocarbon condensate, water, and entrained solids, it generally requires further processing to prevent hydrate formation, to remove water vapor and impurities such as hydrogen sulfide and carbon dioxide, and to separate natural gas liquids from the gas. Whether these operations are performed in the field, in a gas processing plant, or not at all depends on such factors as the type of well, the content of the well fluids, climate conditions, stipulations of the sales contract, and the economics of using field processing equipment. Field handling also includes control of the delivery pressure of the sales gas by the use of pressure-reducing regulators or compressors.

Preventing hydrate formation. Most natural gas contains substantial amounts of water vapor. If the water vapor is cooled below a certain temperature, hydrates can form. A hydrate is a solid crystalline compound formed by hydrocarbons and water under reduced temperature and pressure. It often resembles dirty snow. When gas comes from the wellhead, it is usually at a high temperature and pressure. As the gas cools, it can hold less water in the vapor form. When water vapor condenses into a liquid, hydrate formation becomes a real danger. Hydrates may pack solidly in gathering lines and equipment, blocking the flow of gas. In addition to forming hydrates, water accelerates corrosion of pipelines and equipment. Excess water content is undesirable from the buyer's standpoint. Many pipeline companies will not buy gas containing more than 7 pounds of water per million cubic feet of gas.

To prevent the water vapor in the gas from forming hydrates, two commonly used methods are heating the gas and injecting chemical inhibitors into the flow stream. The most widely used equipment for heating gas is an *indirect heater* because it is simple, economical, and relatively trouble-free. It consists of a heater shell, a removable fire tube and burner assembly, and a removable coil assembly (fig. 5.33). The heater shell is usually filled with water, which completely covers the fire tube and the coil assembly. The fire tube heats the water bath, the water bath heats

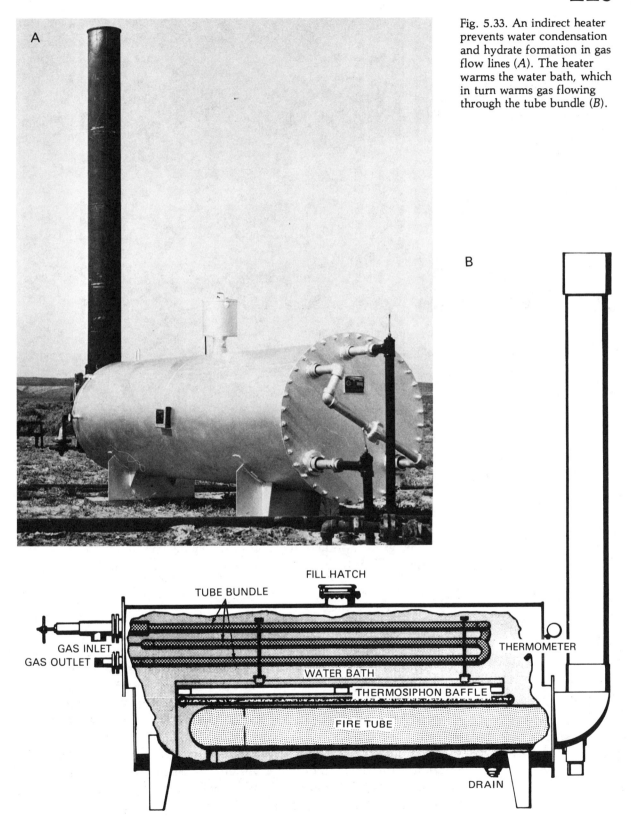

Fig. 5.33. An indirect heater prevents water condensation and hydrate formation in gas flow lines (A). The heater warms the water bath, which in turn warms gas flowing through the tube bundle (B).

Fig. 5.34. Arrows indicate the injection points for introducing glycol or methanol into the gas flow stream to prevent hydrate formation.

Fig. 5.35. This glycol liquid-desiccant dehydrator unit is skid-mounted.

the coil assembly, and the gas is heated as it passes through the coil assembly. Hydrate inhibitors are products such as ammonia, brine, glycol, and methanol that lower the freezing point of water vapor. They are injected into the gas stream at strategic points (fig. 5.34).

Dehydrating. To prevent operating problems and to meet contract specifications, the gas is usually dehydrated. *Dehydration* is the removal of water. Two methods are used: absorption and adsorption. *Absorption* is the removal of water vapor from gas by bubbling the gas through a liquid desiccant (fig. 5.35). A desiccant, whether liquid or solid, is a substance that has a special attraction or affinity for water. The liquid desiccant, often a glycol solution, absorbs the water, and these liquids can then be separated from the gas in a dehydration unit (fig. 5.36). *Adsorption* is the removal of water vapor by passing the vapor-laden gas over a bed of granular solids with an attraction for water (fig. 5.37). The water adheres to the solid desiccant, and the dry

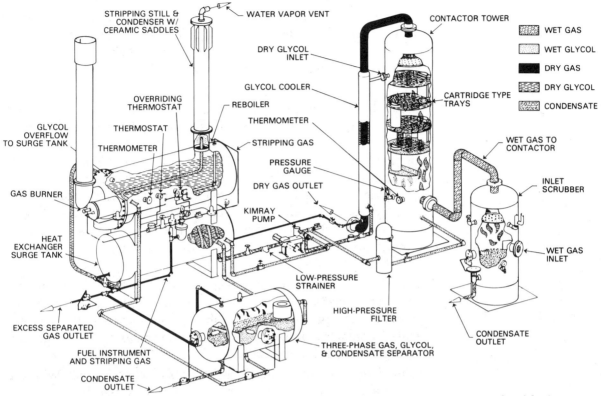

STRIPPING STILL & CONDENSER W/ CERAMIC SADDLES

WATER VAPOR VENT

DRY GLYCOL INLET

GLYCOL COOLER

OVERRIDING THERMOSTAT

THERMOSTAT

THERMOMETER

REBOILER

THERMOMETER

STRIPPING GAS

CONTACTOR TOWER

WET GAS

WET GLYCOL

DRY GAS

DRY GLYCOL

CONDENSATE

CARTRIDGE TYPE TRAYS

GLYCOL OVERFLOW TO SURGE TANK

PRESSURE GAUGE

DRY GAS OUTLET

WET GAS TO CONTACTOR

GAS BURNER

KIMRAY PUMP

INLET SCRUBBER

HEAT EXCHANGER SURGE TANK

LOW-PRESSURE STRAINER

WET GAS INLET

HIGH-PRESSURE FILTER

EXCESS SEPARATED GAS OUTLET

THREE-PHASE GAS, GLYCOL, & CONDENSATE SEPARATOR

CONDENSATE OUTLET

FUEL INSTRUMENT AND STRIPPING GAS

CONDENSATE OUTLET

Fig. 5.36. This dehydration unit uses liquid glycol as its desiccant. (*Courtesy of Sivalls, Inc.*)

Fig. 5.37. This dry-bed dehydration system uses a solid desiccant.

gas passes out of the adsorption tower. After the desiccant bed is dried by heating and vaporizing the water, the bed can be used again. Each of the processes has its own special advantages and disadvantages. Dry-bed adsorption is more likely to be used in a gas processing plant than in the field.

Removing contaminants. Natural gas often contains carbon dioxide and compounds of sulfur such as hydrogen sulfide. These gases are called *acid gases* because they form acids or acidic solutions in the presence of water. Gases with high concentrations of sulfur are also called *sour gases.* If present in significant quantities, acid gases must be removed from natural gas. Both carbon dioxide and hydrogen sulfide are corrosive when water is present, and hydrogen sulfide is very poisonous in relatively small concentrations. Removing these contaminants from natural gas is often called *sweetening* the gas.

Several commercial processes are used to sweeten natural gas; they can be classified as chemical absorption, physical absorption, or adsorption processes. In the absorption processes, the acid gas stream contacts a liquid absorbent that selectively removes certain acid gases from the natural gas.

In *chemical absorption*, the acid gases react chemically with the liquid absorbent. Heat and/or low pressure is used to regenerate the absorbent, that is, to separate the acid gases from the absorbent so that it can be used again. The most widely used sweetening processes in the industry are the amine processes (fig. 5.38). An amine process is a continuous-operation that uses a solution of water and a chemical amine to remove carbon dioxide and several sulfur compounds.

In *physical absorption*, the acid gases are physically dissolved in the liquid absorbent, and the absorbent is regenerated by driving off the acid gases through heating or pressure reduction. Commercial processes of this type include the fluor solvent, Selexol, sulfinol, and Rectisol processes.

In *adsorption*, or dry-bed, processes, the acid gas stream contacts a solid adsorbent that removes sulfur compounds and/or carbon dioxide. The acid gas is vaporized and removed from the adsorbent bed by heating or pressure reduction.

Removing natural gas liquids. Natural gas is a vaporous mixture of methane, ethane, propane, and a certain amount of heavier hydrocarbons such as butane and natural gasoline (ranging from pentanes through nonanes or decanes). In field separators, some of the heavier hydrocarbons liquefy into a condensate that is removed from the natural gas. It is often desirable to liquefy some or all of

the remaining ethane and heavier hydrocarbons and to remove them from the methane gas. The liquefied hydrocarbons, known as *natural gas liquids* (NGLs), have greater market value as liquid fuel or refinery feedstocks than they have as components of fuel gas. Also, if butanes and heavier hydrocarbons are allowed to remain in the natural gas stream, they may condense in the transmission lines during cold weather.

When gas is produced from high-pressure gas wells, the NGLs are usually separated in field facilities. It is generally more economical to process gas from an oilwell in a gas processing plant. The usual equipment for processing gas from a gas well is the low-temperature separation (LTX) unit (fig. 5.39). This equipment partially dehydrates the gas stream as well as removing

Fig. 5.38. This amine sweetening unit uses monoethanolamine to remove carbon dioxide and sulfur compounds. (*Courtesy of Sivalls, Inc.*)

Fig. 5.39. A low-temperature separation unit is used to separate NGLs produced from a gas well.

NGLs. It takes advantage of the fact that the gas stream coming from the wellhead is under high pressure. After passing through a free-water knockout, the compressed gas is cooled and allowed to expand through a choke into a separator vessel. The expansion further cools the gas and condenses the heavier hydrocarbons into a liquid. The hydrocarbon liquid and the hydrates formed from water vapor drop to the bottom of the separator, and the dry gas passes out the top. The gas is warmed and allowed to enter the gas gathering system. The hydrates are melted, the water is sent to a disposal system, and the NGL is stored in a low-pressure tank.

Storing crude oil

Oil, water-cut oil, and water produced by the well move from the wellhead or separator through the treating facilities and finally into stock tanks, or a tank battery (fig. 5.40). The number of tanks in a battery will vary, as will their size, depending on the daily production of the well or wells and the frequency of pipeline runs. The total storage capacity of a tank battery is usually three to seven days' production. Also, since a battery has two or more tanks, one tank can be filling while oil is being run from another.

Before a tank battery is put into operation, each tank is strapped, or calibrated, meaning that the measurements or dimensions of the tank are taken and the amount of oil that can be contained for each height increment of the tank is computed. The capacity in barrels according to the height of the liquid in the tank is prepared in table form, and these data make up what is known as the tank capacity table. This table commonly shows the capacity for each

Fig. 5.40. This central treating station has a free-water knockout, electrostatic treaters, and stock tanks.

¼ inch from the bottom to the top. The strapping and table preparation are usually done by a third party.

Most tanks are constructed of either bolted or welded steel and are equipped with a bottom drain outlet for draining off sediment and water (S&W), still commonly referred to as BS&W (basic sediment and water). It is sometimes necessary for a workman to enter an empty tank to clean out the collection of paraffin and sediment that cannot be removed through the drain outlet.

Oil enters the tank at the top at an inlet opening. The pipeline outlet is usually 1 foot above the bottom of the tank; the space below this outlet provides room for the collection of S&W. The pipeline outlet valve is closed with a metal seal when the tank is being filled and similarly locked in the open position when the tank is being emptied. The application of seals to the tank outlet valves is a method of assuring that the shipper receives the quantity and quality of oil mutually agreed upon by the shipper and the producer.

When measuring or gauging the level of oil in a tank, a steel tape with a plumb bob on the end is lowered into the tank until it just touches the bottom. The highest point at which oil wets the tape shows the level or height of the oil in the tank. By referring to the tank table, the volume of liquid in barrels in the tank is determined. An automatic tank gauge, a newer method of measuring, consists of a steel gauge line contained in a housing; a float on the end of the line rests on the surface of the oil in the tank. The other end of the line, which is coiled and counterbalanced outside the tank, runs through a reading box and shows the height of oil in the tank.

MEASURING AND TESTING OIL AND GAS

The volumes of oil, gas, and salt water produced by each lease are usually measured by the lease operator every 24 hours. This operation is important because in many fields the oil or gas is produced in accordance with daily allowables set by state regulations. These allowables are based on the market demand for oil or gas and the most efficient rates of production for the particular fields. Thus, proper credit must be given to the lease for oil and gas actually delivered from the lease.

Oil sampling

All oil delivered to pipeline companies is subject to their testing. Therefore, in order to assure that the oil will be accepted, the producer should sample and test the oil in the same manner as prescribed by the pipeline company that receives the oil. The procedures for taking samples and making water and sediment tests vary from field to field and company to company and must be agreed on by both the producer and the shipper.

Sampling methods. Many methods exist for sampling oil. Broadly, sampling can be done either automatically or manually. Frequently, oil in lease stock tanks is sampled manually. Perhaps the most common way is by *thief,* or *core, sampling.* In the thief sampling method, a thief (a round tube about 15 inches long) is lowered into a tank to the desired levels (fig. 5.41). A thief has a spring-operated, sliding valve that can be tripped, thus trapping the sample. The thief is designed so that a sample can be obtained within a half inch of the bottom of the tank. A more desirable but more difficult manual sampling method is *bottle sampling.* In one method of bottle sampling, a 1-quart bottle or beaker with a stopper and cord assembly is used (fig. 5.42). The sealed bottle is lowered to the desired depth, the stopper is removed, and the bottle

Fig. 5.41. A device called a thief is used to obtain oil samples from a storage tank.

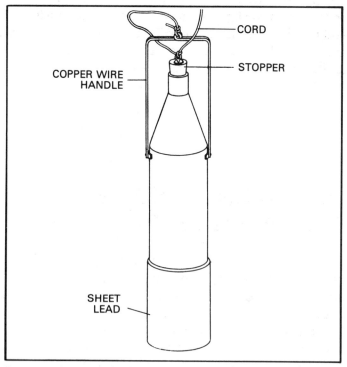

CORD

STOPPER

COPPER WIRE HANDLE

SHEET LEAD

Fig. 5.42. A 1-quart beaker is used in bottle sampling.

or beaker is pulled to the top at a uniform speed. If pulled at the proper speed, the bottle should be about three-fourths full of liquid; if not, the sampler must start over.

Types of samples. Several different types of samples may be taken from a tank of oil, and all parties concerned should agree on which type to take. An average sample is one that consists of proportionate parts from all sections of the tank. A *running sample* is one obtained by lowering an unstoppered beaker or bottle from the top of the oil to the level of the bottom of the outlet connection and returning it to the top of the oil at a uniform rate of speed. A *spot sample* is one obtained at some specific location in the tank by means of a thief, bottle, or beaker. Many types of tank samples can be taken, all relating to specific points in the tank.

Crude oil is bought and sold in this country on a volume basis. This volume must be corrected for any S&W present. Since its volume varies according to temperature, it must also be corrected to the standard base temperature of 60°F. These requirements call for a series of tests, such as temperature, gravity, and S&W content. These elements are all measured and gauged in the presence of witnesses representing the lease and the pipeline. This information, along with the terms of the lease, the producer, the transporter, the number of the tank, the date, and other necessary data are written on the pipeline *run ticket* (fig. 5.43).

Oil measurement and testing

Temperature measurement. The temperature of oil in lease stock tanks generally is close to that of the air surrounding the tanks unless the oil has just recently been produced and is still carrying an elevated temperature from the subsurface or if the oil has been heated in a treater to separate it from S&W or salt water. Temperature is usually measured with a special thermometer that is lowered into the oil on a line and then withdrawn to observe the reading.

Gravity and S&W content measurement. The S&W content and the API gravity (specific gravity, or density, measurement) of oil in stock tanks are measured from samples taken from the tanks. Samples are obtained by using a thief or similar device or by withdrawing oil through sample cocks installed at various levels in the shell of the tank. The requirements for cleanliness of oil vary with individual buyers, but the maximum S&W content in most states is 1 percent. Most pipelines accept all oil from 4 inches below the bottom of the pipeline connection on up. The S&W content of the samples can be determined by a centrifuge test, also

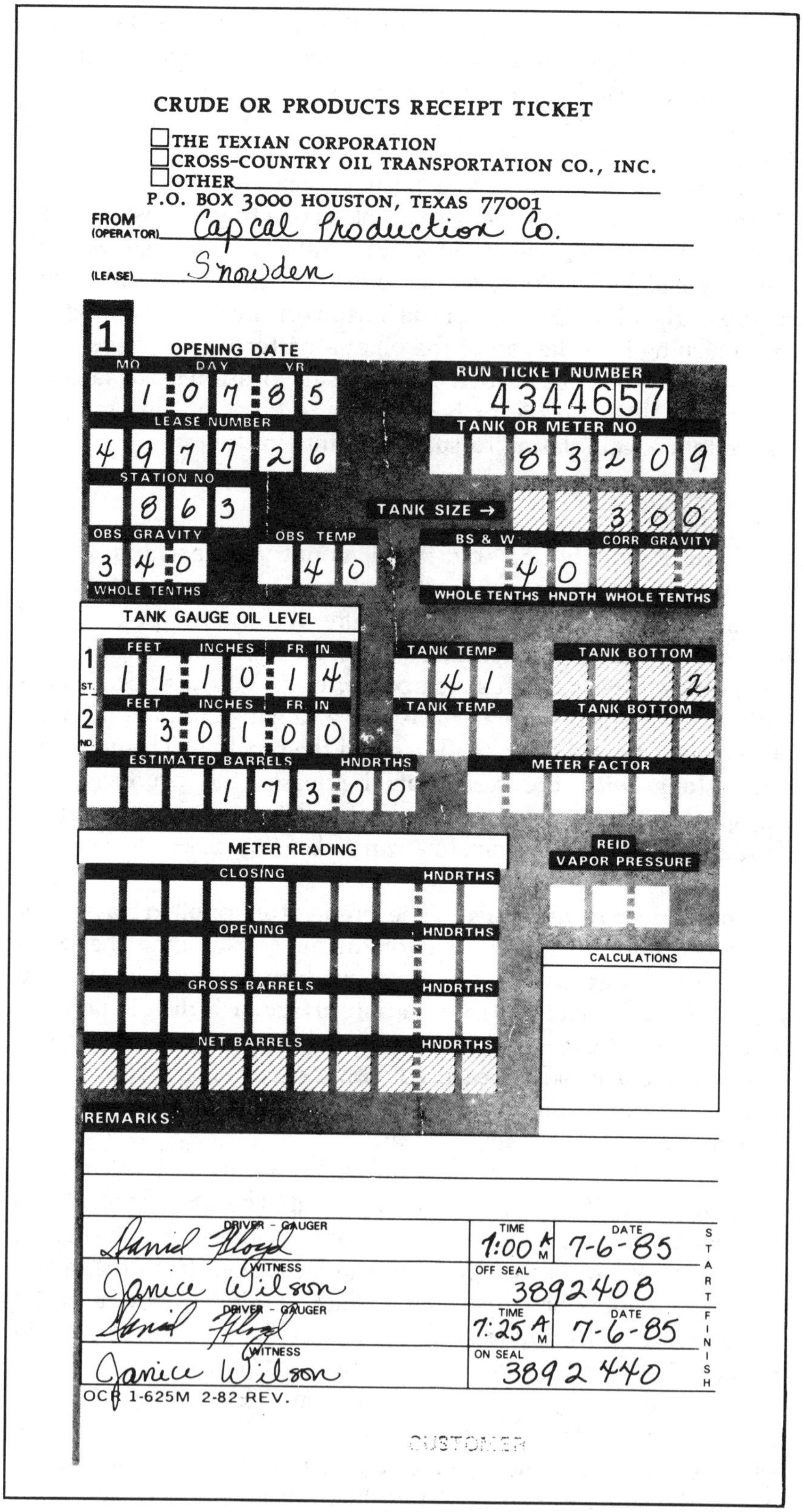

CRUDE OR PRODUCTS RECEIPT TICKET

☐ THE TEXIAN CORPORATION
☐ CROSS-COUNTRY OIL TRANSPORTATION CO., INC.
☐ OTHER _____

P.O. BOX 3000 HOUSTON, TEXAS 77001

FROM
(OPERATOR) _Cap Cal Production Co._

(LEASE) _Snowden_

1 OPENING DATE

MO	DAY	YR
1 0	7 8	5

RUN TICKET NUMBER
4 3 4 4 6 5 7

LEASE NUMBER
4 9 7 7 2 6

TANK OR METER NO.
8 3 2 0 9

STATION NO
8 6 3

TANK SIZE → 3 0 0

OBS GRAVITY	OBS TEMP	BS & W	CORR GRAVITY
3 4 0	4 0	4 0	
WHOLE TENTHS		WHOLE TENTHS HNDTH	WHOLE TENTHS

TANK GAUGE OIL LEVEL

	FEET	INCHES	FR. IN	TANK TEMP	TANK BOTTOM
1 ST.	1	1 0	1 4	4 1	2
2 ND.	3 0	1 0 0			

ESTIMATED BARRELS   HNDRTHS   METER FACTOR
1 7 3 0 0

METER READING

CLOSING   HNDRTHS

OPENING   HNDRTHS

GROSS BARRELS   HNDRTHS

NET BARRELS   HNDRTHS

REID VAPOR PRESSURE

CALCULATIONS

REMARKS:

DRIVER - GAUGER	TIME	DATE	
Daniel Floyd	1:00 AM	7-6-85	S T A R T
WITNESS _Janice Wilson_	OFF SEAL 3892408		
DRIVER - GAUGER _Daniel Floyd_	7:25 AM	7-6-85	F I N I S H
WITNESS _Janice Wilson_	ON SEAL 3892440		

OCB 1-625M 2-82 REV.

CUSTOMER

Fig. 5.43. The run ticket becomes the basis for all payments for oil delivered from the lease.

called a shake-out test (fig. 5.44). The glass container in which the test is made is graduated so that the percentages of S&W can be read directly. The API gravity of the oil is measured by a hydrometer graduated in °API and designed to measure oil at 60°F. Since the gravity, or density, of oil varies with its temperature, gravity readings of oil at temperatures other than 60°F must be corrected. Tables are available that make these corrections easy.

Like oil, gas must also be sampled before it can be sold. Sampling, which is always done in the field, is the way the operator determines the sales potential, composition, heating value, specific gravity, and other physical properties of the gas being produced from a specific well. It also reveals information about the performance of the field and the processing equipment. Because each well has its own unique characteristics — age, equipment downhole and on the surface, production rates, and so forth — sampling procedures vary from well to well. Some wells may require that gas samples be taken frequently from several different points on the

Gas sampling

surface equipment; other wells may have samples taken less frequently and from perhaps only one point. Regardless, sampling is vitally important to both the seller and the buyer, and proper procedures must be followed if a representative sample is to be taken and analyzed.

The best sampling place on the flowing pipeline is the point where a representative sample of the gas moving through the line can be taken. The sample container should, therefore, be connected to a vertical flow line or to the top of a horizontal line. These placements minimize the entry of any oil, condensate, or water that may be present in small quantities in the line. A sample taken on an ell or other location where impurities collect will not be representative. Similarly, if a sample is taken from the side of a vessel containing wall condensation, a probe that extends a good distance into the vessel is used.

Several manual methods are used to sample gas: purging at reduced pressure, fluid displacement, air displacement, and the use of a floating piston and vacuums. Each has its advantages and disadvantages, but they all require that sampling or testing equipment be leak-free and that sample containers are purged of all extraneous gases or vapors. Also, if the sample is to be accurate and useful to both the operator and the buyer, the technician taking the sample must know what he is doing and must follow standard procedures to the letter. Properly taken samples are shipped to a laboratory where they are analyzed.

Although more manual sampling methods are employed, automatic sampling devices connected to the pipeline are also used. These devices allow gas to flow into the sample container over a specified period of time at a constant rate. In some cases, automatic samplers are interfaced with analyzing equipment, but in most instances, field samples are sent to the laboratory for testing.

Gas testing

Gas testing is an important phase of field handling of natural gas. Since the analysis of gas samples determines the liquid hydrocarbon content of the gas, the results of field and laboratory tests may determine how much a seller receives for his gas and whether the gas meets the buyer's contract specifications. Three of the most common methods for gas analysis are compression testing, charcoal testing, and fractional analysis. Compression testing is usually done in the field, while charcoal testing and fractional analysis are conducted in the laboratory.

Compression testing. Compression testing is used extensively on casinghead gas. The sample is compressed in a portable compressor and then cooled in an ice-water bath or in a refrigerated condenser. This is the one form of gas testing that is done completely at the field location, and all equipment used *must* be in top operating condition so that repeatable test data are valid. Compression testing is generally considered to be reliable.

Charcoal testing. In a charcoal test, gas is drawn from the stream and passed over activated charcoal. After the adsorption process is complete, the charcoal is taken to a laboratory where the adsorbed liquid is driven off by heat, condensed, and measured. This test is used when other methods of testing either give unsatisfactory results or are too expensive.

Fractional analysis. Fractional analysis is used when gas composition must be determined. In this procedure, a sample of the gas stream is obtained in a metal container and shipped to a laboratory where the analysis is performed. The analysis will usually reveal not only the composition in percentage of each hydrocarbon present, but also the gallons per thousand cubic feet by component and the heating value of the gas.

Gas metering

Gas metering is the process of measuring the volume of natural gas flowing past a particular point. Since the volume measurement is affected by the gas temperature, specific gravity, and pressure, standards have been established for these variables so that differing pressures, for example, can be taken into account. These standards allow the individual properties of natural gas flowing from a particular well to be converted to what is called standard condition. This standardization of properties is required for metering.

Every gas purchase or sales contract made between major suppliers and consumers is concerned with metering methods, calibration, and equipment. Therefore, gas must be metered whenever there is a change of custody, or ownership. Measurements of natural gas are also used in regulatory reporting, reservoir studies, and material balance studies in plant processing.

A common method of measuring flowing gas is to use an orifice meter installation. Basically, an orifice meter is a device that has an orifice plate (a finely machined plate with an orifice, or hole, in its center) mounted vertically in a gas line. The upstream and downstream tubing or pipe on either side of the orifice plate and its holder is called a *meter run.* When gas flows downstream through the orifice, which is smaller than the upstream tubing, its pressure drops. This pressure drop, called *differential pressure,* is gauged and recorded on a circular chart. Also measured and recorded on the chart is the *static pressure*—the pressure in the line upstream from the orifice plate. By using these two measurements and other known factors, the rate of flow, the volume flow rate, and other data can be calculated.

Recording devices are installed above the orifice tap levels in order to avoid measuring problems caused by the presence of liquids. Other problems in gas measurement are gas pressure and flow pulsations. Pulsations are cyclical pressure surges caused by reciprocating pumps or compressors. Irregular pulsations also occur when valves are opened or closed on the line. Meter installations need to be as far away from pulsation sources as possible in order to avoid inaccurate readings. Pulsation dampeners—which are gas- or liquid-charged, chambered devices that minimize pressure surges—may need to be installed in some cases. Specific gravity and temperature recording devices are placed in such a manner so as not to interfere with meter measurements.

The standards for installing and operating field metering equipment are outlined in recommendations from the American Gas Association's Gas Measurement Committee. The committee was organized to combat the widespread nonconformity of practices that existed among different localities and operators in the fledgling gas industry. The reports and findings of the committee, detailed in AGA Committee Report No. 3, *Orifice Metering of Natural Gas*, are based on impartial research and contain the most accurate information available on large-volume gas measurement. These AGA committee recommendations are the accepted standards for gas measurement and ensure uniform industry-wide practices.

LACT units

The development of lease automatic custody transfer (LACT) units has changed the processes of measuring, sampling, testing, and transferring oil from a tedious, time-consuming business that required many hours and was error-prone into an efficient measuring and recording system that leaves lease personnel free to do other operations (fig. 5.45). Despite the complexities of gauging and testing crude oil in the field, automatic equipment can perform the following tasks:

- Measure the volume of the oil and make a record of volume readings.
- Detect the presence of water in the oil stream and indicate the percentage thereof.
- In cases of excessive water, divert the flow into special storage where it will be held for treatment.
- Isolate individual wells on a lease and test them separately from the commingled flow from other wells.
- Determine and record the temperature of the oil.

Fig. 5.45. A skid-mounted LACT unit is an efficient measuring and recording system.

- Prove the accuracy of flowmeters and provide for their calibration when necessary.
- Take samples from the stream in proportion to the rate of flow and hold these for verification by conventional test procedures (fig. 5.46).
- Measure and record the API gravities of the oil produced.
- Switch well production from a full tank to an empty one and turn the oil from the full tank to the pipeline gathering system.
- Shut in the wells and relay an alarm signal to a remote point in case of any malfunction.

Of course, not all LACT installations are as elaborate as the system just described. Operators use the system designed to best meet their individual needs and budgets.

Fig. 5.46. A sample container on a LACT unit holds oil samples automatically taken from the oil stream flowing through the unit.

WELL SERVICING AND WORKOVER

Well servicing is the maintenance work performed on an oil or a gas well to improve or maintain the production from a formation already producing. A well servicing unit carries the hoisting machinery that is used to pull sucker rods, an operation that is done most frequently for pump changes or rod string repairs (fig. 5.47). The same unit can be used to pull tubing by arranging the equipment for the heavier load. The most common well servicing operations are those related to artificial lift installations, tubing string repairs, and work on other downhole equipment that may be malfunctioning.

Fig. 5.47. Sucker rods (seen hanging) are pulled from the well during a well servicing job.

Routine maintenance work is needed throughout a well's life if economical production is to be maintained. Pump parts wear out and must be replaced periodically, rods break and must be repaired, and gas lift devices must be replaced occasionally. Other remedial work needed to keep a well on steady production may include repair of tubing leaks, replacement of packers that have failed, and sand control.

Part replacement. Well equipment such as downhole pumps, sucker rods, gas lift valves, tubing, and packers must be in good working condition. Rod pumps eventually wear out because of abrasive or corrosive conditions and their reciprocating characteristic. Sucker rods are often highly stressed and may fail because of repeated load reversals. Corrosion, scale, and paraffin deposits may accelerate such failures (fig. 5.48). Tubing—and rods for that matter—will wear due to the reciprocating movement in the well as the string stretches and unstretches to adjust to the changing fluid loads in pumping. Packers and other accessory devices sometimes fail because of the hydraulic and mechanical loads that are imposed on them.

Cleanout operations. Major cleanout operations include sand removal, liner removal, casing repair, cementing, drilling deeper, or sidetracking. These jobs usually require a string of pipe that can be rotated. Also, it is generally necessary to circulate the well—that is, run pump fluid to the bottom and back to the surface. Usually salt water or specially prepared circulating, or workover, fluid is used. Circulating the well (1) removes sand, cuttings, or chips; (2) prevents blowout by maintaining adequate hydrostatic pressure to overcome formation pressure; (3) cools bits and cutters; (4) actuates hydraulic tools; and (5) in open-hole operations, supports the wall of the hole until casing or a liner can be set. These functions require a suitable circulating system: pump, hose and swivel, string of pipe to bottom, and pit or tank to receive fluid returning from the well.

Sand control. Production men have tried for many years to devise a method of keeping sand out of a well. Men who work in hard-rock country may never have sand trouble, but in California and the Gulf Coast, problems occur every day and are handled almost routinely. Loose sand is controlled by plastic squeeze, gravel pack, screen liner, or a combination of these methods. Plastic squeeze involves placing a resinous material into the sand formation and then allowing the plastic to harden. The plastic glues the sand grains together and, if successful, prevents the grains from entering the well. In the gravel-packing process, graded

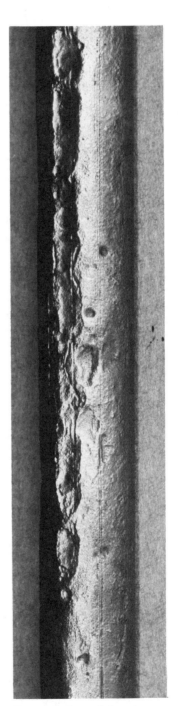

Fig. 5.48. Corroded sucker rods must be replaced to keep the pumping well in good working condition.

gravel is placed outside the casing or liner; this gravel holds back the sand but allows the well fluids to enter the wellbore. One type of screen liner used to control sand is a slotted pipe with the slots cut transversely. The liner also acts to admit oil to the bottom of the well and exclude sand. When production shows sand, it usually indicates that the liner has failed and that remedial work is in order.

Paraffin control. Like sand, paraffin can also cause problems in a producing well. Hydrocarbons in some areas contain heavy waxes, asphalts, and resins, which may solidify in the tubing or other production equipment into a sticky mass called paraffin. Paraffin deposits restrict the flow of well fluids, and left untreated they may completely block flow paths downhole or cause equipment problems on the surface. One way of dealing with these troublesome deposits in the well itself is using a tool called a paraffin scratcher. Lowered into the hole on wireline, the scratcher is a long, cylindrical device that has spiky barbs radiating outward in all directions along its length. Since this tool is usually run while the well is flowing, the paraffin bits cut loose by the barbs are washed up out of the tubing. Another way of removing paraffin is to circulate hot oil in the bore to melt the solidified deposits so that they can flow out of the well. Also, a solvent can be used to dissolve paraffin in the wellbore or formation. When solvents are used, they are squeezed in at very low rates and very low pressures; then the well is closed in for a 24- to 72-hour soaking period to allow the solvent to dissolve the paraffin. Also, special chemicals that may inhibit paraffin formation are being developed.

Workover operations and equipment

Like well servicing, the purpose of a workover is to maintain or increase production of a producing well, but workovers include more extensive repairs, changing the producing zone in a well, cleaning out sand, and recompleting to produce from a different pay zone. To perform these operations, workover rigs, snubbing units, and sometimes coiled-tubing units are used. Wireline units are also frequently used.

Workover rigs. Workover rigs, in many respects, are scaled-down drilling rigs (fig. 5.49). They are equipped to stand the pipe in the derrick, they have some arrangement for turning the pipe string while it is in the hole, and they are furnished with a high-pressure pump. The pump is used to circulate fluid in the well—that is, to force water or other liquid inside the tubing to the bottom of the well by pump pressure and then outside the tubing back to the surface into a tank to complete the circuit.

Fig. 5.49. A typical workover rig is truck-mounted.

Snubbing units. Snubbing units are used primarily on high-pressure wells that require workover operations such as running and pulling production or kill strings; using nitrogen, natural gas, or foam for sand washing; or fishing for lost wireline tools. Snubbing units can also be used for pulling and rerunning damaged joints, collars, and subsurface control equipment and other similar tasks. Usually classified according to their power source, two general types of snubbing units are used: the hydraulic unit and the rig-assisted unit. The hydraulic unit (fig. 5.50) operates with a self-contained hydraulic system and utilizes single or multiple

Fig. 5.50. This snubbing unit operates with a self-contained hydraulic system.

hydraulic cylinders to move the pipe into or out of the well. The rig-assisted unit, on the other hand, uses the rig drawworks in a block-and-pulley arrangement to move the pipe.

Coiled-tubing units. Although they have limited applicability, coiled-tubing units are also used for some workover operations, such as initiating flow, cleaning out sand and paraffin in tubing, and acidizing. Coiled tubing is a continuous string of small-diameter tubing, and its use means that joint connections, which may leak, are not present. Coiled-tubing equipment consists of the tubing roll, storage injector heads to move the pipe into or out of the well, power unit and control assembly, pressure control equipment, and a crane to support the injector head over the wellbore (fig. 5.51). Reduction in rig time is a primary advantage of this type of unit. Rig-up time is usually about 1 hour, and coiled tubing can generally be pulled about three times faster than jointed tubing. A disadvantage is that wireline operations cannot be conducted through coiled tubing because of the coil configuration on the reel and the small diameter of the tubing itself.

Fig. 5.51. This coiled-tubing unit is being used to run tubing into the well under pressure.

Wireline tools. Wireline operations, which are an integral part of both completion and workover, are procedures performed with wireline-suspended tools. A wireline, in essence, is a strong, thin length of wire mounted on a powered reel at the surface of a well. Wireline work is usually performed with a line that varies in size from 0.072 inch to 0.1875 inch in diameter. The 0.1875-inch line is a braided electric line, while the lines from 0.072 to 0.092 inch are solid steel, or slick, lines. Among the more common jobs performed with wireline are depth measurement, logging, perforating, sand and paraffin control, fishing and junk retrieval, and the manipulation of subsurface well pressure and flow controls. These operations are performed by wireline units. On land these units are usually mounted on trucks that are backed up to the well (fig. 5.52). Marine wireline units are mounted on skids and transferred to the wellsite on boats or barges. Regardless of the type, a wireline unit contains a reel of wireline; a power system to pay out and retrieve the line; instruments to indicate weight, line speed, and depth of the tool string; and the string of tools attached to the end of the line to perform the particular job. Pressure containment devices are also common.

Fig. 5.52. Reels of wireline, shown here mounted on the back of a truck, are used to perform many completion and workover jobs.

OFFSHORE AND ARCTIC PRODUCTION

Most offshore activities were initiated in order to recover oil from already defined shore reservoirs or to extend field boundaries. Offshore production was done as early as 1900 in the Pacific Ocean from California beach wharves. Every step away from solid ground toward the sea presented the petroleum industry with new problems in design, methods, and materials. The operator usually tried existing equipment and methods first, and when these proved inadequate, he sought new technology. Progressing from wooden piles and posts in shallow waters, the industry began to employ small fixed platforms accompanied by World War II surplus tender barges for supply and auxiliary services (fig. 5.53). These platform-tender rig combinations were used in waters of 60 feet or less, and after well completion, the tender was moved to another location.

Early offshore platforms

Fig. 5.53. A fixed platform and its tender are being used for drilling and production in shallow waters.

238

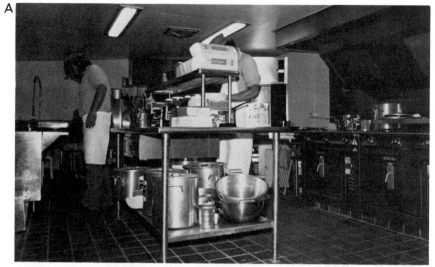

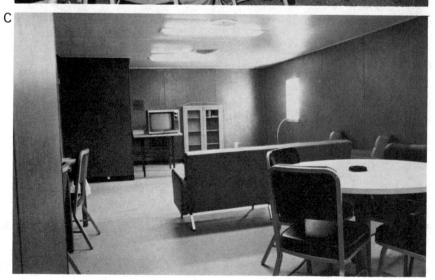

Fig. 5.54. A production platform offers all the "comforts of home" for the crew: galley (*A*), dining facility (*B*), and recreation room (*C*).

The first specifically designed steel structure was installed in 1947 in the Gulf of Mexico at a depth of about 20 feet. In the following decades, more than 2,000 fixed platforms sprang up in the Gulf, in depths ranging up to 1,200 feet. The fabrication and erection techniques perfected in the Gulf influenced platform construction around the world. At water depths of 50 to 300 feet, however, more complex structures were needed. Mooring and anchoring problems in less-sheltered ocean areas caused damage to the platform from the tender, and the costs of separators, loading equipment, and pipelines made it economically mandatory for a large number of wells to be drilled from a central location.

Today's drilling platform (see chapter 4) becomes a production platform when one or more development wells have been drilled and production becomes the primary activity of the platform. The drilling rig may be removed, but it is usually kept on the platform and used to service and repair the producing wells.

In addition to the rig's derrick, the modern platform houses all the usual production equipment, as well as living quarters, offices, galley, and recreation rooms for the crew (fig. 5.54). A platform generally has a helipad, one or more cranes, antipollution devices, and a myriad of safety devices such as firefighting equipment, life rafts, and escape capsules (fig. 5.55).

Modern production platforms

Fig. 5.55. An offshore platform is well-equipped with safety devices. Note the escape capsule at right.

Fig. 5.56. Helicopters and cranes are common sights around an offshore operation.

Fig. 5.57. Work boats are used to transport crew and cargo between shore and platform.

Platforms are serviced by transportation and communications systems. Helicopters are used for emergency evacuation and to transport personnel and supplies to and from the platform (fig. 5.56). Crewboats, supply boats, and other types of work boats also move crew and cargo between the shore and the platform (fig. 5.57). Communications includes remote monitoring of production equipment, closed-circuit television, radio transmissions, and recording devices of all kinds. Weather balloons and satellites give offshore platforms and auxiliary vessels early weather warnings. The use of interactive computer terminals in every phase of operation keeps the offshore drilling and production platforms in touch with the rest of the world.

Offshore completions

A well completion offshore is much like one on land, except that a different technique is used to control the fluids as they flow from the top of the well. The wellhead valves and fittings may be placed above the surface of the water (surface completion), or they may be installed on the seafloor (subsea completion). A well drilled from a platform is usually completed with the Christmas tree on the deck of the platform, while in a subsea completion, the wellhead is located on the seafloor. A *wet* subsea installation is one installed on the wellhead, exposed to the surrounding water (fig. 5.58). A *dry* subsea installation is one housed in a protective dome (fig. 5.59).

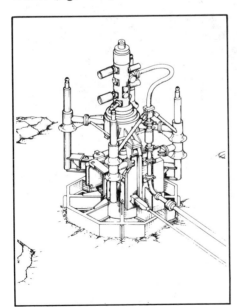

Fig. 5.58. In a wet subsea completion, the valves and equipment used to control the well fluids are placed on the seafloor.

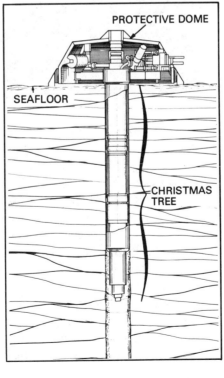

PROTECTIVE DOME

SEAFLOOR

CHRISTMAS TREE

Fig. 5.59. In a dry subsea completion, the Christmas tree is protected from the water.

242

Fig. 5.60. Subsea repairs are often made by workers using a diving bell to reach the seafloor.

Wet subsea repairs and maintenance are handled by divers with special underwater clothing and breathing apparatus. On a dry installation, workers may use a small submarine or diving bell to reach the seafloor (fig. 5.60); they can enter the protective dome and work without special underwater gear.

Handling well fluids offshore

Well fluids produced from a platform generally have the water and other impurities removed before the oil and gas are put into tankers or pipelines for transportation to shore. Therefore, the platform must contain separating and treating equipment, compressors, pumps, dehydrators, special storage tanks, and other fluid-handling equipment. In some subsea completions, oil, gas, and water are routed into flow lines that run from the wells to a production riser, or pipe. Production flows up the riser to a floating buoy. An oil tanker can tie up to the buoy, take on the produced fluids, and transport them to shore for separation and treatment.

A more elaborate subsea system consists of a grouping of satellite wells that supply natural gas or crude oil to a centralized production system (fig. 5.61). These satellite wells, often widely scattered along the reservoir, are attached by flow lines to a central gathering point where storing, separating, and product transfer can take place. Satellite wells allow the reservoir to be drained from widely separated points. Their use is more practical and economical than using only a single well attached to a conventional production platform.

The first successful submerged production system (SPS) was tested in the Gulf of Mexico between 1974 and 1979. The SPS was created to handle all phases of production, from drilling to abandonment. The major components of this system included using a seabed template (a manufactured design pattern with built-in accessories and guides for equipment) through which several wells were drilled and completed. Pipelines connected the template to a convenient conventional platform, and a robot maintenance system (RMS) was used to service the underwater equipment.

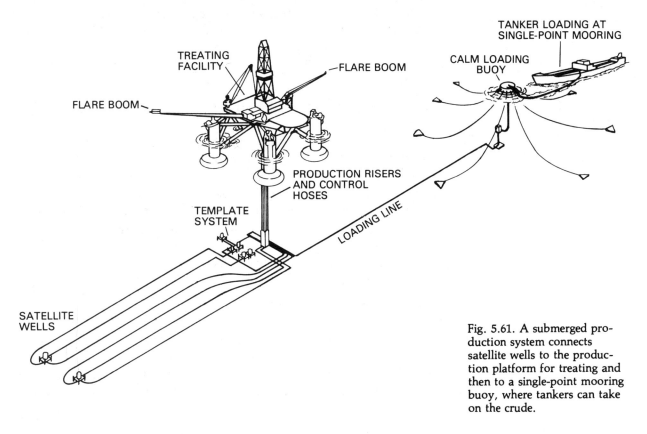

Fig. 5.61. A submerged production system connects satellite wells to the production platform for treating and then to a single-point mooring buoy, where tankers can take on the crude.

The successful use of the SPS in deepwater recovery inspired the technology for an underwater manifold center (UMC) for the North Sea. This revolutionary UMC system allows the exploitation of smaller offshore fields all over the world — once thought too unprofitable to produce in conventional ways. The UMC is connected by pipeline to the South Cormorant platform, more than 4 miles away. The system has several functions, including acting as a template for wells and as a connection for satellite wells drilled some distance away. The UMC collects fluid from the wells and sends it to the platform. Other functions are to aid in enhanced recovery and pressure maintenance methods by the injection of seawater and to control and service individual wells with flow-line equipment.

The technological advances made in platform construction and the use of subsea completions in deepwater environments have greatly expanded the frontiers of oil and gas production throughout the world.

Arctic production considerations

Production problems in the oil-rich Arctic icefields are quite different from those posed by ordinary drilling environments. The frozen north, with its extremely low temperatures that can freeze human flesh in minutes, ice fogs, and overwhelming transportation problems caused by the constant enemy — ice — has challenged the technological skills of petroleum engineers and exploration companies worldwide.

One specific problem related to the completion of an Arctic well involves the presence of permafrost, which is any permanently frozen subsurface formation (fig. 5.62). Casing and wellhead strains are generated in the Arctic because of the partial thawing and refreezing near the surface that is common in permafrost regions. Thawing activity causes deformation of the permafrost and thereby puts pressure on the casing. The strain caused on the casing is directly proportional to the amount of permafrost deformation that occurs. The casing could, in this case, be compared to a tin can being compressed in a vise. As long as the squeezing pressure is less than the can's strain limit, the can does not collapse.

The design of the lease site in the Arctic must also take permafrost into account even if the structure is above the permafrost table. Thaw on a large scale underneath a well can lead to different frost heave and settlement pressures and can therefore adversely affect production. Christmas trees are generally housed and heated, as are other pieces of lease equipment, to facilitate production in below-zero temperatures.

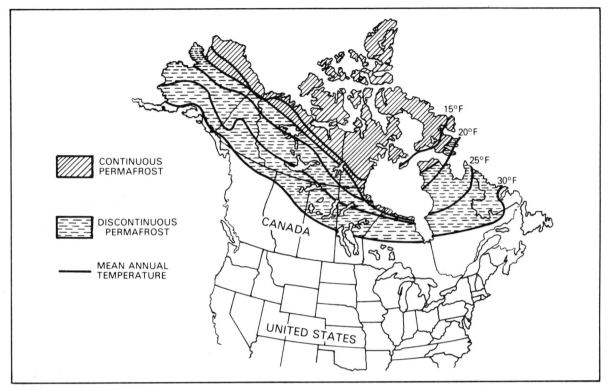

Fig. 5.62. Arctic production must take into account the areas of permafrost, shown here in North America.

The steps between the wellhead and the pipeline or other mode of transportation involve bringing the well fluids to the surface; separating and treating emulsions; disposing of the water and contaminants; and temporarily storing, measuring, sampling, and testing oil and gas. These functions are fairly routine on the average land lease facility and have been solved on the average-depth offshore platform. The challenge of today's engineers is gathering and producing the world's petroleum as the search moves to greater depths in the ocean and more remote regions of the world.

Transportation VI

Moving oil and gas from a field to refining and processing plants and petroleum products from refineries to consumers requires a complex transportation system. Inland waterway barges, railway tank cars, transport trucks, oceangoing tankers, crude oil and products pipelines, and gas transmission pipelines all play an important part in the oil and gas transportation industry—an industry that started with horse-drawn wagons carrying wooden barrels of crude oil to nearby streams. As the industry grew, so did the methods of transportation, and today millions of barrels of crude oil, gasoline, fuel oils, and other petroleum products, as well as billions of cubic feet of natural gas, are moved from the wellhead to the plant, from one refinery to another, from offshore to onshore, and from continent to continent to reach the ultimate consumer.

EARLY METHODS OF TRANSPORTATION

Moving oil from well site to refinery became a necessity when the Drake well began producing 20 barrels of crude oil per day and started an oil boom in the area of Oil Creek, a tributary of the Allegheny River. The first refinery was constructed at Oil City, some 11 miles from the field. Shortly thereafter, a refining complex developed on the Allegheny River at Pittsburgh, about 75 miles downstream, making Pittsburgh a major refining and oil transportation center. Demand for crude oil was pressing, and many competed for the job of hauling the oil to wherever it was needed—to the nearest waterway, railhead, pipeline dump tank, or refinery.

Wagons and water

Because of the proximity of the rivers and streams, one solution to the transportation problem was a massive system of boats, barges, and wagons operated by teamsters. Horse-drawn wagons carried wooden barrels of oil to the streams, where they were loaded on barges and boats of all types and sizes. Local sawmill operators agreed to release the water normally impounded to float logs downstream, and the oil boats (and sometimes the barrels themselves) were floated down the Allegheny River to Pittsburgh.

Until the invention of the steamboat, barges laden with oil were pulled by horses walking on the river banks, or they were floated downstream completely dependent on the river current for power (fig. 6.1). At the height of the operation, as many as a thousand boats were in use, including twenty to thirty steamers, passenger boats, and towboats pushing barges.

Rails and tank cars

After many barrels of oil were lost to the hazards of water travel, railroads were turned to as an alternative solution for moving petroleum and petroleum products. Railroad construction proceeded rapidly. The nearest rail outlet to Oil Creek prior to May 1861 was at Corry, Pennsylvania, some 20 miles north of Titusville. By the following October, tracks had been laid to Titusville, and by 1864 the rail network provided crude oil to Cleveland and New York. Since the railheads were still located some distance from the wells, teamsters transported the crude in their wagons, each carrying from five to seven 360-pound barrels. As many as 2,000 teams a day lumbered through the mud into Titusville carrying oil to the railway for $1 to $5 per barrel. The

Fig. 6.1. Horse-drawn wagons and barges were initially used to move oil to rail and water shipping points.

wooden barrels of oil were then transferred to boxcars for shipment to refineries. But the barrels were expensive, were easily damaged or pilfered, tended to leak, and required too much handling.

These difficulties resulted in the development of the wooden tub tank car by Charles P. Hatch in 1865. In 1866, brothers Amos and James Densmore of Meadville, Pennsylvania, improved on the design and patented the first tank car. Their tank car consisted of two 1,700-gallon wooden tubs, glued and banded with iron hoops and mounted on a flatcar (fig. 6.2). The Densmore tank car was

Fig. 6.2. The Densmore brothers patented this early railway tank car design.

first used to ship oil from Pennsylvania to New York City. In 1869, a horizontal tank car with an expansion dome on top was developed and has remained the industry's basic design. The expansion dome is a space in which any gas that breaks out of the oil can collect. It prevents the buildup of pressure in the tank car that could cause damage.

The first oil pipelines

The high cost of moving oil from the field to shipping points stimulated the development of a more economical and safer method of transportation: *pipelines*. After several unsuccessful attempts to pipe crude oil from the field to the nearest shipping points, in 1865 Samuel Van Syckel succeeded in building a pipeline and starting the first oil pipeline business. His Oil Transportation Association laid a 2-inch wrought-iron line from Pithole Creek (8 miles above Oil City) to the Miller Farm railroad station, 5 miles away. Sometime later he built a second line, and with his two lines he was able to move 2,000 barrels of oil per day at the low charge of $1.00 per barrel—much lower than the teamsters' charge of about $2.50 per barrel for the same distance. The teamsters reacted by tearing up portions of the line but were unsuccessful in deterring pipeliners as they developed crude oil gathering systems and trunk lines.

Crude oil trunk lines

The railroad companies were the first to construct and buy pipelines of their own or to form exclusive arrangements with pipeline transportation companies. In an effort to maintain monopolistic control over the oil transportation business, the railroads attempted to prevent other pipeline companies from crossing their rail lines (fig. 6.3). But public sentiment was against monopolies, and the Pennsylvania and Ohio legislatures passed laws in 1872 that granted common-carrier pipelines the privilege of eminent domain in obtaining their rights-of-way. These acts were the prelude to the development of oil trunk line systems, because main trunk pipelines could connect the oil centers directly to the refineries and provide more economical transportation.

Until 1900, crude oil production was concentrated in Pennsylvania, West Virginia, Ohio, and other eastern states, and refineries were also centered around the eastern population centers. But between 1901 and 1905, oil was discovered in Texas, Kansas, Oklahoma, Louisiana, and California, and the eastern refineries were forced to seek additional crude supplies from the western centers of production. By 1914 the mid-continent oilfields supplied over 60 percent of the crude oil to the Eastern refineries, and by 1940 they were supplying over 85 percent. Pumping stations were installed to move the crude along its way (fig. 6.4).

Fig. 6.3. Denied the right to cross a railroad, one pipeline company used tank wagons to transfer the oil from the pipeline to holding tanks on the other side and then back into the pipeline.

Fig. 6.4. An early pipeline pump station used steam power to build sufficient pressure to move the oil.

Fig. 6.5. As unit trains carrying oil sped east, pipeliners raced to finish the Big Inch crude oil pipeline.

A network of crude oil trunk pipelines was built to connect the source of supply with the refining centers, although a large part of transporting was done by ship as well. The combined system of ships and pipelines worked quite well until World War II when enemy U-boats sank many of the oil tankers destined for wartime industries of the East Coast. So the federal government joined forces with the petroleum industry and constructed the first large-diameter crude oil pipeline, called the *Big Inch* (fig. 6.5), and the refined products line, called the *Little Big Inch*. The Big Inch was 24 inches in diameter, while the Little Big Inch had a 20-inch diameter.

Gathering systems

Early pipelines were not connected directly to the oilwells but received their oil from large dump tanks located at strategic shipping points. Pioneer Henry Harley was the first to build a pipeline to carry oil from the wells to his dump tanks, a job generally handled by the teamsters. Soon after he built two lines from Benninghoff Run to tanks at the Oil Creek railroad loading depot, a longtime dispute between the teamsters and the pipeliners was brought to a head. A bitter fight resulted in much destruction, but the pipeliners persisted.

In 1866 A. W. Smiley and G. E. Coutant formed the Accommodation Pipeline Company and constructed a network of 2-inch pipe some 4 miles across the Pithole field to connect the tanks at the wells with the dump tanks of the pipelines. Later, gathering lines were connected directly into the trunk lines, and elaborate accounting systems were developed to keep track of the owners.

Products pipelines

The first *batching* of refined products (shipping one petroleum product after another through the same pipeline) was performed in 1901 in Pennsylvania, but construction of products pipelines did not begin in earnest until 1930 when the growth of large population centers in the Midwest had created new marketing areas for refined products. Before the use of pipelines became widespread, products were transported with horse-drawn carts and Model T trucks (fig. 6.6).

During the economic depression in the 1930s, the established oil marketers battled more fiercely than ever for a share of the refined products market. Competition demanded that the least expensive method of transportation be employed from the refinery to the market: pipelines designed to transport many grades of refined products. As a result, 3,000 miles of products pipelines were put into operation during 1930 and 1931. In some instances, older crude lines were cleaned and the direction of flow reversed to move products from the East Coast to the Midwest.

Fig. 6.6. Petroleum products were transported with horse-drawn carts and Model T trucks before products pipelines became widespread. (*Courtesy of Humanities Research Center, The University of Texas at Austin*)

Fig. 6.7. Using bamboo pipes, the Chinese were among the first to pipe natural gas. (*Courtesy of American Gas Association*)

Gas transmission pipelines

Although people knew about the uses of natural gas 3,000 years ago when the Chinese piped natural gas through hollow bamboo poles (fig. 6.7), widespread use of natural gas did not develop until it became easily available at competitive prices.

Aaron Hart, a gunsmith, drilled the first American natural gas well in 1821. He had drilled only 17 feet when a hissing sound indicated that he had struck not water but something unexpected—gas. Hart promptly experimented with the gas to see what could be done with it. Before long he had strung together a few hollowed logs to serve as a gas pipeline. With this primitive system, he managed to light some buildings nearby, and this was the beginnning of the natural gas industry—an industry that depended totally on an efficient method of transportation.

From then on, the problem that took longest to solve was finding a practical means of transmitting gas from the fields where it was produced to the cities and factories where it could be used. By the beginning of the twentieth century, natural gas was available as a fuel for homes and industries, but costs were high because the delivery system was still inefficient. Constructing the

Fig. 6.8. True horsepower was used in early-day construction of gas pipelines and compressor stations. (*Courtesy of El Paso Natural Gas Co.*)

first natural gas pipeline was backbreaking work, especially since heavy machinery had not yet replaced horses and mules (fig. 6.8). Until the 1920s, construction was largely performed by hand labor, including clearing right-of-way, ditching, screwing ends of pipe together, lowering in, and backfilling (fig. 6.9).

Fig. 6.9. Hand tongs were used by early pipeline construction workers.

The first large-scale natural gas transmission lines came about as a direct result of World War II. After the war, private industry bought the crude oil and products pipelines—the Big Inch and the Little Big Inch—and converted them to natural gas transmission lines, thereby solving the problem of supplying gas to the markets at rates competitive with other fuels.

Ships at sea

During the 1860s a few sailing ships fitted with iron tanks had begun hauling crude across the Atlantic; however, bulk transportation did not replace barrels for some time, because the shifting of a load of oil in a big tank tended to throw a ship out of ballast. Also, loading and unloading large tanks was a problem. Petroleum, in barrels or in bulk, was a dangerous cargo with the threat of fire and explosion ever present. Barrels often leaked because the glue used to hold them together dissolved from exposure to crude or refined oil. In bulk containers, gases often accumulated—gases that were very susceptible to explosion.

When refining was promoted by European industrialists, oceangoing ships were designed to carry bulk crude from the producing countries (the United States and Russia) to Europe. In 1886 the *Gluckauf*, hauling refined oil, made its maiden voyage from New York to Bremen, Germany. Designed by Wilhelm Riedemann, a German importer, the *Gluckauf* was to be the first successful oceangoing tanker using the ship's hull as part of the storage compartments (fig. 6.10).

New designs that stored the oil in separate steel compartments with a pipe for loading and unloading and a pipe to vent gas accumulations, along with conversion from sail to steam power, revolutionized the tanker as a means of transporting petroleum and its products. By the turn of the century, companies were shipping products from their Gulf Coast plants to the East Coast and abroad. In 1901, at the time of the Spindletop strike, Sun Oil

Fig. 6.10. The *Gluckhauf* was the first successful oceangoing tanker that carried oil in storage tanks built into its hull.

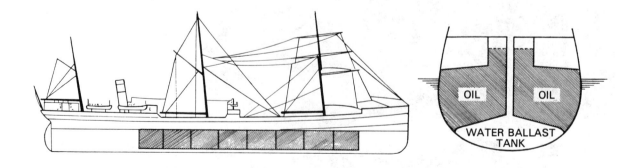

bought a cargo vessel, the *S. S. Paraguay*, and converted it into a tanker to ship oil from the Gulf Coast ports. Other companies followed suit and bought or leased tankers to ship their oil.

After World War I, steam-powered tankships were developed with cargo capacities of 9,000 deadweight tons (dwt) each and with double bottoms. (Deadweight tonnage is a ship's load carrying capacity, expressed in long tons. A long ton is 2,240 pounds.) Diesel engines began to replace steam engines as the power source. In World War II, fleets of U. S. tankers, carrying some 34,000 dwt, played a key role in supplying petroleum for wartime industries on the East Coast, as well as products from the Western Hemisphere to the war zones. These tankers became prime targets for enemy submarines, forcing the United States to expand its pipeline network— a less vulnerable mode of transportation.

When automotive vehicles came into use at the end of the nineteenth century, tank trucks, patterned after the tank wagons, were used to transport crude oil from the wellhead to the pipeline or other shipping outlet (fig. 6.11). With the requirements of World

Tank trucks

Fig. 6.11. An early tank truck hauls crude oil from the well's wooden storage tank. (*Courtesy of Goldbeck Collection, Humanities Research Center, The University of Texas at Austin*)

War I, better truck designs emerged, overcoming the handicaps of inadequate motive power, small size, and unsafe features. Welding replaced riveting as a means of fabricating the tanks, and soon after World War II, aluminum, stainless steel, and plastic tanks replaced many of the steel ones. The transport of chemicals and corrosive commodities and the development of new welding methods were largely responsible for this change in tank construction.

As gathering lines took over the job of transporting crude, tank trucks were used primarily for gasoline and fuel oil. During and after World War II, liquefied petroleum gas (LPG) became a major commodity for tank truck transportation. Tank trucks, specially designed to contain the pressures of liquefied gas, transported LPG not only to dealers throughout the country but from the dealers to their many rural customers.

Design improvements in tank trucks, as in all other methods of transporting petroleum and products, continued to accommodate the needs of the industry.

BARGE TOWS

Petroleum and petroleum products are barged wherever there is a waterway and a need for the products. Barges carrying gasoline, asphalt, crude oil, chemicals, and industrial fuel are shuttled back and forth on a network of lakes, rivers, channels, and intracoastal canals. Levees, dikes, and other flood control devices and navigational aids have made the inland waterway system a safe and efficient method of moving huge amounts of crude oil and products to coastal tankers and refineries (fig. 6.12).

Barges Oil barges were the first to travel the Gulf Intracoastal Waterway, which runs the length of the Gulf Coast from Brownsville, Texas, at the Mexican border more than 1,000 miles to Carrabelle, Florida. This waterway has contributed to making the Gulf Coast the nation's leading petrochemical region. Barges hauling huge spherical chemical tanks are a common site on the canal. Barging is considered one of the safest methods of moving chemicals, since barrier islands, peninsulas, and inland routes in the area protect barge tows from the rough waters, squalls, and storms common in ocean waters.

Petroleum and fuel oil are still the largest cargoes being barged on the Gulf waterway. In 1981, nearly 12 million tons of crude petroleum and nearly 11 million tons of residual fuel oil traveled the canal. Basic chemicals, gasoline, and distillate fuel oil were next in size of shipments on the canal.

Fig. 6.12. This tide gate at Freeport, Texas, is one of the flood control devices that make the Gulf Intracoastal Waterway safe for navigation. (*Photo by Greg White; courtesy of* Texas Highways)

Fig. 6.13. The draft marks on the side of this barge docked at a refinery show that it is virtually empty. Note the bow thruster, which aids the towboat in steering the barge. (*Courtesy of Valero Refining Co.*)

An empty barge rides high in the water, showing its draft marks on its side (fig. 6.13). As it is filled, the barge rides deeper and deeper in the water, and tonnage is estimated by reading the marks on the hull. Barges have no motive power of their own. They depend on tugboats, towboats, and water currents to move.

Tugboats

Tugboats are strongly built, keeled boats, designed to push or pull other vessels in harbors, on inland waterways, and in coastal waters. Although comparatively small, a tugboat is powerful and can often be seen pushing a vessel many times its size. Tugboats are used to help maneuver barges into line as they are laced together with ratchets and steel cable to make a barge tow or to hold a barge in place at the dock for loading and unloading. (fig. 6.14).

Fig. 6.14. A tugboat holds a barge in place at a loading dock (left), while a towboat pushes a loaded barge out of the harbor. Note the keeled hull and bow of the tugboat and the square bow of the towboat. (*Courtesy of Saber Energy, Inc.*)

262

Towboats Unlike the tugboat, whose hull and bow are keeled, the towboat is a relatively flat-bottomed boat with a square bow (front). Affixed to the bow are large upright *knees*, which are used for attaching and pushing barges (fig. 6.15). Sometimes a control barge called a *bow thruster* (see fig. 6.13) is used between the towboat and the barge to give more control in steering the barge down the waterway. Diesel towboats are used to push a string of barges carrying tons of petroleum products over great distances at low cost. A towboat may range in power from 600 to 2,400 horsepower. A large towboat, sometimes as tall as a four-story building, can push as many as forty barges in a string, carrying some 18,000 tons of cargo. This string can "make tow" at about 6 miles per hour or more, depending on the waterway—its current, crosscurrents, sandbars, wind, ice, and other traffic. On the river, the string going with the current has the right-of-way, so those barges going against the river current must stop and wait for oncoming traffic to clear the waterway before they can proceed.

Fig. 6.15. A towboat has large upright knees on its bow to facilitate pushing. (*Courtesy of Valero Refining Co.*)

KNEES

The towboat crew (typically, a captain, a ship's mate, a chief engineer, an assistant engineer, a tankerman, a pilot, a cook, and deckhands) build the tow as they stop along the way to pick up barges. Oil barges "may be strung out for more than 1,000 feet in barges 54 feet wide and almost 300 feet long, lashed tightly together so they can be steered as a single unit."[1] Such a barge tow may be loaded with as much as 60,000 barrels of oil. A river tow may be as large as two side-by-side strings, each stretching some 800 feet. With oil, chemicals, and other petroleum products on board, no one is inclined to take chances, especially when going through control gates, drawbridges, or swing bridges. The barges have no lights, and smoking is strictly prohibited on them. Visitors are not allowed on the tow or the barges.

A towboat crew may be on a trip up the Mississippi River for as long as a month. A trip from Tulsa, Oklahoma, down through the rivers and canal to plants in Nederland, Texas, may take from 16 to 18 days. After picking up its string of barges, a towboat seldom stops. Fuel and supplies are furnished by service boats that come alongside the tow in midstream.

In one year one company shipped nearly 66 million barrels of products by barge. The economic attractiveness of barging was explained by the company's manager of barging in marine affreightment: "In many instances barging is the key logistical link between our manufacturing facilities and our distribution and marketing location throughout the United States. . . . We're getting more and more involved in many aspects of the barging business, from design to ownership, particularly for specialty barges. It's been a steady and positive evolution."[2]

RAILWAY SYSTEMS

Compared to the amount of petroleum and products shipped by other means, the railroads handle only a small volume; however, rail is a vital link in the total petroleum transportation network. In 1984 two major rail lines — the Santa Fe and the Southern Pacific — carried an average of 90,000 tons of lubricating oils, 64,440 tons of petroleum refining products, and 67,560 tons of residual fuel oil between major cities in the United States. Approximately 170,000 tank cars are in regular service on the 320,000 miles of railroad track across the United States, transporting petroleum feedstocks to refineries and refined products to consumers.

1. Tommie Pinkard, "Gulf Intracoastal Waterway: Texas' Big Ditch," *Texas Highways,* August 1984, p. 22.
2. John P. Abbott, "We Just about Cover the Map," *Shell News,* February 1985, p. 21.

Petroleum products transported by rail

The nation's $100 billion chemical industry relies heavily on its fleet of 62,000 or more tank cars to safely transport millions of gallons of chemicals used in virtually all manufacturing processes. Many of these chemicals are shipped an average of 2,000 miles from refineries along the Gulf Coast and in the Southwest to the Northeast and Midwest manufacturing centers for conversion into plastics, synthetic fibers, and paints.

In addition to petrochemicals, major hazardous materials are shipped by rail. Approximately 23,000 pressurized tank cars, 14 percent of the nation's entire tank car fleet, are devoted to moving 70 percent of the major hazardous materials transported across the nation and in Alaska and Mexico. For example, 40,000 to 50,000 carloads of anhydrous ammonia are carried annually from petroleum refineries to farm areas for use as fertilizer.

Probably the hazardous material carried in greatest quantity by rail is liquefied petroleum gas (LPG). About 100,000 carloads of LPG go by rail each year, although most LPG now moves through pipelines.

Tank cars are used to ship crude oil from waterways or ocean tankers to small inland refiners (fig. 6.16). Most refineries have rail lines running through their plants with loading and unloading

Fig. 6.16. This general-purpose tank car is capable of carrying 10,000 gallons of crude oil by rail. (*Courtesy of Union Tank Car Co.*)

depots. Crude oil produced in many states west of the refinery centers is shipped to the East Coast by rail. Two factors contribute to this use of the rails: (1) most pipelines run from east to west, carrying *products* from refineries along the East Coast to consumers in the west, and (2) local government regulations prevent the building of new refineries on the West Coast.

Government regulations

Since most petroleum derivatives are considered hazardous materials, the Federal Railroad Administration of the Department of Transportation (DOT) controls their transport. Rules are found in the *Code of Federal Regulations, Title 49, Transportation.* Diamond-shaped placards appearing on cars indicate hazardous materials, and Form 8620, *Instructions for Handling Hazardous Materials,* issued by the U. S. government, tells carriers how they must handle these products. Tank car design is also regulated by the DOT.

Tank car design and manufacture

Although tank cars seem identical on the outside, they are tailored to carry specific products (fig. 6.17). Special linings and insulation, along with metals unaffected by corrosive liquids, are necessary for each type of product carried.

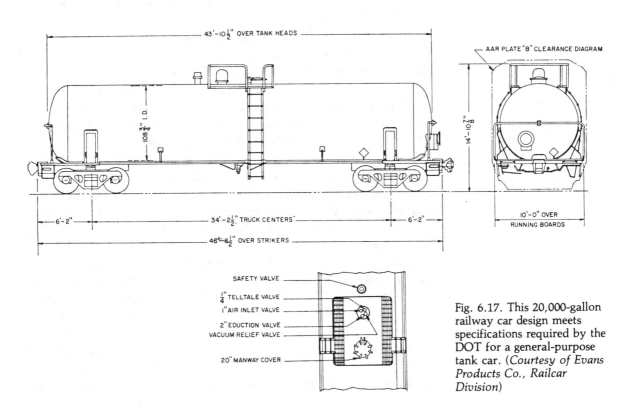

Fig. 6.17. This 20,000-gallon railway car design meets specifications required by the DOT for a general-purpose tank car. (*Courtesy of Evans Products Co., Railcar Division*)

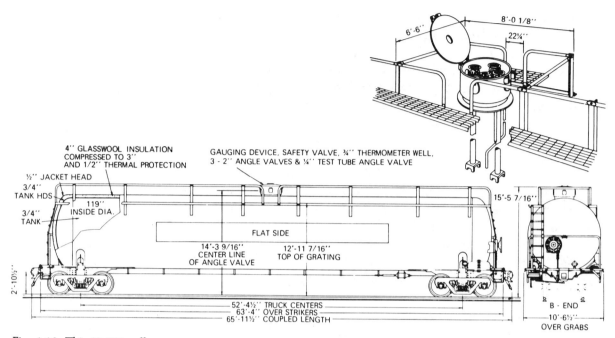

Fig. 6.18. This 33,500-gallon LPG tank car requires special insulation and safety valves to meet DOT requirements.

The huge cylindrical cars range in length from 24 to 64 feet and are up to 15½ feet high and 10½ feet wide. A general-purpose car carries 20,000 to 23,000 gallons of liquid compared to an LPG tank car, which carries about 33,500 gallons (fig. 6.18).

All tank cars are built in accordance with regulations of industry and governmental agencies that establish specifications for each tank car design before it is constructed. In addition to meeting DOT requirements, each tank car design must be reviewed and certified in writing by the Association of American Railroads (AAR) before it can be built and placed in service.

Other than the tank itself, most of the components used in building tank cars are common to all other railroad cars. For example, all tank car wheels, axles, brake systems, bearings, and couplers are built by specialty suppliers of all rail car builders. These specialty builders are also regulated by government and industry agencies.

A tank car is monitored throughout its operating life, primarily by the railroads that operate the cars. Safety inspections are required as well as regularly scheduled tests of pressure rating and checks of all valves for leaks. With the help of the Railroad Tank Car Safety Research and Test Project, tank car manufacturers have improved the safety of tank cars. Three design changes resulted from this project: (1) a new coupler that is less likely to disengage during derailments, (2) head shields mounted on both ends of the car to give added protection against punctures during derailments, and (3) thermal protection provided through either special steel jackets or thermal coating of the tank. By 1981, these modifications were in effect, and the industry spent some $200 million to make pressurized tank cars safer.

Tank car engineers and manufacturers continue to strive for the strongest metal alloy, the optimum tank capacity for each material, improved safety features, and the best valves and fittings.

A unique design patented by General American Transportation Corporation (GATX) is the TankTrain®, a system of interconnected tank cars for transporting bulk liquids. The concept of pumping liquid through interconnected tank cars was first successfully tested by GATX in 1971, and in 1983 the first TankTrain system designed to carry crude oil began operation.

A TankTrain consists of a string or strings of interconnected cars with each string being loaded or unloaded through a single connection, unlike the individual loading and unloading of conventional tank cars (fig. 6.19). Each TankTrain car, with elbow

Tank car strings

Fig. 6.19. Conventional tank cars are loaded and unloaded individually (right), while interconnected tank cars may be handled from one loading rack. (*Courtesy of General American Transportation Corp.*)

pipes, protective housings, and tank isolation valves located at its top ends, is connected to the car in front of it by a large flexible hose. A string of cars is loaded through a connection to the first car. During loading, displaced vapors are collected in the last car and processed through a flare or scrubbing unit or piped to storage for recycling.

At the unloading site, a reverse procedure is used to unload an entire car string, using compressed air or inert gas, depending on the flammability of the liquid being transported (fig. 6.20). A pump is often used to increase the rate of flow.

Two basic loading and unloading systems—*closed* and *simplified*—are capable of handling petroleum products with a large range of viscosities and volatilities (fig. 6.21). The closed system is used when products have flammable or toxic vapors. The simplified system handles products with low volatility and no vapor hazard.

Fig. 6.20. The last car of this thirty-car string is attached to a compressed air supply (pipe in foreground), which is pumped into the last car at 60 psi, forcing the entire lading toward the front car and into a storage facility. (*Courtesy of General American Transportation Corp.*)

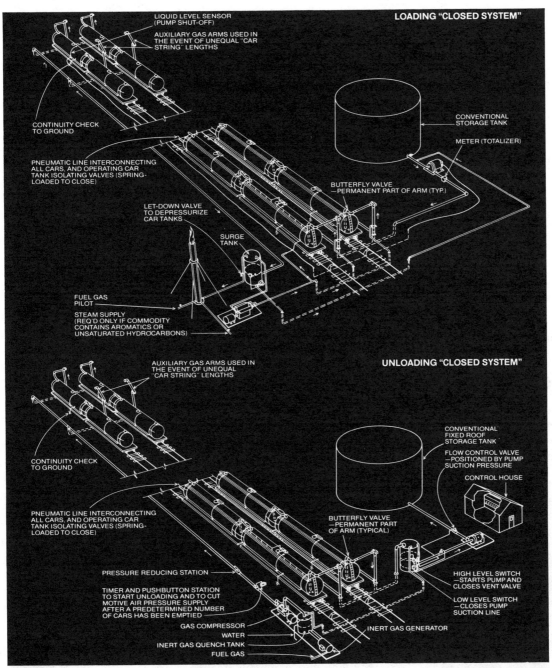

Fig. 6.21. This closed system for loading and unloading maintains an average flow rate of 3,000 gallons per minute for general petroleum products at a maximum pressure of 60 psi in the tank cars. (*Courtesy of General American Transportation Corp.*)

Before the advent of the TankTrain, transporting crude oil from a production field to a refinery over a long distance by rail was too costly and time-consuming. But one petroleum company's system has proved successful in moving crude on a regular basis from its California oilfield to its refining complex some 250 miles away. The system consists of 72 insulated tank cars, holding almost 550 barrels of warm (140° F) crude oil each, making a single shipment of almost 40,000 barrels of oil. Five 3,600-horsepower diesel-electric locomotives lead the train with six more locomotives interspersed throughout, providing 43,200 horsepower to push and pull the fully loaded, 12,000-ton train over its mountainous path.

To handle such high-volume shipments efficiently, the company required some changes on the part of the railroad. Loading and unloading facilities were modified, heavy-duty roadbeds were laid, and new sections of track were installed to facilitate unloading six strings of twelve cars each simultaneously. A two-man crew, situated at the refinery storage end of the train, and one operator can unload the complete train in approximately 5 hours (fig. 6.22). At the other end of the train, nitrogen—an inert gas—is used to pressure the heavy crude oil out of the tank cars.

Fig. 6.22. Interconnected hoses and valves allow the entire train to be unloaded by a two-man crew in a short time. (*Courtesy of General American Transportation Corp.*)

The advantages of simplified loading and unloading, coupled with safety and environmental acceptability, make this "pipeline on wheels" particularly applicable to the transportation of crude oil and other liquid petroleum products.

MOTOR TRANSPORTATION

The design and engineering skill that has gone into the development of tank truck transports and the expansion of the interstate highway system have made motor transportation important to the petroleum industry. Also contributing to the growth of the petroleum trucking industry are the large number of petroleum products being produced and the varied locations of production, refining, and distribution centers.

Many types of motor vehicles are used to transport petroleum, refined products, and gas. A *tank truck* is a self-propelled vehicle that carries a tank on its own chassis. A *truck-tractor* (simply called *tractor*) is usually diesel-powered and is designed to pull a *tank semitrailer* (often called *semi* or *trailer*). In combination, the *truck-tractor semitrailer rig* is generally known as a *tractor-trailer* but is also called a *semi* or an *eighteen-wheeler.* The semitrailer has part of its weight resting on the tractor and part on its own wheels. A *full trailer* is like a semitrailer except that its full weight is on the trailer wheels, and it is designed to be pulled by a truck or behind a tractor-trailer.

Types of vehicles

To carry petroleum products, all vehicles except the tractor have cylindrical, horizontal tanks. The type of tank used depends on the products it is designed to carry.

For purposes of this discussion, the term *truck* is used to mean any of the types of motor carriers just described.

A lot of crude oil is moved by truck, especially from new fields where pipeline gathering lines have not been constructed. Large diesel-powered oil tank trucks can go almost anywhere to serve small lease tank batteries. For example, of the 68,968,945 barrels of oil produced in Texas during April 1985, 16,025,111 barrels were transported from the field to a central facility by motor vehicles. From the central facility, the oil might go into a pipeline and on to a refinery. Of this same total, 52,573,964 barrels were shipped by pipeline from the field, with the balance falling into an "other disposition" category.

Crude oil trucks

Crude oil is carried in a general-purpose tank. This tank needs no special insulation, compartments, or multiple outlets since it carries only one product—crude oil.

Nonpressurized refined products transport

Some refined products can be transported in general-purpose tanks, but others require tanks designed for special purposes. The general-purpose products tank is often lined with epoxy resin, divided into compartments, and fitted with a sump to ensure drainage. Each compartment has its own outlet, enabling the trucker to carry more than one product at a time. Gasoline, kerosine, and diesel are examples of products carried in general-purpose tank trucks (fig. 6.23). Constructed of steel, stainless steel, or aluminum, they are designed to operate at atmospheric pressure. Federal regulations require that trucks transporting hazardous materials display signs or placards stating that fact.

Special-purpose tanks must be used for highly corrosive products and products requiring special equipment. For example, heavy oils that have to be heated for loading and transporting require tanks equipped with flame tubes or steam coils, are usually insulated, and are made of nonconductive materials.

An asphalt tank also contains heating equipment and is insulated, since it may carry products of temperatures up to 400° F.

Fig. 6.23. Tank tractor-trailers regularly line up at refineries to pick up refined products. (*Courtesy of Valero Refining Co.*)

Fig. 6.24. At this terminal, a computerized meter lets truckers load up with gasoline, day or night. (*Courtesy of* PhilNews)

Corrosives such as caustics and acids require tanks made of noncorrosive materials and with special linings and seals.

Some refineries pipe their gasoline, kerosine, and diesel to remote terminals equipped with truck racks and meters that facilitate self-service loading by large tanker trucks (fig. 6.24). Authorized truck drivers can pull into the terminals and take on fuel at the computer-controlled truck loading rack anytime of the day or night. Billing and record-keeping are automated, allowing truckers to load up to 10,000 gallons of fuel in minutes and be on their way without even seeing a refinery or marketing employee.

Fig. 6.25. A liquefied petroleum gas transport truck is used to carry propane to various LPG storage sites. (*Courtesy of Martin Gas Sales, Kilgore, Texas*)

LPG transport

Some products require refrigeration and compression tanks. Liquefied petroleum gases (butane and propane), ethylene, and similar products must be transported in tanks designed for pressures between 100 and 500 pounds per square inch (psi). These tanks are stronger, usually made of high-tensile steel, and not compartmentalized (fig. 6.25). Pressure and very low temperatures are used to liquefy the gases before they are loaded into the tanks, and they are kept refrigerated by insulation and controlled vapor release. Air compressors or pump units are used to discharge gas products from the tanks.

LPG is not only picked up by trucks at the refinery but also dispensed from remote storage systems. One company's propane flows through an 8-inch pipeline from storage tanks some 100 miles away directly into the propane trucks via a propane loading facility. This system eliminates area storage problems and makes tanking up easy for the drivers. The system includes computerization and special meters that measure product flow and volume, computer controls that automatically print bills of lading, and a security system that ensures safe operations.

CNG tube trailers

Until the mid-1970s, most of the natural gas produced at lease sites that did not have gas gathering lines was flared or vented for lack of transportation. Many of these wells were too isolated or remote, or they were marginal producers of gas. In 1979, the Pressure Transport System™, pioneered and patented by Texas Gas Transport Company, provided an economical mode of transportation for natural gas and casinghead gas (gas associated with a well primarily producing oil). Special equipment at the well site dehydrates, compresses, and loads the gas into diesel tractor-tube trailer rigs (fig. 6.26). Each trailer holds from 140 to 180 thousand cubic feet (Mcf) of compressed natural gas (CNG) at 2,400 psi,

Fig. 6.26. Natural gas from a nearby well is compressed to 2,400 psi by a two-stage compressor (right); then it is piped to the load station (left) and loaded into the tube trailer. A diesel tractor pulls the DOT-classified vessel of CNG to its destination. (*Courtesy of Pressure Transport, Inc.*)

depending on the Btu content of the gas. (Btu, short for *British thermal unit*, is a measure of heat energy. One Btu is equal to the amount of heat needed to raise the temperature of 1 pound of water 1 degree Fahrenheit.) By compressing the gas to high pressures, a large volume of gas can be put into a relatively small space.

A standard two-trailer system can move up to 2 million cubic feet (MMcf) of gas per day. After a loaded tube trailer has been delivered to the discharge site, the gas is unloaded into the receiving pipeline or end user's system through a receiving station. The system is designed to operate 24 hours a day, allowing the well to flow continuously into an empty trailer while the full trailer is being transported and emptied.

All tractor-tube trailers operate under a DOT permit, and many safety devices are incorporated because of the extremely high pressures involved. The system is generally recommended for short-haul usage but has been used to transport gas some 60 miles or more. During the transportation process, the CNG remains in its gaseous form for transporting through traditional gas distribution systems.

State regulations vary widely on the operation of motor vehicles carrying petroleum and petroleum products. In addition to tank specifications, drivers and safety equipment must meet state and federal regulations. Driver qualifications, driving rules, equipment and accessories, accident reporting, maintenance requirements, and weights and sizes are some of the things controlled by the government. The Interstate Commerce Commission governs those carriers engaged in interstate or foreign commerce. Many states have adopted ICC regulations to govern intrastate commerce, and the DOT sets the standards for safe operation.

Government regulations

OCEANGOING TANKERS

After World War II, oceangoing tankers became the major method of transporting world oil. Bigger and better crude carriers were built to meet the demand for oil by the industrialized nations. Although the global transportation of oil has slackened somewhat, modern tankers still transport over two-thirds of the millions of barrels of petroleum produced in many areas of the world. Crude oil tankers range in size from the ultralarge crude carriers with a carrying capacity of some 500,000 deadweight tonnage (dwt) to the smaller tankers of 20,000 dwt or less. Tankers are also designed to carry refined products, natural gas liquids, and liquefied natural gas.

Supertankers   In 1973 the first half-million-ton tanker was launched, and in 1979 — the peak year of worldwide petroleum trade — some 190 million dwt of petroleum products were transported by supertankers. Many companies were operating *very large crude carriers* (VLCCs) and *ultralarge crude carriers* (ULCCs), although ports and waterways that could handle these mammoth vessels were limited.

A VLCC has a 100,000-dwt to 250,000-dwt range, while a ULCC ranges from 250,000 up to and over 500,000 dwt. A typical supertanker is twice as wide as and hundreds of feet longer than the largest passenger ship and can hold some 2 million barrels of oil. As one person explained it, a VLCC transports enough petroleum "to fill a 50-mile line of tank trucks or to heat the homes and power the cars in a city of 85,000 people for a year."[3]

VLCCs and ULCCs are designed for the greatest capacity, safety, and fuel efficiency possible. Navigational aids, electronic control systems, automated power plants, satellite communications, and radar equipment allow for maximum security and a minimum crew, making this type of marine transportation an economical one for long-distance hauls.

During their prime, supertankers filled a specific need: they transported huge amounts of crude from new oilfields to the new refineries being built to supply an increasingly energy-hungry world. "The global tanker fleet doubled in size between 1947 and 1957, again in 1966, and once again by 1973."[4] By the early 1980s, a worldwide oil slump caused a surplus of supertankers, especially the very large ones. The oil shipping industry depression put about

3. "How to Sail in an Oil Surplus," *Shield*, Vol. X, No. 2, 1985, p. 2.
4. "Chevron Shipping — Battened Down and Making Way through Rough Seas," *Chevron World*, Summer 1985, p. 16.

one-third of the world's supertankers out of the transportation business.

Companies having marine transports have had to cope with this worldwide slump in oil demand. Some companies have downsized their surplus supertankers, some have scrapped them rather than bear the cost of maintaining them, others have mothballed all or part of their fleets, and still others have utilized them as offshore storage tanks.

Of course, supertankers are still filling a transportation need. In 1984, Marathon's *MT Garyville*, a 140,000-dwt crude oil carrier, traveled 5,700 miles from Scotland's east coast to the Louisiana Gulf Coast (fig. 6.27). On this sixteen-day voyage, the tanker carried 1 million barrels of oil beneath her main deck.

Fig. 6.27. The tanker *MT Garyville* takes on its first load of Brae Field crude oil at a terminal off Scotland. (*Courtesy of Marathon Oil Co.*)

Average-size tankers

Global supertanker fleets are gradually being replaced by smaller tankers ranging in size from 30,000 to 80,000 dwt. The construction of new pipelines, the deepening and widening of canals and waterways that accommodate average-size tankers, the discovery of new fields that eliminated the necessity of many long hauls, and a use conservation program have all contributed to the popularity of smaller tankers.

U. S. tankers are constructed and operated under strict regulations. Using a communications network of telephone, telex, and satellite systems, a company can locate any of its tankers at any time. On board the ships, strict safety measures are followed, minimizing loss of personnel and product. Automatic collision avoidance systems track approaching ships as far away as 50 miles, alerting a tanker to an obstruction on its course.

Responsible shippers adhere to these environmental and safety regulations, although many countries have no building restrictions or rules for operation. In spite of the notoriety of oil spills, the percentage of crude lost to spills each year is small when compared to the total amount transported by tanker.

NGL and LNG tankers

Tankers equipped with pressurized, refrigerated, and insulated tanks are used to transport natural gas liquids (NGL) and liquefied natural gas (LNG). NGLs are those hydrocarbon liquids that are gaseous in the reservoir but recoverable as liquids in field facilities or in gas processing plants. Gas liquids that are not highly volatile are treated and shipped as natural gasoline or added to the crude oil. If separated before shipment, the other NGLs—butane, propane, and ethane—must be transported in tankers designed to handle the high pressures and low temperatures needed to keep the gases in a liquid state. NGL is usually transported as a mixture and

Fig. 6.28. This cryogenic tanker carries 450,000 barrels of liquefied natural gas per trip from Alaska's Kenai Peninsula to Tokyo. (*Courtesy of Marathon Oil Co.*)

separated into LPG, natural gasoline, and ethane after it reaches the refinery.

Natural gas is liquefied at the source before it is transported by special LNG tankers. Liquefied natural gas is regasified and generally fed into a gas pipeline system when it reaches its destination. One company that furnishes natural gas to Tokyo liquefies the gas near its source by lowering its temperature to $-259°$ F. As the gas liquefies, it shrinks to $\frac{1}{600}$ of its original volume. The company uses cryogenic (extremely low temperature) LNG tankers, each holding 450,000 barrels of LNG, or the equivalent of 1.5 Bcf of natural gas (fig. 6.28). The LNG cargo is kept in its liquefied state during a 3,000-mile journey and is converted back to its gaseous state as it enters natural gas pipelines in Japan.

The technologies of icebreakers and oil tankers have been combined to develop icebreaking tankers used for shipping oil and gas in the Arctic, Baltic, and other ice-laden seas. Icebreaking tanker design is unique and is based on ship-ice interaction. Design engineers record loadings on different ships in controlled ice conditions and perform continuous measurements on ships in regular arctic service in order to build economically operable icebreaking tankers. The tankers must have hull strengths that prevent spillage but avoid the high cost and excess weight of very thick steel hull plating.

Although U. S. companies are studying the feasibility of operating icebreaking tankers, they have not yet utilized them for petroleum transportation. Icebreakers in operation were designed by Wärtsilä Shipbuilding Division in Helsinki and have been used since 1977 by Neste Oy, a government-owned oil company in Finland (Fig. 6.29). Neste uses the *Lunni* and her sister ships to distribute oil products from the refineries in the south of Finland to the towns on the Finnish coast, and to transport crude oil from the Baltic area to the refineries in Finland.

Icebreaking tankers

Fig. 6.29. Hull strength is of vital importance to this icebreaking oil tanker used by Neste Oy of Finland. (*Courtesy of Wärtsilä Shipbuilding Division*)

280

Loading/offloading
facilities

Because very few ports are large enough to accommodate VLCCs and ULCCs, oil entrepôts and methods of discharging at sea have been developed. (An *entrepôt* is a transshipment center for the collection and distribution of goods, as a *depot* is a center for storing goods.) Most ports capable of handling supertankers have offshore mooring facilities, and underwater pipelines are used to transfer oil from ship to ship or from ship to barge. In the United States, these shuttle tankers and barges then take the oil or products to shore or inland through the 25,000 miles of major waterways maintained by the U. S. Army Corps of Engineers. Many of these waterways are deep-draft canals and channels, deep enough and wide enough to handle small tankers (fig. 6.30).

The only U. S. mainland deepwater port equipped to handle supertankers is the Louisiana Offshore Oil Port (LOOP), and its capability is limited to 150,000 dwt. Oil discharged at LOOP is shipped by pipeline to company refineries, with computers controlling and supervising deliveries. Tankers of various sizes take on North Slope crude oil from the Trans-Alaska Pipeline at the Valdez terminal in southern Alaska (fig. 6.31).

Floating production, storage, and offloading (FPSO) units are used in offshore areas to facilitate the transfer of crude oil from production platforms to shuttle tankers. Well fluids are first transferred by subsea pipeline to a moored tanker outfitted with storage and treating equipment. Then the treated crude, which is sold by meter transfer, is picked up by shuttle tankers that dock at the FPSO unit for loading.

After its journey from the oilfield or from a supertanker to its destination, a shuttle tanker may offload its cargo at the refinery's loading/offloading docks (fig. 6.32). A products ship, which may carry more than one type of cargo, is often unloaded at different destinations; therefore, the ship must be loaded so that the trim is not upset when the first product is discharged.

Fig. 6.30. A tanker shuttles down the Sabine Ship Channel on the Gulf Intracoastal Waterway after unloading its cargo at an inland plant. (*Photo by Greg White; courtesy of* Texas Highways)

Fig. 6.31. The *ARCO Fairbanks*, a 120,000-dwt VLCC, docks at Alaska's Port Valdez terminal at the southern end of the Trans-Alaska Pipeline. (*Courtesy of Alyeska Pipeline Service Co.*)

Fig. 6.32. Crude oil and feedstocks are shipped in and products are shipped out from this refinery dock used for small- to average-size tankers. Note the loading/offloading arms, which swing out to reach a tanker's hold. (*Courtesy of Valero Refining Co.*)

282

CRUDE OIL PIPELINES

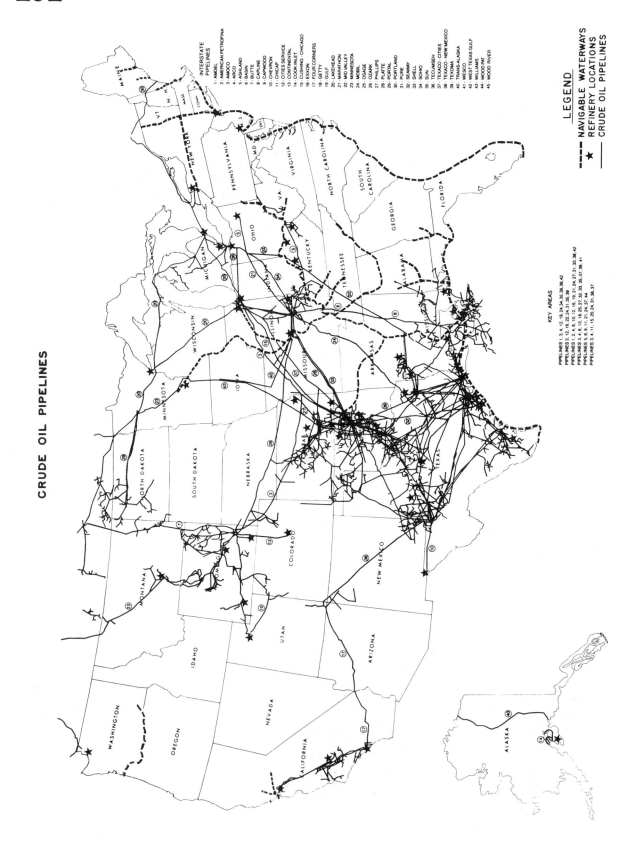

LEGEND

NAVIGABLE WATERWAYS
★ REFINERY LOCATIONS
CRUDE OIL PIPELINES

INTERSTATE PIPELINES

1. AMDEL
2. AMERICAN PETROFINA
3. AMOCO
4. ARCO
5. ASHLAND
6. BASIN
7. BUTTE
8. CAPLINE
9. CAPWOOD
10. CHEVRON
11. CHICAP
12. CITIES SERVICE
13. CONTINENTAL
14. COOK INLET
15. CUSHING - CHICAGO
16. EXXON
17. FOUR CORNERS
18. GETTY
19. GULF
20. LAKEHEAD
21. MARATHON
22. MID VALLEY
23. MINNESOTA
24. MOBIL
25. OSAGE
26. OZARK
27. PHILLIPS
28. PLATTE
29. PORTAL
30. PORTLAND
31. PURE
32. SEAWAY
33. SHELL
34. SOHIO
35. SUN
36. TECUMSEH
37. TEXACO - CITIES
38. TEXACO - NEW MEXICO
39. TEXOMA
40. TRANS-ALASKA
41. WESCO
42. WEST TEXAS GULF
43. WILLIAMS
44. WOOD PAT
45. WOOD RIVER

KEY AREAS

PIPELINES 1, 3, 4, 12, 19, 24, 34, 35, 38, 39, 42
PIPELINES 3, 12, 16, 22, 24, 31, 35, 39
PIPELINES 1, 2, 4, 6, 10, 12, 16, 19, 21, 24, 27, 31, 33, 38, 42
PIPELINES 3, 4, 6, 13, 18, 25, 27, 32, 33, 35, 37, 39, 41
PIPELINES 5, 8, 9, 11, 21, 24, 37, 44
PIPELINES 3, 4, 11, 15, 20, 24, 31, 36, 37

CRUDE OIL PIPELINES

The construction of oil pipelines increased tremendously during and after World War II. From prewar deliveries of less than 50,000 barrels, daily pipeline shipments eastward soared to 754,000 barrels, with the companies east of the Rockies increasing their total daily movements about 60 percent between 1941 and 1945. Companies discovered that large-diameter trunk lines proved to have operating flexibility previously considered available only in smaller lines and that the larger lines reduced transportation costs per barrel considerably. Pipelines devoted to the transportation of liquids increased from 124,000 miles in 1948 to more than 222,000 miles in 1976.

Since 1976, actual mileage has declined but total quantities shipped has steadily risen because of larger-diameter lines and the *looping of long lines*, that is, laying additional lines alongside existing pipelines (fig. 6.33). A major contributor to the increased quantities of oil transported is the Trans-Alaska Pipeline System (TAPS). This 48-inch-diameter pipeline, which runs from Prudhoe Bay to Port Valdez—a distance of some 801 miles, can deliver 2 million barrels of oil per day to the Valdez terminal. Sophisticated handling facilities, including four tanker berths, permit the safe and efficient loading of tankers at rates up to almost 110,000 barrels of oil per hour.

State and federal regulations

In an industry where crude oil is shipped regularly, accurate recording of shipments sent and received can mean the difference between profit and loss and even between legal and illegal operation. Some states require the pipeline companies to keep track of individual and connected shippers to see that each company is complying with the established monthly proration schedules. Monthly forms are filed with the appropriate state agency to indicate the quantity of oil taken from each lease.

Fig. 6.33. Shown as solid black lines, the main interstate pipelines of the United States connect key production areas with refineries (stars). Also shown on this 1983 map are the major inland waterways (dashed lines) used for transporting petroleum and petroleum products. (*Courtesy of Association of Oil Pipe Lines*)

The federal government also imposes stringent controls on various aspects of the pipeline industry. Oil pipelines are classified as common carriers and therefore operate under the U. S. Department of Transportation. At one time, they were under the jurisdiction of the Interstate Commerce Commission.

Field gathering systems

An oilfield may have several hundred wells, with flow lines that carry crude oil from the wells to the lease tanks (fig. 6.34). Various tanks and treating vessels are often a part of the field gathering system, but lines can take crude oil away without having tanks or other ground equipment except for printing meters to register the number of barrels taken.

After the oil is gathered from the field, it is usually treated, measured, and tested before entering the pipeline—a major network of small and large arteries installed by the pipeline company to give oil producers an outlet to market.

Fig. 6.34. Photographed from an altitude of 27,000 feet, this southwestern oilfield is a maze of wells (light spots) and flow lines (white lines).

To move the oil into and through the pipeline, pressure is supplied by pumps at a pump station. Functionally, pump stations are gathering stations, trunk-line stations, or a combination of both. A gathering station is located in or near an oilfield and receives oil through a pipeline gathering system from the producers' tanks. From the gathering station, oil is relayed to a trunk-line station, which is located on the main pipeline, or trunk line. Of much greater capacity, the trunk-line station relays the oil to refineries or shipping terminals. Since the pressure gradually drops as the oil is moved through the line, booster pump stations are spaced along the trunk line as needed. Tank farms along the line serve as receiving and holding stations. The oil that enters and leaves the tank farms and the pump stations is regulated by manifolds.

Gathering station. The gathering station may include one or more pumps and may move several thousand barrels of oil daily from various producers' tanks and gathering station tanks. Pumping units usually consist of electric-drive reciprocating or centrifugal pumps. Internal-combustion engines have been almost totally displaced by electric motors as prime movers. The method of operating a gathering station depends on its size and function. At tanks where custody transfer occurs, the pumps are activated by a gauger, who measures the volume and draws samples for analysis before pumping begins (fig. 6.35). On leases where LACT units are installed, pumps are started automatically or by remote

Fig. 6.35. A battery of suction booster pumps activated by a gauger pull oil from the tanks and move it to the suction of the trunk-line pumps.

control. After the oil is received at the gathering station, it is pumped either to another gathering station closer to the trunk line or to a trunk-line station directly.

Trunk-line station. A trunk-line, or mainline, station is located on a main petroleum-carrying artery. It may serve as a gathering station if near the source, but usually a trunk-line station relays or boosts the oil to the next pump station down the line. Pumps used in trunk-line stations today are typically high-speed centrifugal units having a direct connection to electric motors (fig. 6.36). Internal-combustion engines and gas turbines are used to a lesser degree because of increases in both fuel and maintenance costs. In a large station, the pumps are usually placed in series so that each pump handles the total flow, imparting an increment of pressure to the oil. A sufficient number of pumps are used to achieve the required pressure, and the oil is sent down the pipeline to the next station. Booster stations are located along the trunk line so that the work load performed by each station is approximately equal. This division of load is called *hydraulic balancing.* The work load is determined by the volume and velocity of the fluid being pumped (bbl/hr/mile), the density and viscosity of the fluid, and friction of the fluid on the inside surface of the pipe.

Station tank farm. A tank farm is to a pipeline what a railway yard is to a railroad. It is a place where oil in transit may be temporarily sidetracked for sorting, measuring, rerouting, or simply holding during repairs on a line or station. A tank farm may also be a receiving station where oil comes in from the producing fields

Fig. 6.36. A routine check of temperature and lubrication keeps this large centrifugal pump operating at a trunk-line station.

or from other carriers to be injected into the pipeline transportation system (fig. 6.37). Trunk lines bring crude oil belonging to many different shippers to the tank farm. By segregating shipments in the tanks, the pipeline company can make deliveries to various owners at refineries or terminals.

Pump station manifold. Tank farm pipelines converge at a station manifold (fig. 6.38). A manifold is an accessory system of piping that serves to divide the flow into several parts, to combine

Fig. 6.37. This pipeline junction has a tank farm, pump station, and converging pipelines. Note the positions of the floating roofs on the tanks.

Fig. 6.38. The manifold at a pump station and tank farm permits routing of the oil stream to its destination.

several flows into one, or to reroute the flow to any one of several possible destinations. The manifold permits the switching of the oil stream to the proper destination. It can be very complex as in a main station or simple as in a booster station. A manifold might connect a mainline, a field gathering line, several tanks, and one or more pumps. This network of lines and valves, operated by the tank farm and station personnel, can (1) pump oil through the trunk line with tanks inactive, (2) receive production from the field into any tank, (3) receive from a trunk line into any tank, (4) transfer from one tank to another, (5) pump from any tank into the trunk line, (6) isolate all pumps and tanks while the trunk-line station upstream pumps through, or (7) inject oil from any tank into oil being pumped through the trunk line.

Control of oil movements

Pipeline shipping of crude oil is based on large shipments, high throughput, and efficient operation to reduce costs to a minimum for the benefit of the shippers. Hence, a single shipment, or batch, must consist of a minimum volume of several thousand barrels. Any single pipeline may handle several grades of oil for each of several shippers. The shipment is scheduled by the pipeline scheduler to leave a stated point of origin on a set date and to be delivered to a destination point on a set date. The shipper must have tank space to receive the grade of oil at the destination.

A typical shipment might be accumulated over a period of several days. Each day a part of the shipment is delivered from the leases to pipeline company tanks to be held until the required volume is ready for delivery to the trunk line. The pipeline takes custody at the lease where the oil is produced and holds it on account for the shipper who either produced it or bought it.

Suppose transit time is ten days and ten different shippers regularly use a pipeline. At any given time each of these shippers might have several shipments in the line, and the pipeline company might have in the line and in its tanks more than a million barrels of oil, all of which belong to the shippers. As oil is pumped in at the upstream end, an equal volume is pumped out at the downstream end. Each shipper, therefore, is continually receiving deliveries and putting more oil in the line. Thus, the shipper always has a balance of oil in the system. While his oil is in transit, a shipper may trade it to another shipper for oil that was shipped earlier. Then the pipeline changes the route and delivers it to the new owner at a different destination. This buying, selling, trading, and rerouting of shipments is routine in pipeline operation.

Just as a railway freight shipment may travel on more than one railroad in transit, so a shipment of crude oil may travel on more than one pipeline system between its origin and destination. Requests for shipping space in one section of a line may be greater than those in other sections of the same line because of transfers to another pipeline system at some point miles downstream from the originating station. The second line must be prepared to receive the diverted shipment, and the originating line must pump from tankage at the diversion point while the diverted shipment is being withdrawn from the line. Otherwise, oil movements on the downstream sections of the originating line will come to a standstill.

Scheduling. To handle all of the operations involved in pipelining, a coordinator or scheduler works with the shippers to set up a mutually agreed-upon schedule. On line with many pump stations and pumping units, computer programs determine the stations to be operated and the exact pumping unit combinations to be used to obtain the desired flow rate at minimum power cost. The movement of crude oil is scheduled according to the projected dates of both accumulation from producing areas and connecting carrier sources. Though tanks are common items at points of origin and at terminals, the fact that modern pipeline systems are designed with a minimum of operating tankage may pose a major problem in pipeline scheduling. The sequencing of batches to arrive at final destination or intermediate delivery points for barge or tanker loading requires considerable skill on the part of the scheduler. Computers are used on the modern pipeline to develop schedules, to project the schedules to completion, to print out the pertinent information of delivery times and dates, and to inject and switch instructions needed to implement the schedules (fig. 6.39).

Fig. 6.39. Schedules show point of origin, amount and grade of oil to be pumped, batch number, starting and ending time and place, and pumping rate of each batch. (*Courtesy of American Petroleum Institute and Continental Oil Co.*)

Dispatching. After the schedule has been determined, a pipeline dispatcher, sometimes called a line operator or oil movement controller, communicates with other points on the pipeline system. Using a computer console, he controls the pumping stations hundreds of miles away and monitors line pressures, rates of flow into and out of the pipeline system, and gravities of the various crude oil streams. In executing the schedule, the dispatcher operates motor valves, starts and stops pumps, and performs other operations in the system through computer-assisted remote control. As the dispatcher executes the schedules, the computer records shipping information—the quantity and quality of oil tendered, its origin and destination, and shipping and arrival dates. The dispatcher is also responsible for the security of the lines. He makes sure that the liquid is not being lost because of a leak or theft.

Testing and measuring. Along with the scheduler who devises the plan and the dispatcher who executes it, an oil measurements group has the job of preserving the quality and quantity of material tendered to the pipeline by the shipper. This group checks methods used by field personnel to measure the quantity of oil by meters and tank gauges. Testing procedures and sampling devices are constantly monitored.

Accounting. The run ticket is the basic accounting document in the buying and selling of millions of barrels of crude oil each day. Every time custody of oil changes hands, a run ticket is made and signed by the receiver and the deliveryman (see fig. 5.43). This ticket serves as a record and a receipt for all oil handled by the pipeline. In pipeline operations, control is essential to ensure the proper handling of oil movements. Records are maintained and used many months after the oil has been delivered to its destination.

PRODUCTS PIPELINES

Just as crude oil pipelines perform a *gathering* function, products pipelines perform a *distribution* service, with refineries in the middle. A products pipeline transports refined products derived from crude oil, such as gasoline, diesel fuel, kerosine, jet fuel, heating oil, and other liquid hydrocarbons *not* produced by liquefaction of gas. Thus, products pipelines usually begin at or near refineries and end at terminals in areas of high market demand for the products (fig. 6.40).

Most of the nation's products pipeline systems originate in and around the Gulf Coast area, since 50 percent of the country's refining companies and 60 percent of the petrochemical industry are concentrated there.

Fig. 6.40. A products pipeline terminal at New York City's John F. Kennedy International Airport supplies an area of high market demand. Shown are (left to right) a storage tank, prover loop, and filter vessels. (*Courtesy of Buckeye Pipe Line Co.*)

A products pipeline operates much like a crude oil pipeline in reverse. Intake, movement, and delivery of products is generally directed by remote control from the pipeline company's headquarters. A dispatcher at a computer terminal in the main control center controls all movements in the company's mainlines (fig. 6.41). A backup control center, located away from the main control center, takes over operations in case of emergency.

Control of products movement

Fig. 6.41. A dispatcher controls product movement and monitors other operations from the main computer console. (*Courtesy of Colonial Pipeline Co.*)

Fig. 6.42. Tricolor video displays enable an operator to view any aspect of the modern products pipeline pumping station. (*Courtesy of Marathon Pipeline Co.*)

Located on the mainline are booster stations (usually unmanned) and junction stations, generally operated by the main control center. Instrumentation at the center keeps the operator informed of suction and discharge pump pressures and destination and time of arrival of each shipment en route. A booster station usually consists of a central hydraulic system that operates block valves and electrically operated pumps that keep the pressure needed to move products along.

Operators on duty at a pipeline's pumping station use computer consoles to monitor all aspects of station operation (fig. 6.42). Operations are initiated at input stations, and local deliveries and transfers are handled at delivery sites by station operators. Electricity supplies the motive power throughout the system.

Some products reach their destination through stublines. Stublines branch off from the mainlines and carry products to nearby areas. For example, one company covers a thirteen-state area with 2,885 miles of mainlines, 2,190 miles of stublines, and 173 miles of delivery lines (fig. 6.43). In 1983 this company

293

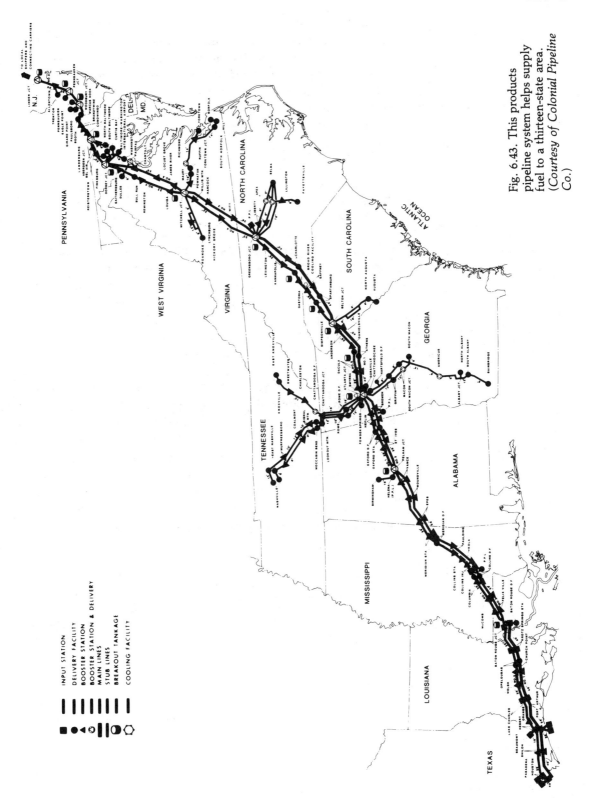

Fig. 6.43. This products pipeline system helps supply fuel to a thirteen-state area. (*Courtesy of Colonial Pipeline Co.*)

294

typically transported more than 1.6 million barrels per day of refined petroleum products an average distance of 1,100 miles.

Breakout tanks, located at mainline junction stations, are used to hold products temporarily until they can be relayed to local shippers' tanks, shippers' tanks on stublines, or terminals farther up the mainline. Manifolds may be found in conjunction with breakout tanks or at the company's facilities.

Batching

A pipeline company takes care of many shippers and their various products, so batching is a very important aspect of products pipelining. One company, for instance, handles as many as thirty different grades of products. To operate such a process, the company transports gasolines and distillate fuel oils in batches (fig. 6.44). The sequence of products varies according to the types of products being shipped. The company operates on a ten-day cycle, meaning that every shipper is able to repeat shipments of any product grade at least every ten days. The shipper's exact place in the cycle is determined by the kind of product he plans to ship and the products of other shippers.

Verification of arrival of a shipment is established by sampling the incoming batch of products. The gravity, color, and appearance of the product usually verifies the change from one shipment to another.

Sometimes batches of products are separated in the line by batch separators, such as spheres; however, commingling of products at interfaces between shipments is usually not a problem. Spheres, as well as line scrapers of various kinds, are used to remove water and sediment from the pipeline, thereby preserving product quality (fig. 6.45).

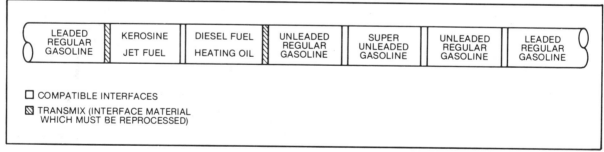

Fig. 6.44. A products pipeline runs batches of different products in sequence similar to this typical cycle. (*Courtesy of Association of Oil Pipe Lines*)

Fig. 6.45. A sphere is used to clear water and sediment from a pipeline. (*Courtesy of Association of Oil Pipe Lines*)

Federal and state regulations stipulate the type and efficiency of equipment to be utilized by the industry to protect the air, soil, and water from becoming contaminated by hydrocarbon products. These regulations differ by states, although products pipelines are considered common carriers and, like crude oil pipelines, operate under the Department of Transportation. Because the products carried are considered hazardous, the Occupational Safety and Health Administration (OSHA) and the Environmental Protection Agency (EPA) are also involved in their handling.

A petroleum products pipeline is considered a closed system; however, limited exposures are associated with product sampling, routine station operations, and occasional leaks and spills. Because petroleum liquids are flammable and volatile, products pipeline companies are aware of potential hazards and have taken every step to protect their employees, the public, and the environment.

Environmental and safety regulations

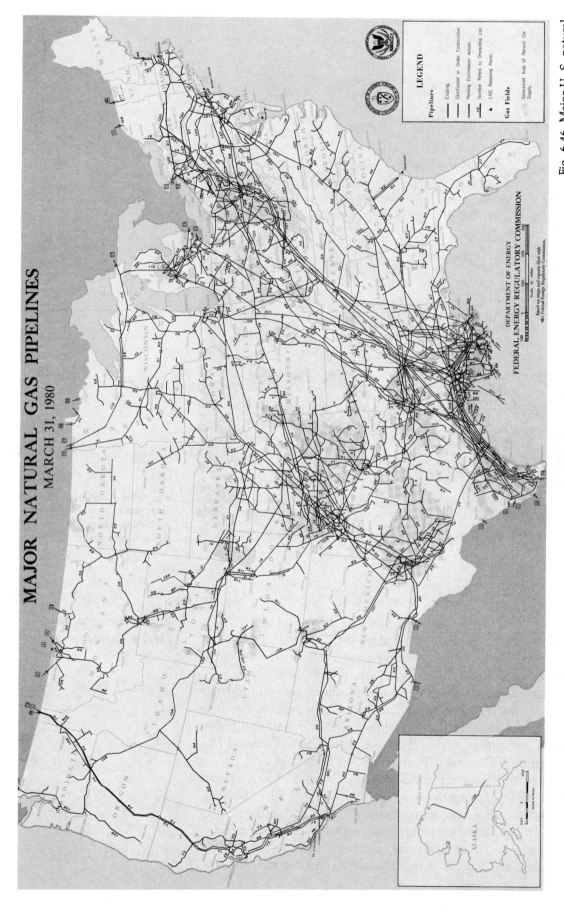

Fig. 6.46. Major U. S. natural gas pipelines connect gas supply areas.

NATURAL GAS PIPELINES

Gas pipelines and oil pipelines historically have been separate industries. A gas pipeline uses pressure from *compressors* as its driving force, while an oil pipeline uses *pump* pressure. Different agencies govern their operations. Before the establishment of the U. S. Department of Energy, gas pipelines were regarded as a utility and operated under the regulation of the Federal Power Commission (FPC), unlike oil pipelines that operated under ICC regulations. At present, the sale and transportation of gas are controlled by the Federal Energy Regulatory Commission (FERC), a branch of the U. S. Department of Energy, while oil and products pipelines are controlled by the Department of Transportation.

Since 1950, the natural gas industry has grown tremendously as technological advances have cut the cost of installing pipe. From 1950 to 1975, the number of miles of gas pipeline tripled. Of the more than a million miles of U. S. gas pipelines in use today, about 270,000 miles are transmission trunk lines (fig. 6.46), with the remaining mileage consisting largely of distribution, main, and gathering lines. Table 6.1 shows the number of miles of pipeline and the average number of gas customers by state. More than 1,000 compressor stations are associated with the trunk lines, using over 12 million horsepower to drive the compressors (fig. 6.47). In addition to the trunk-line stations, another 1,000 compressor stations for field gathering and other uses require 2.5 million horsepower from prime movers.

Modern transmission systems

The method used for collecting, conditioning, and transmitting gas to its primary destination depends on the individual situation, but within any gas gathering system, specialized equipment is required for (1) conditioning the gas so it will flow safely and steadily and (2) controlling, measuring, and recording its flow through the pipeline.

Conditioning equipment such as separators, heaters, dehydrators, and compressors is located at the wellhead where the gas first reaches the surface or at other locations in the field. Each piece of equipment has its particular function related to the safe and steady flow of gas from the wellhead to the pipeline and finally to its destination often thousands of miles away.

Conditioning equipment

Separators isolate components flowing in the line. Heaters prevent hydrate formation by raising the temperature of the gas to a safe level until it reaches the next stage of conditioning, and dehydrators remove all but a minute amount of water vapor from the gas.

TABLE 6.1
Transmission, Distribution, and Field Gathering Pipelines
and Natural Gas Customers by State

State	Pipelines (miles)	Gas Customers (thousands)
Alabama	19,644	674.1
Alaska	1,212	45.5
Arizona	16,817	585.2
Arkansas	18,387	491.9
California	80,837	7,410.7
Colorado	23,488	872.7
Connecticut	6,798	412.2
Delaware	1,339	82.7
District of Columbia	1,162	147.8
Florida	12,651	441.9
Georgia	25,147	1,053.6
Hawaii	587	32.7
Idaho	4,435	110.0
Illinois	53,501	3,317.8
Indiana	27,762	1,266.8
Iowa	18,042	756.2
Kansas	37,318	735.5
Kentucky	20,110	630.1
Louisiana	42,256	1,006.7
Maine	389	15.3
Maryland	8,764	783.0
Massachusetts	16,915	1,093.5
Michigan	44,486	2,484.4
Minnesota	17,232	836.5
Mississippi	18,101	400.3
Missouri	21,404	1,230.3
Montana	8,538	184.9
Nebraska	14,149	469.7
Nevada	3,658	175.9
New Hampshire	1,162	52.2
New Jersey	23,106	1,850.1
New Mexico	25,373	337.7
New York	41,681	3,943.3
North Carolina	12,090	385.2
North Dakota	2,950	88.6
Ohio	52,981	2,768.1
Oklahoma	38,578	834.2
Oregon	10,070	283.9
Pennsylvania	52,527	2,362.2
Rhode Island	2,425	173.2
South Carolina	9,973	286.2
South Dakota	2,622	105.6
Tennessee	16,502	515.0
Texas	120,880	3,260.6
Utah	8,504	386.5
Vermont	404	15.7
Virginia	11,345	533.4
Washington	11,883	380.6
West Virginia	22,331	393.8
Wisconsin	22,197	1,060.2
Wyoming	8,457	79.9
Total U. S.	1,064,190	47,844.1

Source: American Gas Association, Gas Facts, Arlington, Virginia.

Fig. 6.47. This compressor plant speeds gas to consumers by pipeline.

Large compressors are used to compress the gas up to or in excess of a hundred times the normal atmospheric pressure. Usually large reciprocating compressors driven by gas engines are used, but centrifugal units driven by gas turbines or electric motors are also utilized (fig. 6.48). Large compressor stations along the

Fig. 6.48. This 9,200-horsepower gas turbine engine is typical of those used in gas pipeline compressor stations. (*Courtesy of American Petroleum Institute and Cities Service Co.*)

300

pipeline are often fueled by natural gas from the pipeline; the high pressure of the gas is lowered to a usable pressure by regulators.

Controlling and measuring flow

Compressed gas exhibits a complex behavior pattern; in order to operate a gas pipeline properly, most of the variables of the gas must be measured and controlled. Orifice meters are commonly used to measure the volume of natural gas, although turbine meters have become popular. Turbine meters are ideally suited for easy transmission of measured data via electronic means and are more accurate over a wide range of flow rates.

Modern gas pipelines are also equipped with devices and instruments that measure and record density, dew point, temperature, and pressure. Recording gravitometers are used to measure and record the specific gravity of natural gas, which is needed to calculate gas flow. Pertinent information is gathered by electronic data recorders and processors and is sent to an office for analysis.

Automation

Many gas pipelines have become so automated that they are capable of operating under command of a computer system that coordinates the operation of valves, prime movers, and conditioning equipment. The computer receives input from each part of the system, including the conditioning and measuring equipment. In case of malfunction at any point in the process, the computer searches its programs for possible corrective actions and simultaneously sounds alarms at the appropriate control points (fig. 6.49).

Fig. 6.49. A computerized control center keeps tab on deliveries throughout the gas pipeline system. (*Courtesy of Tennessee Gas Pipeline Co., a Division of Tenneco Inc.*)

An interesting aspect of natural gas pipelines is the introduction of odorants into the gas system. Natural gas is almost odorless as it comes from the well or processing facility. If the gas is destined for use as a fuel in homes or industry, a compound called *mercaptan* is used to odorize the gas so that it can easily be detected when its presence in the atmosphere reaches a concentration of 1 percent. Gas and air mixed in this concentration are not hazardous, but a mixture containing 5 percent gas is explosive.

The odorant makes leaks or other unburned discharges of the gas quite evident long before a real hazard exists. Odorants injected into the gas burned in homes or industry do not create odors while burning, nor do they leave troublesome residue. Odorants are usually not introduced into gas sent to petrochemical plants where the gas is used as a feedstock for producing other commodities such as plastics, since the mercaptan will frequently interfere with the chemical process.

PIPELINE CONSTRUCTION ON LAND

Before construction begins on a modern pipeline, months and sometimes years of engineering studies and surveys of potential reservoirs and markets precede the final decision to build the line. Routes are surveyed by aerial photography and surface mapping. During the surface mapping, a right-of-way must be secured from each owner of the property that the pipeline will cross. Before the pipeline superintendent goes out to begin construction, he has in his possession several guidelines by which he determines all operations: maps, a line list, and contract specifications. His maps show in accurate detail all surface features affecting construction. The line list sets out in sequence the names of all landowners, the number of linear rods of line to be built on each piece of property, and any written limitations or restrictions. Included are copies of permits for crossing roads, rivers, and other public property. Specifications in the contract cover every planned operation and as many eventualities as possible.

The spread—that is, the equipment and crew needed to build a pipeline—must be assembled by the pipeline contractor. A spread may be composed of 250 to 300 men in an average operation and up to 500 in a very large operation. The amount of construction equipment depends on the size of the pipeline to be built and the difficulty of the terrain. Stream crossings, marshes, bogs, heavily

Assembling the spread

timbered forests, steep slopes, or rocky ground can require different pieces of machinery. The pipeline contractor might rent rather than buy the necessary machines since outfitting a big-inch (large-diameter) pipeline spread can be a multimillion-dollar operation. A spread can move at a rate of 3 miles a day with a distance of sometimes 10 or 15 miles separating the beginning of the work crew from the end.

Clearing right-of-way

A right-of-way is the legal document granting a right of passage over another person's land; by common usage it has come to mean a cleared strip of land from 50 to 75 feet wide, depending on the size of the pipe and the type of terrain. The clearing crews open fences and build gates, cattle guards, and bridges as the first segment of the spread moves up on the job. Bulldozers are standard equipment for clearing timbered areas (fig. 6.50). Salable timber cut by clearing crews is stacked; the rest is cut and burned. Grading and completion of a roadway capable of supporting all vehicles follow. The road must be large enough for the largest sideboom tractor and other necessary equipment. In rocky terrain, a machine equipped with a ripper that extends several feet into the ground is often used to loosen rocks for removal before the ditching operation begins.

Fig. 6.50. A bulldozer is used to clear a pipeline right-of-way.

Fig. 6.51. In this above-ground section of the Trans-Alaska Pipeline, 15 miles south of Livengood, Alaska, the roadbed along the right-of-way supports working vehicles as well as the pipeline itself. (*Courtesy of Alyeska Pipeline Service Co.*)

Construction of pipelines in arctic climates involves routes that cross permafrost (permanently frozen ground), forests, snow-covered mountains, tundra, rivers, and streams. During the construction of the Trans-Alaska Pipeline, for example, many miles of diverse terrain were traversed by the pipeline. On the North Slope, the route required a special gravel and insulation workpad to prevent degradation or melting of the subsurface permafrost. The gravel road base acted as an insulator, protecting the permafrost as vehicles, equipment, and crew members of the spread built the above-ground portion of the pipeline (fig. 6.51). Parts of the line were laid in a zig-zag pattern to allow for expansion and contraction of the pipe during temperature changes.

Fig. 6.52. A wheel ditcher leaves a neat spoil bank (left) that aids in rapid backfilling.

Ditching   Digging a ditch in which to bury the pipe requires various kinds of equipment. For loose dirt or stable soil, the wheel ditcher is most commonly used (fig. 6.52). The excavated dirt, or *spoil*, is picked up by rotating toothed buckets and piled to the side of the ditch for later use as backfill. Rocky terrain requires the use of pneumatically drilled holes and dynamite, a rock ditcher, or a bulldozer with ripper attachment. Sometimes backhoes are used to dig through rock and to clear blasted rock out of the trench.

The ditch must be at least 12 inches wider than the pipe diameter and deep enough to ensure that it will not interfere with plowing and other normal land use. The Department of Transportation requires a minimum of 30 inches of cover across farm land in normal soil and 36 inches in built-up areas.

In arctic areas where permafrost conditions exist, pipeline contractors may elect not to use a ditch for the pipeline. The pipeline, or sections of it, may be built above the ground on specially designed horizontal crossbeams mounted between pairs of vertical support members (VSMs) (fig. 6.53A). Approximately 400 miles of the Trans-Alaska Pipeline is constructed in this manner. When the above-ground technique is used, the pipeline is surrounded by

Fig. 6.53. In above-ground construction, VSMs hold insulated pipe at various heights (A), while special finned radiators and heat pipes draw heat away from the ground (B). (*Courtesy of Alyeska Pipeline Service Co.*)

insulation panels of steel-jacketed fiberglass with weatherproof expansion joints. Finned radiators atop the VSMs improve heat transfer between the atmosphere and the 2-inch-diameter heat pipes to which they are attached (fig. 6.53B). The heat pipes maintain soil stability in permafrost areas by drawing heat from the ground. In situations where ditching is done in unstable permafrost soil, the buried section of pipe may be refrigerated to maintain the integrity of the subsurface soils in the area. About 7 miles of the Trans-Alaska Pipeline is constructed in this special burial mode.

Fig. 6.54. Pipe joints are strung along the right-of-way so that they are accessible but do not obstruct spread equipment and personnel.

Pipe is purchased, transported, and strung along the right-of-way by the pipeline contractor or its subcontractor (fig. 6.54). On some jobs the pipe is cleaned, primed, and coated before it is delivered to the right-of-way. This pipe requires special handling to prevent damage to the coating as well as the pipe ends. Special curved aluminum plates at the end of wire rope slings are placed inside the pipe ends so that contact with the coating and the beveled ends of the pipe is kept to a minimum (fig. 6.55). As pipe is lifted off the trailer, the weight of the load must be evenly distributed to prevent buckling.

On jobs where the terrain is relatively flat and the road system permits it, the pipe may be double jointed prior to being strung along the right-of-way. *Double jointing* is the welding of two sections of pipe to form one piece approximately 80 feet long. A spreader bar is generally used between the two lifting lines to prevent buckling of these long joints.

Fig. 6.55. A stringing side boom unloads coated pipe and places it along the right-of-way. (*Courtesy of Sheehan Pipe Line Construction Co.*)

Bending pipe

When the ditching and stringing have been completed, engineers measure the contours of the ditch to determine how many degrees of bend must be put in the pipe so that it will conform to the bottom of the ditch. Often as many as seven out of every ten pieces of pipe must be bent to fit the ditch. Special equipment allows the pipe to be bent on location (fig. 6.56). A bending shoe—an attachment that works off the side-boom tractor and winch—can be used for small-diameter pipe. With large-diameter pipe a bending mandrel is used, which keeps the pipe from buckling or wrinkling when it is bent.

Cleaning, aligning, and welding pipe

Before the pipe is welded in place, the ends must be absolutely clean of all dirt, scale, and coating. Line-up clamps are used to align the ends of the pipe prior to welding. They ensure that the two ends match around the entire circumference and the proper spacing is maintained between the ends. Clamps are used outside on small-diameter pipe and inside on large-diameter pipe. Side-boom tractors and winches lower the pipe into place and hold it while skids are put under it and the initial welds are made (fig. 6.57). The welds are then inspected visually and by X ray before the pipe is coated and wrapped.

Coating and wrapping pipe

Three types of coatings used on pipe are fusion-bonded epoxy, enamel, and tape. Fusion-bonded epoxy coatings are usually applied at the mill because they require pipe heated to around 400° F or higher for application. Epoxy coatings do not require an outer wrap, offer strong resistance to handling damage, and can withstand high operating temperatures such as those found in gas transmission lines.

Enamel and tape coatings are applied to pipe while it is being laid (called *line travel applied coating*). When enamels are used, a self-propelled cleaning and priming machine moves along the top of the pipe, followed by a coating and wrapping machine. Coal tar enamels are particularly effective for line travel applied coating. The heated liquid enamel is applied by the coating machine, and then the coated pipe is wrapped with fiberglass, felt, and kraft paper to protect it while it hardens and while the pipe is lowered in. Enamel coatings are most effective when the pipeline's operating temperature range is between 30° F and 180° F.

When line travel tape is used, cleaning, priming, coating, and wrapping processes are performed with one machine (fig. 6.58). Rolls of tape (polyethylene, polyvinyl, or other adhesive) mounted on spindles are wrapped in overlapping segments around the pipe by the taping machine as it moves along the line.

Fig. 6.56. A pipe-bending machine shapes pipe joints to conform to the contours of the terrain.

Fig. 6.57. Welders work on opposite sides of the pipe as they make the initial weld on two joints held together by an internal line-up clamp.

Fig. 6.58. When line travel tape is used, the cleaning and priming and the coating and wrapping processes are performed with one machine.

310

Fig. 6.59. Side-boom tractors with correctly spaced slings lower pipe into the ditch.

Lowering in and backfilling

If the pipe is lowered into the ditch in conjunction with the coating and wrapping operation, two steps are eliminated: lowering of the pipe to skids and picking it up later for final placement (fig. 6.59). When the special burial mode is used in arctic environments, the pipe is lowered into the ditch enclosed in a sheath of insulation. Refrigerant lines are laid along both sides of the pipe, and special padding is placed in the ditch to minimize melting in the perma-frost layer. Side booms with nonmetallic slings, correctly spaced, are used to place the pipe safely into the ditch.

After pipe has been lowered into the ditch, adequate fill material is provided underneath the pipe as well as above it. An auger fitted to the front of a bulldozer is often used to distribute the backfill evenly in the ditch. Backfilling requires breaking up rocky or frozen soil before returning it to the ditch to prevent damage to the pipe coating. Uneven distribution of frozen soil may leave sections of the pipe unsupported after the fill has thawed, causing cracks and leaks.

Specialty and tie-in crews

As the spread moves along, portions of the cross-country line are bypassed. These unfinished sections of the line are completed by specialty crews and tied in to the main pipeline by the tie-in crew—a miniature spread that joins the segments into one con-tinuous line.

Highway, railroad, and inland waterway crossings require special construction techniques and equipment, as do swamps and marshes. For example, a boring machine may be used to go under a highway, railroad, or preexisting pipeline. In some areas, the use

Fig. 6.60. A boring machine pushes casing under a road-bed as its cutting head rotates just inside the front of the casing. Pipe will be installed inside the casing.

of casing to protect the pipe is mandatory (fig. 6.60). Some operations require six or more side-boom tractors to maneuver a crossing. Creek crossing and steep terrain also require the skillful maneuvering of side-boom operators as they move long sections of pipe into position.

One way to lay pipe across a river is to bury it under the riverbed. A directional drilling rig is used to drill a controlled horizontal pilot hole in an inverted arc under the riverbed. Guided by computers, a pilot string of pipe is pulled through first, followed by a work string. Finally the actual hole is bored and the pipeline pulled through (fig. 6.61). Water crossings may also be made by

Fig. 6.61. A pullback assembly, composed of fly cutter, barrel reamer, and swivel, is used in controlled directional drilling.

312

Fig. 6.62. Flotation devices support sections of pipe during a river crossing.

attaching flotation devices to the pipe, then walking or pulling the pipe across (fig. 6.62). Pipe is then lowered into an underwater ditch, which has been dredged out with a dredging unit.

Special sections are tied in to the adjoining cross-country pipeline by the tie-in crew. Tie-in involves welding pipe ends together, testing, and fabrication when necessary. Fabricated sections are special valve connections and fittings made up ahead of time and joined to the main pipe where needed.

As the pipeline is completed, special care is taken to restore the right-of-way as nearly as possible to its original condition (fig. 6.63). Of course, the right-of-way must be kept clear of trees and brush, allowing visual inspection for leaks.

Fig. 6.63. After construction is completed, a neatly restored right-of-way allows this land to produce crops once again.

OFFSHORE PIPELINE CONSTRUCTION

Pipelining offshore in shallower depths has been well established for quite some time, and modern technology has made it possible to lay pipe in deepsea areas. In all offshore pipeline construction, various types of barges play a key role.

The pipeline industry has also researched and developed techniques for building offshore pipelines in ice-laden arctic waters. Equipment that can detect the presence of ice features such as ice islands and pressure ridges that cause scouring is being used to reduce potential pipeline damage from contact with ice. To protect the pipeline against ice scouring, it must be buried so that the depth of the trench on the seafloor to the top of the pipe exceeds the deepest scour marks in the area.

Thawing may cause excess stress on pipelines constructed in permafrost near the shore. To minimize thawing in the permafrost layer and to keep the stress on the pipeline at an acceptable level, several methods have been proposed for use: (1) placing pipelines on special beds of granular material to keep an insulating layer between the permafrost and the pipeline, (2) artificially reducing the temperature of the oil at the processing facility, (3) raising the pipeline above the surface of the water on piles or artificial berms, and (4) circulating refrigerant through cooling tubes surrounding pipeline enclosed in a jacket.

Offshore lines are laid by pipe-laying barges, or simply *lay barges* (fig. 6.64). A lay barge is a complete seagoing plant that allows the pipeline to be assembled and laid continuously along the selected

Conventional lay barge

Fig. 6.64. The main deck of a lay barge, as seen on this center-slot vessel, may be thought of as an offshore pipeline spread. (*Courtesy of McDermott, Inc.*)

Fig. 6.65. Welding, coating, and inspection are handled at stations on the lay barge.

route either on top of the ocean floor or in dredged trenches on the seafloor. When used in conjunction with supporting tugs and an anchoring system, the lay barge can be self-sufficient for months at a time.

An eight-point mooring system of wire rope and anchors holds the lay barge on a precise heading to prevent buckling of the pipe as it is laid. The system also propels the barge as anchor lines are reeled in and out. As the barge progresses to the end of the lines, the mooring system is moved ahead by anchor-handling tugs.

The line pipe is brought by pipe barge from yards on shore and stored aboard the lay barge for addition to the lengthening pipeline. Welding, coating, and inspection are handled at stations, or work areas, along the length of the lay barge deck (fig. 6.65). The completed pipeline is lowered into the water by way of an inclined ramp and a *stinger* attached to the ramp to guide the pipeline to the seafloor at the proper angle (fig. 6.66). The pipeline is coated with a high-density cement to overcome buoyancy so that it can be sunk into place. To prevent cracking of this concrete coating or buckling of the pipe when lowered, especially in deep water, an attachment called a *pontoon* is sometimes added to the stinger. The pontoon is control-flooded to lower the line toward the seafloor at an angle that will not overstress the pipeline. Also, a tensioner is sometimes used in deepwater lay operations. The tensioner maintains an upward force in the pipe so that the pipe does not buckle under its own weight.

The method used to keep the pipeline on the bottom and out of danger varies with the topography of the seafloor. On rocky bottoms, the line is fastened to the seafloor with pipeline anchors.

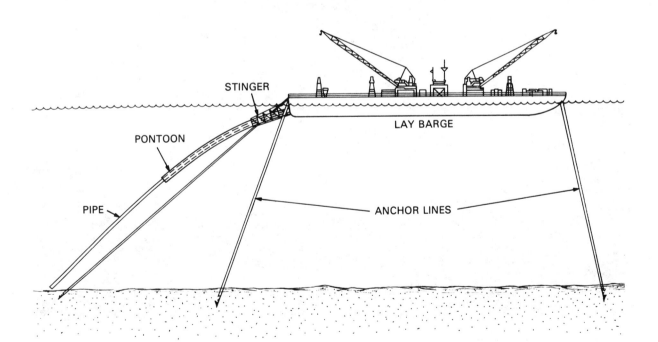

Fig. 6.66. The stinger guides the pipeline safely to the seafloor while anchor lines move the vessel along.

Bury barge   Subsea pipelines that might be harmed by fishing or other marine operations are laid in a seafloor trench. On softer bottoms, a bury barge, or pipe-trenching barge, with a sled is used for this operation (fig. 6.67). The sled, which is attached to the submerged line, moves along the line on the seafloor and uses a high-pressure jetting action to form a trench in which the pipeline will rest. Usually, the pipe is not covered at the time it is laid but is eventually covered by migrating fill material.

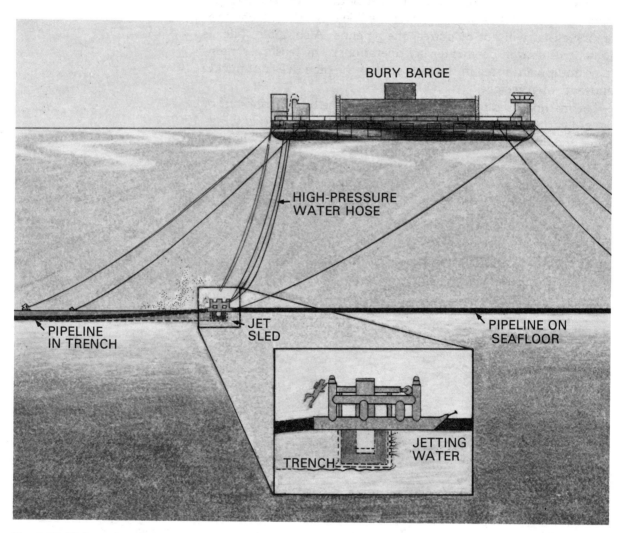

Fig. 6.67. High-pressure jets of water from a jet sled dig a trench for a pipeline on the seafloor. The sled is operated from a bury barge as it moves forward by means of anchors.

Designed for use in extremely rough and unusually deep water is a huge lay barge that might be called a superbarge (fig. 6.68). It is 650 feet long, 140 feet wide, and 50 feet deep. The barge's center-line, elevated pipe ramp, and three pipe tensioners permit the laying of 80-foot length, double-jointed pipe in deep water, without the use of a pipe-laying pontoon.

The superbarge houses 350 people and can store as much as 20,000 tons of pipe. It has seven work stations for double-jointed pipe or nine work stations for 40-foot automatic or manual welding joints. It also includes a pipe-beveling station, two 125-ton gantry cranes, twelve 60,000-pound anchors, and a heliport.

Superbarge

Fig. 6.68. Designed for use in rough, deep water, this super-barge houses 350 people and can store as much as 20,000 tons of pipe. (*Courtesy of Brown & Root, Inc.*)

318

Semisubmersible barge

Coming to the forefront in offshore pipelining is the use of a semisubmersible lay barge, which operates in a manner similar to the semisubmersible drilling rig. The semisubmersible lay barge is designed to minimize the effects of wave and wind actions when operating in rough waters. For example, a semisubmersible barge crew can lay pipe in the North Sea during 35-foot waves and 70-mile-per-hour winds—an impossible task with a conventional lay barge. The stability comes from the submerged portion of the vessel, which lowers the center of gravity and steadies the vessel against high winds and waves.

Reel vessel

During the past several years, increasing use has been made of reel pipe-laying barges and ships (fig. 6.69). In the early development of the reel vessel, pipe was limited to small gathering line of 1½ to 2 inches in diameter. Currently, lines of 8-inch diameter are being laid from the reels.

On reel vessels the pipe is welded and spooled onto the giant reels at onshore facilities. This pipe must be of sufficient flexibility to withstand reeling and unreeling without buckling or cracking.

The spooled line, ranging from several hundred feet to a couple of miles in length (depending on the pipe diameter), is unwound by moving the barge forward. The line is unspooled cleanly and in a straight path. Giant unspooling devices located at the rear of the barge, where the line extends to the water, keep the line straightened and unwrinkled. When one spool of line ends, the end is held

Fig. 6.69. A giant spool unreels continuous lengths of line pipe from a reel lay barge.

above the water surface while a new spool of line is joined by welding. Then the submerged line is dropped, and unspooling activity continues with the new line.

Economics and safety generally determine which method of transportation is used for shipping crude oil, petroleum products, and natural gas. Of the four methods used by the industry—pipelines, water carriers, motor carriers, and railroads—pipelines have been found to be the safest by the U. S. National Transportation Safety Board. Pipelines are also the most economical, provided that the line is available and can operate at full capacity. However, pipelines require a great deal of capital to build and operate, and once laid, they are immobile.

Since 1980, water carriers have had a slight edge over pipelines in *crude oil* transportation (table 6.2). This jump in water-carrier use is partially due to increased shipments of crude by tanker from

TABLE 6.2
DOMESTIC TRANSPORTATION OF CRUDE PETROLEUM AND PETROLEUM PRODUCTS, 1973–1983
(In billions of ton-miles[a])

MODE OF TRANSPORTATION	1973	1974	1975	1976	1977	1978	1979	1980	1981	1982	1983
Pipelines											
Crude	302.0	303.0	288.0	303.0	326.6	359.5	372.2	362.6	333.1	335.1	332.4[b]
Products	205.0	203.0	219.0	212.0	219.4	226.3	236.1	225.6	230.6	230.6	223.7[b]
Total	507.0	506.0	507.0	515.0	546.0	585.8	608.0	588.2	563.7	565.7	556.1[b]
Water Carriers											
Crude	58.8	53.0	40.6	37.8	63.1	261.3	265.5	387.4	404.9	432.7	471.2
Products	238.0	244.0	257.4	269.1	270.2	269.3	257.4	230.4	212.3	184.2	159.3
Total	296.8	297.0	298.0	306.9	333.3	530.6	522.9	617.8	617.2	616.9	630.5
Motor Carriers[c]											
Crude	1.3	1.3	1.4	2.1	2.0	2.0	2.3	2.5	2.2	2.0	2.0
Products	23.7	27.7	26.2	30.4	27.6	28.6	27.8	24.3	22.7	20.7	22.2
Total	25.0	29.0	27.6	32.5	29.6	30.6	30.1	26.8	24.9	22.7	24.2
Railroads											
Crude	1.1	1.6	1.5	0.9	0.8	0.7	0.6	0.5	0.5	0.4	0.5
Products	13.7	14.1	12.6	12.4	13.7	12.5	12.9	12.0	12.1	12.5	11.3
Total	14.8	15.7	14.1	13.3	14.5	13.2	13.5	12.5	12.6	12.9	11.8
All Modes											
Crude	363.2	358.9	331.5	343.8	392.5	623.5	640.6	753.0	740.7	770.2	806.1
Products	480.4	488.8	515.2	523.9	530.9	536.7	534.2	492.3	477.7	448.0	416.5
Total	843.6	847.7	846.7	867.7	923.4	1,160.2	1,174.5	1,245.3	1,218.4	1,218.2	1,222.6

[a]A ton-mile is a standard measure for moving one ton of product one mile.
[b]Preliminary figures.
[c]Amounts carried by motor carriers are estimates.

SOURCE: Association of Oil Pipe Lines, Washington, D. C.

320

Fig. 6.70. Transportation of all kinds is evident at this refinery in Louisiana, located near a river and ship channel. A portion of the facility's crude oil feedstock is delivered by barge; products move out by railcar and tank trucks; and pipelines fill and empty the many storage tanks with feedstocks coming in and products going out. (*Courtesy of Cities Service Co.*)

Valdez, Alaska, to U. S. ports on the Gulf Coast, East Coast, and Caribbean. Most *petroleum products*, however, are shipped by pipeline, with water carriers running a close second. The decrease in domestic petroleum products carried by water since 1981 is partly the result of an increase in imports of products from foreign refineries. *Natural gas* is almost totally shipped by pipeline, although LNG tankers and pressure transport trucks are used in special cases.

Regardless of the volume carried, each transportation method plays a key role in the functioning of a refining or processing plant—bringing feedstocks in and taking refined products out (fig. 6.70). Transportation is also vital to the marketing and distribution of consumer products made from petroleum.

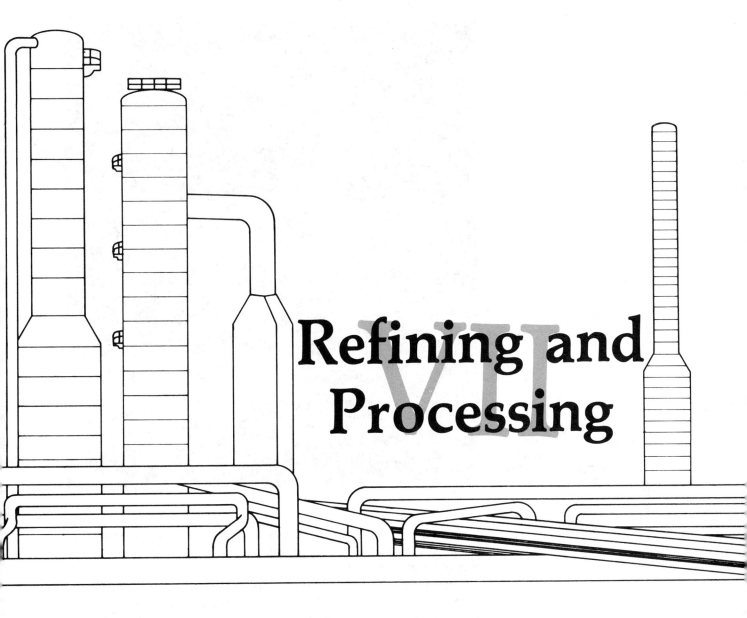

Refining and Processing VII

Collected crude oil and gas are of little use in their raw state; their value lies in what is created from them: fuels, lubricating oils, waxes, asphalt, and petrochemicals. The refining world is highly technical. To the layman, a refinery appears to be a strange conglomeration of illuminated towers and walls at night (fig. 7.1) and an equally strange maze of pipes and tanks during the day. In reality, a refinery is an organized and coordinated arrangement of manufacturing processes designed to produce physical and chemical changes in crude oil and natural gas (fig. 7.2). These changes result in salable products of the quality and quantity desired by the market. A refinery also includes nonprocessing facilities required to store crude oil and products, maintain equipment, and ensure continuous operation.

Fig. 7.1. A refinery unit lighted against the backdrop of a Louisiana sky at dusk is an impressive sight. (*Courtesy of Cities Service Co.*)

As crude oil comes from the well, it contains hydrocarbon compounds and relatively small quantities of other materials such as oxygen, nitrogen, sulfur, salt, and water—plus trace amounts of certain metals. In the refinery, nonhydrocarbon substances are removed from the crude oil. The oil is broken down into various components. Some of the components are chemically changed to give them more desirable qualities. The resulting substances are blended into useful products.

Likewise, natural gas is a mixture of hydrocarbon gases, along with impurities such as water, nitrogen, and carbon dioxide. Objectionable impurities are removed and desired products are separated by various processing systems in field facilities, refineries, and gas processing plants.

Most refinery products are ready for use by consumers and industry. A portion, however, goes to petrochemical plants where it is changed into chemical feedstocks used to manufacture an almost limitless number of products ranging from fertilizers to plastics.

Fig. 7.2. A refinery is an organized and coordinated arrangement of processes linked together with miles of pipe carrying feedstocks in and products out. (*Courtesy of Coastal States Gas Corp.*)

324

OIL AND GAS HYDROCARBON STRUCTURES

Crude oil and natural gas are often collectively called petroleum; however, *petroleum* refers to crude oil, the heavier constituents that occur naturally in liquid form in the reservoir, while *gas* refers to the lighter constituents that occur naturally in gaseous form. A third, less common constituent is referred to as condensate. Condensates are in a gaseous state in the reservoir but emerge from the oilwell as liquid petroleum.

The color, viscosity, density, and other qualities of oil and gas depend on the chemical structures of the substances they contain. They contain many different hydrocarbons, each with its unique chemical structure and characteristics. The refinery or processing plant does not deal with every one of these compounds separately, however; it refines or processes groups of hydrocarbons with similar chemical structures and physical traits.

To understand how unrefined oil and unprocessed gas are converted into useful products, it is necessary to be familiar with four types of hydrocarbons: paraffins, olefins, naphthenes, and aromatics. All of these compounds are arrangements of carbon and hydrogen atoms, and sometimes they include other elements such as oxygen, sulfur, and trace metals.

Paraffins

Natural gas and a substantial portion of crude oil are composed of paraffin hydrocarbons (fig. 7.3A). The simplest is *methane*, the main ingredient of natural gas. It has one carbon atom attached to four hydrogen atoms. The other hydrocarbons in this group are chains of carbon atoms with attached hydrogen atoms. Examples are *ethane* and *propane*, also components of natural gas. Each carbon atom can bond or attach to four other atoms, and each hydrogen atom can bond with one other atom.

Paraffins are given names ending in *ane*. Chemically, they are very stable compounds. The paraffin group is large, varying in number of carbon atoms from one to ninety or more. The number of carbon and hydrogen atoms in each hydrocarbon can be expressed by a formula; the formula for ethane is C_2H_6, indicating two carbon atoms and six hydrogen atoms.

Hydrocarbons with very few carbon atoms (C_1 to C_4) are light in weight and are gases under normal atmospheric pressure. Hydrocarbons with more carbon atoms are heavier and are either liquid or solid under atmospheric pressure. As the number of carbon atoms increases, the boiling point also increases—more heat must be applied to change the hydrocarbon from a liquid or solid to a gaseous state.

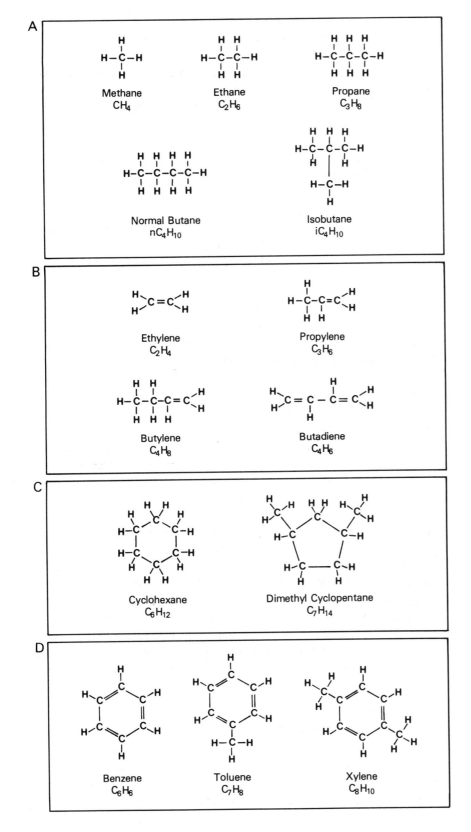

Fig. 7.3. Four types of hydrocarbon structures are paraffins (A), olefins (B), naphthenes (C), and aromatics (D).

Paraffins with four or more carbon atoms can have more than one arrangement of atoms. All the carbon atoms can form a single straight chain, or some of the carbon atoms can form branches attached to the main chain. *Butane*, for example, has two forms: normal, or n-butane (straight chain), and iso, or i-butane (branched chain). Isobutane is known as an *isomer* of normal butane because i-butane has exactly the same formula but a different arrangement of atoms. This difference is important because isobutane boils at a different temperature from normal butane and causes different chemical reactions in the refinery. The larger the paraffin molecule, the greater is the possible number of isomers.

Other paraffins include *pentane* (C_5H_{12}), *hexane* (C_6H_{14}), *heptane* (C_7H_{16}), *octane* (C_8H_{18}), *nonane* (C_9H_{20}), and *decane* ($C_{10}H_{22}$). Because paraffins have the greatest possible number of hydrogen atoms, they are called *saturated* hydrocarbons.

Olefins

Like paraffins, olefins are chains of carbon atoms with attached hydrogen atoms (Fig. 7.3B). However, olefin chains do not have the greatest possible number of hydrogen atoms. If two hydrogen atoms are missing, two carbon atoms in the chain will form a double bond to make up for the deficiency. An example is *ethylene*, an important petrochemical. Its carbon atoms are capable of bonding to six hydrogen atoms, but there are only four hydrogen atoms, so the two carbon atoms form a double bond. If four hydrogen atoms are missing, the carbon chain will have two double bonds; an example is *butadiene*, another petrochemical. Olefins with one double bond have names ending in *ylene* or *ene*, and olefins with two double bonds have names ending in *adiene*.

Other common olefins are *propylene* and *butylene*. Like paraffins, olefins can have isomers with branched chains or with the double bonds at different places in the chain.

Olefins are known as *unsaturated* hydrocarbons because of their hydrogen deficiency. Because the carbon atoms of olefins are always seeking to attach their "empty" bonds to other atoms, olefins are unstable and are easily used to make new chemical compounds. They do not occur naturally in crude oil but are formed in the refinery by the breakdown of larger hydrocarbon molecules. Olefins are very useful in creating certain refinery products and petrochemicals.

Naphthenes

The carbon atoms of naphthenes form rings rather than chains, so naphthenes are called ring compounds or cycloparaffins (fig. 7.3C). Hydrocarbons in this group have names that begin with the prefix *cyclo* to indicate the ring structure. An example is *cyclohexane*, a hydrocarbon often occurring in natural gasoline.

The carbon rings of naphthenes are saturated with hydrogen atoms. Naphthenes are chemically stable. They occur naturally in crude oil and have properties similar to the paraffins.

Aromatic hydrocarbons are compounds that contain a ring of six carbon atoms with alternating double and single bonds and six attached hydrogen atoms (fig. 7.3D). This type of structure is known as a benzene ring. The most important aromatics in refinery production are the BTXs—*benzene, toluene,* and *xylene.* Aromatics occur naturally in crude oil. They are also created by refinery and petrochemical plant processes.

Aromatics

Anywhere from 2 to 50 percent of a crude oil may be composed of compounds containing oxygen, nitrogen, sulfur, and metals. Oxygen content can be as high as 2 percent, and nitrogen content as high as 0.8 percent. Sulfur content ranges from traces to more than 5 percent. If a crude contains appreciable quantities of sulfur or sulfur compounds, it is called a *sour* crude; if it contains little or no sulfur, it is called a *sweet* crude. Trace metals contained in crude oil include sodium, magnesium, calcium, strontium, copper, silver, gold, aluminum, tin, lead, vanadium, chromium, manganese, iron, cobalt, nickel, platinum, uranium, boron, silicon, and phosphorus. Of these metals, nickel and vanadium occur in the greatest quantities.

Other elements

CRUDE OIL REFINING

The processes used by a refinery depend partly on the content and quality of the crude oil it receives, partly on consumer demand, and partly on existing plant facilities and the economics involved in changing them.

Today, many processes are used to refine crude oil. Some are quite complicated, and changes are made as researchers develop new methods of arranging the molecules of crude oil to produce more useful and profitable products. Processes are chosen, arranged, and interrelated according to the market the refinery serves and the products it manufactures.

Every refinery process begins with the separation of crude oil into different components or fractions by distillation (fig. 7.4). The fractions are further treated to convert them into salable products. Various cracking methods may be used to break down large molecules. Molecules may be rearranged by reforming, alkylation, polymerization, and isomerization. The mixtures of new compounds created by these chemical conversion processes are then

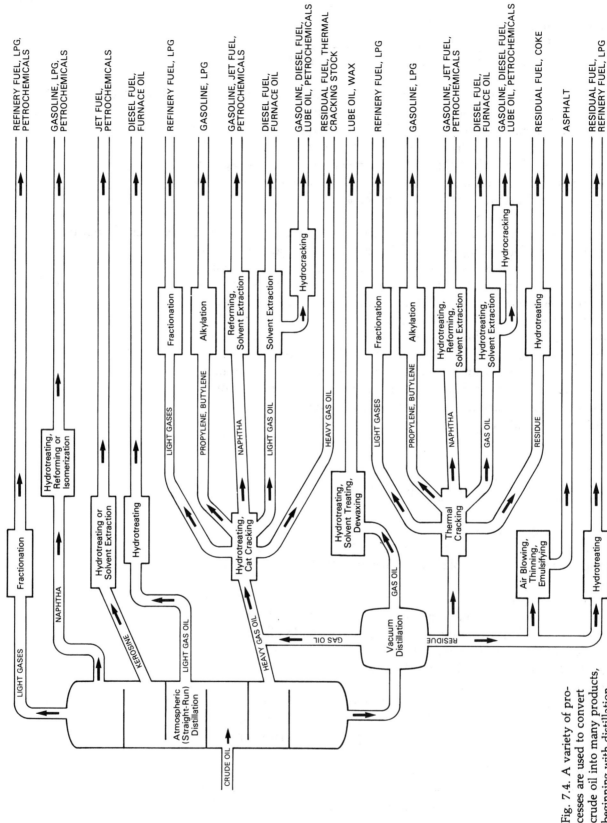

Fig. 7.4. A variety of processes are used to convert crude oil into many products, beginning with distillation.

separated into individual products using methods such as distillation and solvent extraction. Impurities are removed by various methods of dehydration, desalting, sulfur removal, and hydrotreating. Compounds are blended into products that have a certain combination of desirable qualities. By combining these processes, refineries are able to convert virtually all of the crude oil barrel into salable products in relative amounts that meet market demand.

Crude oil is a mixture of many hydrocarbons ranging from light gases to heavy pitch. Crude oil varies in the type and amount of components in its hydrocarbon mixture as well as in its nonhydrocarbon content. These differences are reflected in its qualities. Some crude oils are light, flow easily, and make good gasoline. Others are heavy and thick and are excellent for producing asphalt. Some contain large amounts of contaminants such as sulfur; others do not.

To evaluate the content and properties of crude oil, refining companies conduct assays (fig. 7.5). These assays, or evaluations, serve several purposes:

1. to detect content changes over time in an oil whose properties are known;

Evaluating crude oils

Fig. 7.5. A lab technician monitors a crude assay—in this case, a simulation of vacuum distillation conducted in a refinery laboratory. (*Courtesy of Valero Refining Co.*)

2. to decide whether a new crude oil will yield desirable amounts and qualities of products;
3. to plan the processing of a crude oil in existing or new plants; and
4. to compare the crude oil with oils of competitors and assess its market value.

The completeness of the crude oil assay depends on the purpose. A few routine tests may be enough to monitor changes in well-known crude oils, but planning a refinery operation to process a large quantity of a new crude oil requires a full-scale evaluation.

A complete assay is a detailed description of the chemical and physical properties of the whole crude oil as well as the amount and quality of certain portions of the crude that will be separately processed in the refinery. The whole crude may be evaluated for properties such as API gravity (density), viscosity (resistance to flow), and sulfur content. Specific *fractions* or *cuts* — groups of hydrocarbons with the same boiling-point range and similar properties — are evaluated for qualities important to the products they will yield.

Some refineries purchase feedstock instead of whole crude oil; that is, initial distillation is done by one plant, and selected fractions are purchased by the second plant for further refining. Regardless of their feedstock, most refineries conduct complete crude oil assays.

Based on the composition of its fractions, a crude oil can be classified in one of three groups: paraffin-base, asphalt-base, and mixed-base. *Paraffin-base* crude oil contains mostly paraffin hydrocarbons in its heavier fractions. It is a good source of paraffin wax, quality motor lube oils, and high-grade kerosine. *Asphalt-base* crude oil contains mostly asphaltic hydrocarbons in its heavier fractions. Asphalt is a dark solid or semisolid containing carbon, hydrogen, oxygen, sulfur, and sometimes nitrogen. This type of crude is particularly suitable for making high-quality gasoline and roofing and paving materials. The heavier fractions of some crude oils contain considerable amounts of both paraffins and asphalt. They are called *mixed-base* crude oils. Virtually all products can be obtained from them, although at lower yields than from the other two classes.

Responding to market demands

Refinery processes have developed also in response to changing market demands for certain products. Before the automobile, the prime product of refineries was kerosine for lamps and home heating. With the advent of the internal-combustion engine, the main job of refineries became the production of gasoline. Crude oil was separated into natural gasoline and other components by

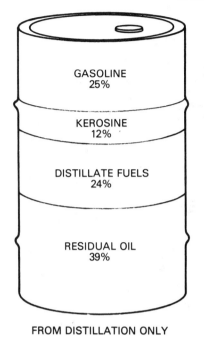

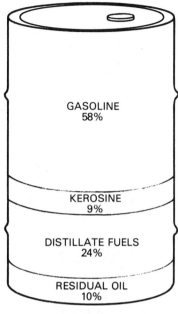

FROM DISTILLATION ONLY   FROM 1983 REFINING PROCESSES

Fig. 7.6. An important job of refineries today is the conversion of oil at the "bottom of the barrel" into gasoline and other more marketable products.

distillation, a physical separation method using the same boiling and condensing processes as a whiskey still. However, the quantity of natural gasoline available from distillation was soon insufficient to satisfy consumer demands (fig. 7.6).

At the same time, refineries had more heavy oil components than they needed to make lubricating oils and other types of fuel. So, refineries began to look for ways of chemically converting a part of the heavier oils into synthetic gasoline. Two types of processes were developed to produce more and better quality gasoline: (1) breaking down large, heavy hydrocarbon molecules and (2) reshaping or rebuilding hydrocarbon molecules.

Separating crude oil

Because crude oil is a mixture of hydrocarbons with different boiling temperatures, it can be separated by distillation into groups of hydrocarbons that boil between two specified boiling points. This process is known as fractionation—separation into fractions. Two types of distillation are performed: atmospheric and vacuum.

332

Fig. 7.7. As heated crude oil is piped into this distillation tower, heavy liquids fall to the bottom, while lighter hydrocarbons are drawn off the center and top. Note lines coming from the various levels of the crude oil tower. (*Courtesy of Diamond Shamrock Refining and Marketing Co.*)

Atmospheric distillation. The initial distillation of crude oil is conducted in a distilling column at or near atmospheric pressure (fig. 7.7). Distillation is quite simple in principle and may be compared to turning water into steam in a teakettle. When water is heated, it vaporizes into steam and escapes from the kettle. The rising steam cools and condenses into water again. In the refinery, crude oil is heated until it partly vaporizes, and the resulting mixture of vapor and liquid is piped into the distilling column. The liquid falls to the bottom of the column and is removed. The vapor

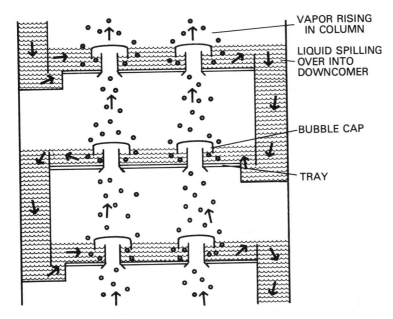

VAPOR RISING IN COLUMN

LIQUID SPILLING OVER INTO DOWNCOMER

BUBBLE CAP

TRAY

Fig. 7.8. Vapor passes through bubble caps and settles on trays in the distilling column.

rises, passing through perforated bubble caps on a series of trays (fig. 7.8). The cooling vapor turns to liquid that settles on the trays. Heavier hydrocarbons condense more quickly and settle on lower trays, and lighter hydrocarbons remain in the vapor state longer and condense on higher trays. Liquid fractions are drawn from the trays and removed from the distilling column. The lightest hydrocarbons remain gaseous and pass out through the top of the column.

The fractions or distillates produced by atmospheric distillation can be used in their new state, blended with other substances, or further processed to make useful products (table 7.1). Gases are

TABLE 7.1
SOME FRACTIONS OBTAINED FROM ATMOSPHERIC DISTILLATION

Fraction	Carbon Atoms	Boiling Range, °F	State under Atmospheric Pressure
Gas	1–4	−258–31	Gas
Naphtha	4–12	31–400	Liquid
Kerosine	10–14	356–500	Liquid
Light gas oil	15–22	500–700	Liquid, solid
Heavy gas oil	19–35	640–875	Solid
Residuum	36–90	875+	Solid

DATA SOURCES: "Petroleum Processing," *McGraw-Hill Encyclopedia of Science and Technology*, and Table 2.1, Cram and Hammond, *Organic Chemistry*.

sent to a gas plant to be fractionated into individual products. Light distillates include naphtha and kerosine. Naphtha is used to make gasoline and petrochemicals. It is called straight-run or virgin naphtha because it is a product of the original distillation of crude oil rather than of chemical conversion processes such as cracking. (This use of the term *naphtha* should not be confused with certain commercial solvents popularly known as naphthas.) Kerosine is used to make jet fuel. Middle distillates include light and heavy gas oils. Light gas oils are made into diesel fuels and furnace oils. Jet, diesel, and furnace fuels are sometimes called distillate fuels. Heavy gas oils are often cracked to produce naphtha and other products. Atmospheric residue is the unvaporized portion of crude oil and is usually sent to a vacuum distillation column for further fractionation.

Fig. 7.9. Heated residual oil is piped to a vacuum distillation tower such as this one. (*Courtesy of Diamond Shamrock Refining and Marketing Co.*)

Vacuum distillation. Heavy hydrocarbons with boiling points of 900°F and higher cannot be separated into fractions by distillation at atmospheric pressure. The high temperatures required to vaporize them would instead cause the molecules to crack into smaller molecules. To prevent cracking, the heated residual oil from the atmospheric distillation column is piped to a vacuum distillation vessel (fig. 7.9). Steam keeps the oil hot, and low pressure allows the hydrocarbons to vaporize at temperatures below their cracking points and to be separated into fractions. Light and heavy vacuum gas oil are separated from the *bottoms*, the heaviest residue.

Gas oil will undergo cracking or solvent extraction to produce products ranging from gasoline and distillate fuels to lubricating oils. Depending on its content, the heavy residue will be cracked or otherwise processed to make fuels, asphalt, lubricating oils, wax, coke, or residual oils.

Asphalt, a material for road pavings and roof coatings, is commonly produced by vacuum distillation (fig. 7.10). The stability and binding qualities of asphalt come from its large molecules composed of interlocking aromatic rings. To perform well, asphalt must be soft enough to pour and mix with sand and gravel during construction and hard enough when dry to resist penetration.

Fig. 7.10. Residuals are made into liquid asphalt and other products in a vacuum distillation unit. The heater preheats the charge before it goes to the vacuum distillation column. (*Courtesy of Valero Refining Co.*)

Hard and soft grades of asphalt can be made by raising or lowering the temperature at which the vacuum distillation tower is operated. The two grades can then be blended to produce asphalts with varying hardness to meet the specifications for different uses. A second way to modify hardness is to blow hot air into asphalt in the presence of a catalyst to cause oxidation and hardening. Asphalt made in this way is known as blown asphalt. Asphalt is heated to soften it for pouring and mixing. To reduce the amount of heat needed, asphalt can be diluted with a thinner or cutback, or it can be mixed with an emulsifier and water to form a water emulsion. After the asphalt is poured, the thinner or water evaporates, allowing the asphalt to harden.

Breaking down molecules

Cracking processes break down heavier hydrocarbon molecules into lighter products such as gasoline and distillate fuels. They are used to decrease the amount of heavier hydrocarbons from the bottom of the crude oil barrel and to convert them to products in greater market demand. These processes include catalytic cracking, thermal cracking, and hydrocracking.

Catalytic cracking. Catalytic cracking is used to chemically convert gas oils obtained from atmospheric and vacuum distillation to gasoline (fig. 7.11). The gas oil undergoes a chemical breakdown under controlled heat and pressure in the presence of a catalyst — a substance that promotes the reaction without itself being chemically changed. The catalyst may be in the form of small pellets or powder. Today, powder is most commonly used. Because the catalyst behaves like a fluid when mixed with the gas oil, the method is known as *fluid catalytic cracking*. Besides cracked

Fig. 7.11. Three catalytic cracking units perform a vital process in breaking down gas oils. (*Courtesy of Cities Service Co.*)

ELEVATOR

HOC REACTOR

HOC REGENERATOR

CATALYST STORAGE HOPPERS

AIR EXCHANGER COOLERS (FIN FANS)

Fig. 7.12. This heavy oil cracker, designed and engineered by the M. W. Kellogg Company, is used to produce unleaded gasoline and other light product yields from heavy crude oil. (*Courtesy M. W. Kellogg Co. and Valero Refining Co.*)

gasoline, the reaction yields light gases, unsaturated olefin compounds, cracked gas oils, a liquid residue called cycle oil, and a solid carbon residue known as coke. The cycle oil is usually recycled through the reaction chamber to cause further breakdown. The coke coats the catalyst and must be removed by burning so that the catalyst can be used again. The other hydrocarbon products of the cracking reaction are sent to a fractionator to be separated, and most of them undergo further processing elsewhere in the refinery before they are ready for sale.

A heavy oil cracker (HOC) is now being used by a few refineries to upgrade residuum into unleaded gasoline and other high-demand products (fig. 7.12). This large fluid catalytic cracking unit consists of a main fractionator, catalyst storage hoppers, vapor lines, heaters, and a bank of air coolers. The fractionator, which is so large that it requires an elevator for access, has a regenerator at the bottom and a reactor on top.

Thermal cracking. Thermal cracking uses heat to break down the residue, or bottoms, from vacuum distillation and sometimes the heavy gas oils resulting from catalytic cracking. The lighter hydrocarbons produced by thermal cracking can be made into distillate fuels and gasoline. The heavy residue is converted into residual oil or into coke, which is used in the manufacture of electrodes, graphite, and carbides. Two common types of thermal cracking are viscosity breaking, or visbreaking, and coking.

In *visbreaking*, a relatively mild form of cracking, the feed is heated in a furnace for a short period of time. To prevent coke from forming, the reaction is performed under high pressure and the cracked products are immediately quenched or cooled. The pressure is lowered, allowing the lighter hydrocarbons to flash or vaporize. They are sent to a fractionator to be separated. Part of the heavy residue is recycled back to the reaction vessel for further breakdown.

In *coking*, a more severe form of cracking, the feed is heated and sent to insulated vessels called coke drums, where it remains under high heat for an additional period of time. The lighter cracked products vaporize and leave through the top of the vessel, and the heavier hydrocarbons form solid coke. The coke is periodically broken up by high-pressure jetted water and removed from the drum.

The products of thermal cracking include both saturated and unsaturated light gases, naphtha, gas oil, and residual oil or coke. The lighter products are of low quality and are usually upgraded by further processing before being sold. Cracked gases are converted to gasoline by alkylation. Naphtha is upgraded to high-quality gasoline by reforming. Gas oil can be used as distillate fuel or can be converted to gasoline and other products by hydrocracking. Coke is heated to drive off water and is crushed.

Hydrocracking. Whenever gas oils produced by catalytic and thermal cracking are not needed to make distillate fuels, they can be converted to high-grade gasoline by hydrocracking. *Hydrocracking* is catalytic cracking in the presence of hydrogen. The extra hydrogen saturates, or hydrogenates, the chemical bonds of the cracked hydrocarbons and creates isomers with desirable characteristics. Hydrocracking is also a treating process, because the hydrogen combines with contaminants such as sulfur and nitrogen, allowing them to be removed.

Gas-oil feed is mixed with hydrogen, heated, and sent to a reactor vessel with a fixed-bed catalyst, where cracking and hydrogenation take place. The hydrogen is separated from the cracked products and recycled to the reactor. The products are sent to a fractionator to be separated. Residue from the first reaction is mixed with hydrogen, reheated, and sent to a second reactor for further cracking under higher temperatures and pressures. In addition to a large quantity of *hydrocrackate*—cracked naphtha for making gasoline—hydrocracking yields light gases useful for refinery fuel or alkylation and components for high-quality fuel oils, lube oils, and petrochemical feedstocks.

Some of the hydrogen for the hydrocracking process is supplied as a by-product of reforming, but additional hydrogen must usually be made by *steam methane reforming*, a series of chemical reactions that extract hydrogen from steam and methane.

Several chemical processes may be used to rearrange molecules or to build new molecules into high-quality gasoline, jet fuel, and petrochemicals. Commonly used processes are alkylation, isomerization, and reforming.

Building desired molecules

Alkylation. Olefins such as propylene and butylene are produced as by-products of catalytic and thermal cracking. In the refining industry, *alkylation* refers to the chemical combining of these light molecules with isobutane to form larger branched-chain molecules (isoparaffins) that make high-octane gasoline. This process is accomplished with an alkylation unit (fig. 7.13).

Fig. 7.13. This alkylation unit consists of a depropanizer, a main fractionator, and a reboiler heater. Hydrofluoric acid used in the unit is stored in the tank next to the heater. (*Courtesy of Phillips Petroleum Co. and Valero Refining Co.*)

340

Fig. 7.14. Normal butane, isobutane, and propane, often obtained as by-products of other refining processes, are stored in these LPG storage tanks until recycled as feed or sold as products. Their spherical shape gives them strength to contain the high pressure necessary for storing liquefied petroleum gas. (*Courtesy of Valero Refining Co.*)

Olefins and isobutane are mixed with an acid catalyst and cooled. They react to form gasoline components known as *alkylates*. In addition, the reaction produces some normal butane, isobutane, and propane (fig. 7.14). The acid catalyst is allowed to settle out of the liquid and is drawn off to be reused. The rest of the liquid is treated with a caustic to neutralize remaining traces of acid, and the products are separated in a series of distillation columns. The isobutane is recycled as feed, and the butane and propane can be sold as liquid petroleum gas.

Isomerization. In the refining industry, *isomerization* refers to chemical rearrangement of straight-chain hydrocarbons (paraffins), so that they contain branches attached to the main chain (isoparaffins). Isomerization units accomplish two things: (1) they create extra isobutane feed for alkylation, and (2) they improve the performance of straight-run pentanes and hexanes as gasoline blending stock.

The isobutane needed as feed for alkylation is obtained as a by-product of catalytic cracking and hydrocracking. If a refinery does not have a hydrocracker, however, more isobutane will probably be needed. It can be made from normal butane by isomerization. Normal butane is mixed with a little hydrogen and chloride and allowed to react in the presence of a catalyst to form isobutane. Other reaction products include normal butane and a small amount of lighter gases. The reaction products are separated in a fractionator. The lighter gases are used as refinery fuel, and the normal butane is recycled as feed.

Pentanes and hexanes are the lighter components of gasoline. Isomerization can be used to improve gasoline quality by converting these hydrocarbons to their higher-octane isomers. The process is the same as for butane isomerization: the hydrocarbons react with hydrogen and chloride in the presence of a catalyst, the reaction products are separated, the light gases are used as refinery fuel, and the normal pentane and hexane are recycled as feed.

Reforming. Reforming is a process for upgrading naphthas into high-octane gasoline and petrochemical feedstocks. The feed can be naphthas from straight-run distillation, thermal cracking, or hydrocracking. Naphthas are hydrocarbon mixtures containing many paraffins (chain molecules) and naphthenes (ring molecules). Reforming converts a portion of these compounds to isoparaffins and aromatics, which make a higher-octane gasoline. Aromatic compounds are also an important feedstock for the petrochemical industry.

Reforming uses heat, pressure, and a catalyst to bring about the desired chemical reactions (fig. 7.15). Of the several ways of exposing the naphtha to the catalyst, the most common method is to send the feed through a series of reactors with fixed catalyst beds.

A

B

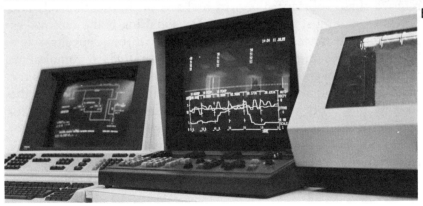

Fig. 7.15. At this refinery, flow rates, temperatures, pressures, and other variables are controlled by a board operator who uses custom consoles, preformatted for the various refining processes (A). These consoles are interfaced with a host computer that allows other stations throughout the plant to access the information (B). (*Courtesy of Valero Refining Co.*)

The feed is mixed with hydrogen, which helps to prevent the formation of coke on the catalyst bed by combining with free carbon to form carbon monoxide and carbon dioxide. Each reactor in the series has a different temperature and pressure in order to promote a specific chemical reaction. Reaction products include hydrogen, light gases, and *reformate*—a liquid mixture of hydrocarbons. Hydrogen is separated from the other products by cooling, part is recycled as feed, and the remainder is used as feed for hydrocracking. The gases are separated from the reformate in a fractionator and used as refinery fuel or as feed for other plant operations.

Extracting with solvents

Solvent extraction is the use of a solvent to selectively dissolve a particular compound and remove it from a mixture of other hydrocarbon compounds. The method is especially useful when the compounds of a mixture have the same boiling point and cannot be separated by distillation. Solvent extraction has many applications in the refining industry. It is used most often to improve the burning qualities of fuels, to refine lubricating oils and wax, and to recover aromatic compounds for use as gasoline components or petrochemicals.

Improving fuels. Kerosines and gas oils used to make jet, diesel, and burner fuels often contain aromatic compounds, sulfur, nitrogen, metallic compounds, and other substances that cause undesirable smoking when they are burned. By solvent extraction these substances can be removed, improving the burning quality of the fuels. Solvent extraction also improves the cetane number of diesel fuel, a measure of its ability to self-ignite.

Making lubricating oils and waxes. Hydrocarbons used to make lubricating oils and waxes often contain substances that interfere with desirable product qualities. Solvent extraction methods such as deasphalting, treating, dewaxing, and deoiling can be used to remove these substances and to make good lubricating oils and waxes from almost any crude oil. Deasphalting is the removal of asphaltic substances that tend to form carbon deposits when lubricating oils are heated. It is an alternative to removal of asphalt by vacuum distillation. Solvent treating or refining is the removal of aromatics, naphthenes, and olefins that decrease the stability of a lubricating oil's viscosity under temperature changes and cause the formation of gummy deposits. Dewaxing removes waxes so that a lubricating oil will flow more readily at low temperatures. Wax that has been separated from lubricating oil is deoiled to remove traces of oil and compounds with a low melting point.

BTX recovery. A third use of solvent extraction is the recovery of the aromatic compounds benzene, toluene, and the xylenes—commonly known as BTX recovery. The aromatics are separated from reformed naphtha and used as gasoline blending stock or sold as petrochemical feedstocks.

Although there are many types of solvent extraction, the basic steps for all of them are the same. A solvent is selected that will dissolve only the extract, the substance that is to be separated from the *raffinate* (the remaining mixture). In a contacting or treating tower, the solvent is thoroughly mixed with the feed so that all of the extract is contacted and dissolved by the solvent. The extract solution is separated from the raffinate by settling, precipitation, cooling, and filtering, or by other means. Next, the solvent is separated from the extract by distillation or evaporation and is recycled to the contacting tower. The raffinate contains traces of solvent, which must also be removed by distillation or other means. In the case of BTX recovery, the extract consists of a mixture of aromatic compounds. The final step is to separate this mixture into benzene, toluene, and xylenes by distillation or fractional crystallization.

Treating

Treating is the removal of contaminants present in crude oil or their conversion into harmless compounds. Treating is often necessary in order to improve the qualities of finished refinery products, prevent the destruction of expensive catalysts, minimize corrosion of refining equipment, or prevent environmental pollution. Sometimes treating may occur as a benefit of a refinery process also performed for other purposes; examples are hydrocracking and solvent extraction. Other treating processes used include dehydration and desalting, hydrotreating, and sulfur recovery.

Dehydration and desalting. Salts break down during refinery processing and cause fouling and corrosion of equipment. They must be removed, along with water, at the outset of refinery processing before crude distillation takes place. When salts are suspended in water, the water and salt can be removed by heating the oil and allowing the water-salt solution to settle out. However, if the oil-water mixture is a stable emulsion that resists separation, chemicals or an electric current must be used to break the emulsion and allow the water-salt solution to separate from the oil.

Hydrotreating. Hydrotreating is a method of removing sulfur, nitrogen, and metals from crude oil fractions. It is used to treat straight-run naphthas, kerosines, and gas oils; thermal-cracked fractions; and residual fuel oils. Hydrotreating is similar to hydrocracking. The feed is mixed with a stream of hydrogen, heated, and sent to a reactor where a series of chemical reactions takes place in the presence of a fixed-bed catalyst. The hydrogen combines with nitrogen and sulfur to form ammonia and hydrogen sulfide, and metals are deposited on the catalyst. The reaction conditions for hydrotreating are less severe than for hydrocracking, because the purpose is to remove contaminants rather than to crack large hydrocarbon molecules. However, a certain amount of cracking and olefin saturation does take place. Heavier oils, in particular, have to undergo more chemical breakdown of molecules in order to free contaminants from the molecules that contain them. In the case of kerosine hydrotreating, molecular changes are beneficial because they improve the burning qualities of diesel fuel. The reaction mixture is separated from the remaining hydrogen and sent to a fractionator where the oil is separated from the contaminant and hydrocarbon gases.

Sulfur recovery. Hydrogen sulfide, the highly poisonous gas sometimes contained in natural gas or created by hydrotreating, can be converted into elemental sulfur, a useful product. Sulfur can be converted into ammonium thiosulfate, a useful fertilizer. The most widely used and reliable process for sulfur recovery is the Claus process. The hydrogen sulfide is sent to a furnace where it is heated and exposed to a controlled amount of oxygen. Most of the hydrogen sulfide reacts chemically with the oxygen to form sulfur and water, but some of the hydrogen sulfide forms sulfur dioxide. The sulfur dioxide is sent to a series of two or three catalytic converters where it is converted into elemental sulfur and water. The vaporous sulfur is cooled and sent through a condenser to change it into a liquid state. The liquid sulfur is drained into a heated tank and stored until it is ready to be pumped into tank trucks or cars for delivery.

Other methods. A number of other chemical and physical methods have been used for treating. They include washing oil with caustic or acid solutions, exposing oil to adsorbent clays, and oxidating oil with catalysts. Although many of these methods have been replaced by hydrotreating, certain methods may still be used.

Blending and using additives

After crude oil has been refined into useful products by various distillation, chemical-conversion, extraction, and treating processes, specific qualities of many of these products can be further improved by blending hydrocarbon components with various characteristics or by adding small amounts of other substances to modify qualities (table 7.2).

TABLE 7.2
REFINERY PROCESSES

Feed	Process	Primary Outputs
Crude oil	Atmospheric distillation	Straight-run fractions: light gases, naphtha, kerosine, gas oils, residue
Atmospheric residue	Vacuum distillation	Catalytic cracking and thermal cracking stocks Lube oil and wax Asphalt Residual fuel
Atmospheric and vacuum gas oils	Catalytic cracking	Alkylation, reforming, and hydrocracking stocks Residual fuel
Vacuum residue	Thermal cracking	Alkylation, reforming, and hydrocracking stocks Residual fuel Coke
Cracked gas oils	Hydrocracking	Gasoline Distillate fuels Lube oil
Cracked olefins and isobutane	Alkylation	Gasoline
C_5–C_6 gases	Isomerization	Gasoline Alkylation stock
Atmospheric and cracked naphthas	Reforming	Gasoline
Kerosines, straight-run gas oils	Solvent extraction	Distillate fuels
Vacuum gas oils	Solvent extraction	Lube oil and wax
Reformed naphtha	Solvent extraction	Aromatic petrochemicals
Crude oil	Dehydration, desalting	Crude oil free of water and salt
Straight-run middle distillates, thermal-cracked fractions, residual oil	Hydrotreating	Oil free of sulfur, nitrogen, and metals Upgraded distillate fuels
Hydrogen sulfide and other sulfur compounds	Sulfur recovery	Elemental sulfur
Fuel oils, asphalts, lube oils, and other products	Blending, additives	Products with improved performance qualities

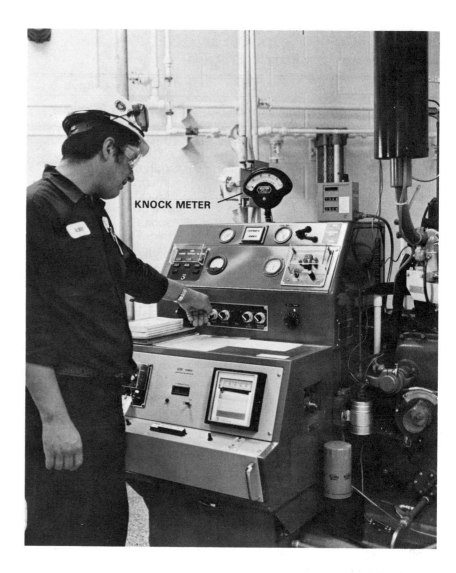

Fig. 7.16. In this gasoline blending department, the blender operator sets the control panel on the research method machine to obtain the octane reading. A similar machine is used to check the octane by the motor method. The average of the two readings constitutes the octane seen at the customer's pumps. (*Courtesy of Valero Refining Co.*)

Blending is particularly important in the refining of gasoline, diesel fuel, furnace oils, and residual fuels. The gasolines produced by straight-run distillation, catalytic cracking, hydrocracking, reforming, alkylation, and gas processing have differing molecular contents and performance qualities. They must be blended into various grades of products that will perform well under varying weather conditions, altitudes, and engine compressions to meet market demands. A gasoline has to vaporize enough to ignite in a cold engine but not enough to cause vapor-lock stalling in hot weather. It has to have an octane number that prevents it from knocking or self-igniting (fig. 7.16). In modern refineries, computers are used to make the complex calculations needed to plan production of the required amounts and types of gasoline components (fig. 7.17).

Diesel fuel is blended from straight-run light gas oils, cracked gas oils, and kerosine. It must have a cetane number that allows it to self-ignite readily, just the opposite of the antiknock quality needed in gasoline. Several grades are sold; the premium grades make better cold starts and smoke less when burning.

Furnace oils are blended from straight-run and cracked gas oils and are used for residential and commercial heating. Residual fuel oils are made from the residue of vacuum distillation and thermal cracking and are used for industrial and ship boilers. These oils must be able to flow at the temperatures at which they are commonly used and must not vaporize enough to form mixtures with air that could start an accidental fire. Residual fuels are blended with heavy gas oils to make them pour more easily and to reduce sulfur content to acceptable levels (if they have not been hydrotreated).

Fig. 7.17. Computers are used to calculate the proper blend to meet the customer's specifications, to open and close valves of lines leading to equipment on the plant site, and to automatically gauge storage tanks on the site. (*Courtesy of Valero Refining Co.*)

348

Protecting the environment

Much antipollution legislation has been aimed at the petroleum industry, especially refineries. Thus, protecting the land, water, and air is a vital function in today's refineries, requiring environmental specialists on the job at every plant. Billions of dollars are also spent by refineries to build, operate, and maintain antipollution facilities.

Refineries use a lot of water for cooling and other purposes. When this water is discharged, it must be pure enough to meet the demands of local and state water quality boards, the U. S. Environmental Protection Agency (EPA), and other regulatory agencies. Chemical treatment tanks, holding ponds, water fauna, and bio-microorganisms are all used to remove contaminants, such as hydrogen sulfide, ammonia, phenols, and salts, from the water (fig. 7.18).

Air quality is also a concern of refinery environmental personnel. Floating roofs on storage tanks that eliminate space for vapor buildup, periodic seal checks along the tank walls, vapor collection systems, scrubbers to remove sulfur from combustion gases, and electrostatic precipitators to collect particulate emission from cat crackers are some of the ways that refineries take care of fumes, emissions, and evaporation of compounds detrimental to the atmosphere.

Fig. 7.18. A bio-reactor uses microorganisms to remove impurities from the water. When not "fed" oil, grease, ammonia, and other contaminants, the tiny bacteria stop "working." (*Courtesy of Valero Refining Co.*)

GAS PROCESSING

As late as the 1930s, natural gas leaving the wellhead had to reach a market nearby or be flared. Natural gas was flared and blown to the air in large volumes in the United States. Flaring is still a common practice in remotely located oilfields, when gas cannot be reinjected into the reservoir or used for local fuel. With the advent of gas pipelines (commonly called transmission lines), gas transport trucks, and field processing facilities for gas, gas production in the United States and elsewhere has become a thriving industry.

Field processing of natural gas is necessary to make the gas marketable. It includes the removal of water, impurities, and excess hydrocarbon liquids as required by the sales contract. It also includes the control of delivery pressure. Gas processing plants can be designed to accomplish the same functions as field facilities when it is economical to gather the gas from several wells to a central point. Often, plants include such processes as dehydration and hydrogen sulfide removal. In addition, gas processing plants are designed to separate hydrocarbon mixtures or individual hydrocarbons from natural gas and to recover elemental sulfur and carbon dioxide.

In general, the larger the gas processing plant, the more economical it is to operate. However, large plants must be located near fields that provide a substantial volume of natural gas. In recent years, portable skid-mounted plants have been developed to provide efficient, relatively inexpensive gas processing for smaller fields. In addition, gas processing facilities are required in refineries to process not only saturated gas resulting from crude oil distillation but also unsaturated gas produced from the cracking and reforming operations of the refinery.

Saturated gas includes methane through nonane or decane. Unsaturated gas includes ethylene, propylene, and butylene. Refinery gas processing provides fuel gas (methane, ethane, and ethylene) to power refinery operations. The facilities also separate individual natural gas liquids (NGLs), which may be used to make various fuel products or may be sent to the alkylation unit for further processing.

Four processes are used in gas processing plants to recover NGL mixtures: oil absorption, straight refrigeration, cryogenic recovery, and dry-bed adsorption. The term *absorption* means taking in molecules and distributing them throughout, as a sponge sucking up a liquid. *Adsorption* is the adhesion of a thin layer of molecules to the surface of a solid body or liquid, as dust particles stick to the surface of flypaper.

Recovering NGL mixtures

Oil absorption. Oil absorption is a process used by older gas processing plants and by many refinery gas plants (fig. 7.19). It may be used with or without refrigeration. Without refrigeration, oil absorption recovers about 70 percent of the propane and all of the butane and heavier hydrocarbons. With refrigeration, the process can recover about 50 to 75 percent of the ethane and all of the propane and heavier hydrocarbons.

Fig. 7.19. This gas processing plant uses the oil absorption process.

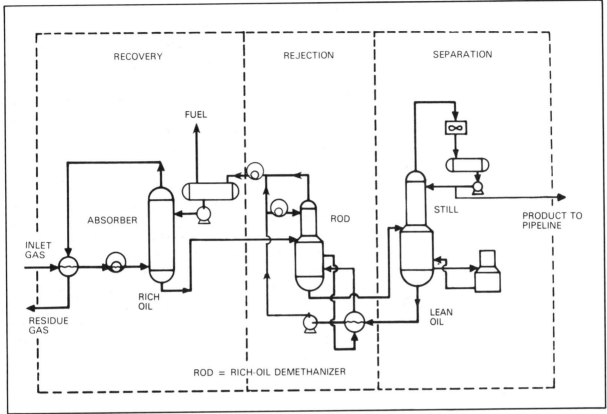

Fig. 7.20. The oil absorption process consists of NGL recovery, methane rejection, and separation of the absorbent.

The oil absorption process has three steps: NGL recovery, methane rejection, and separation of the absorbent from the NGLs (fig. 7.20). Usually, a combination of refrigeration and oil absorption is used in the recovery step. The inlet gas and the absorbent are cooled to temperatures between 0° F and −40° F using a refrigerant such as ammonia, Freon, or propane. Some of the methane is liquefied along with the desired NGLs. Since methane is not a desirable component, it is removed from the NGLs in the rejection step. Most plants use a rich-oil demethanizer (ROD) to separate the methane. Then, a fractionator is used to separate the NGLs from the absorbent. Exposed to steam or hot oil, the NGLs vaporize and flow out the top of the fractionator. The absorbent remains a liquid and flows out the bottom.

Straight refrigeration. Straight refrigeration is often used in smaller gas processing plants, particularly portable, skid-mounted plants that produce from 10 to 20 million cubic feet of product per day. Several designs are available; all of them use refrigerants such as Freon or propane to cool the natural gas and separate the NGLs.

Cryogenic recovery. Cryogenic recovery processes are used in most of the newer gas processing plants. They use high pressures and extremely low temperatures to recover most of the ethane (90 to 95 percent) as well as all of the propane and the heavier hydrocarbons.

Because cryogenic processes use temperatures lower than −150°F, they require specialized equipment and techniques. These systems use plate-fin heat exchangers to achieve low temperatures. Also, construction materials that do not become brittle in low temperatures and cold boxes that enclose and insulate the processing equipment are necessary. Control valves and pumps have extended shafts so that they can be located within the cold box but operated by external motors and works. The inlet gas requires special treating before it can enter the cryogenic equipment. Contaminants such as salt water, wax, dirt, scale, iron sulfide, and oil carried over from wellhead separators and compressors must be removed because they clog the heat exchangers and stop the flow of fluids. Water must be reduced to less than one part per million.

Two types of cryogenic recovery are used: the expander process and cascade refrigeration. The most commonly used recovery method is the *expander process* (fig. 7.21). After being treated, the dry, clean inlet gas is cooled in the heat exchangers. Under high pressure, the gas enters the separator chamber of an expander-compressor, where it expands and cools further. As the inlet gas cools, the NGLs condense and are separated from the methane vapor. The expander-compressor is similar to a centrifugal pump

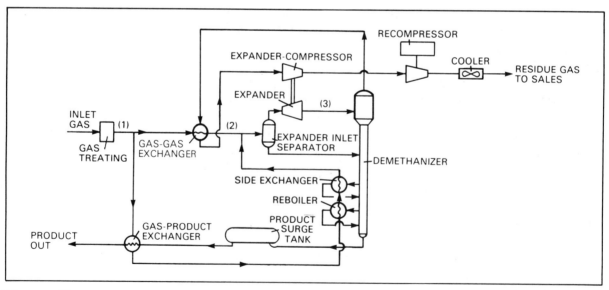

Fig. 7.21. The expander process is one type of cryogenic recovery of NGLs.

Fig. 7.22. A medium-sized expander-compressor will generate up to 1,500 psi of pressure for use in a cryogenic natural gas plant. (*Courtesy of Rotoflow Corp.*)

run backwards (fig. 7.22). When the inlet gas enters the expander separator, the expander wheel begins to spin. This energy is transferred to a centrifugal compressor attached to the expander shaft. The compressor is used to compress methane separated from the NGLs. As the compressor performs work, it causes additional cooling of the inlet gas entering the separator.

Because some methane is dissolved in the NGLs, the liquid is sent to a demethanizer to remove the methane. In the demethanizer, the liquid is heated, and some of it vaporizes and rises through the column. As the vapor contacts cold liquid descending from above, heavier hydrocarbons condense to liquid again, but the methane vapor continues to rise and leaves from the top. The methane is compressed to sales-gas pressure, cooled, and delivered to its destination. The NGLs are piped to a fractionation facility to be separated into individual components.

Cascade refrigeration is a recovery process that uses two-stage refrigeration to separate methane from the NGLs. In the first stage, the treated inlet gas is cooled in heat exchangers by residue gas (methane that has already been separated) and then by a refrigeration system using propane as the refrigerant. In the second stage, the inlet gas is further cooled by residue gas and by a refrigeration system using ethane or ethylene as the refrigerant. The inlet gas, now at a temperature of $-120°F$ or lower, enters a separator where the NGLs condense and are separated from the methane vapor. Because some of the methane remains dissolved in the liquid, the liquid is sent to a demethanizer for further separation.

Fig. 7.23. At this natural gas liquids fractionation center, mixed liquid streams enter the refinery and are fractionated into natural gasoline, normal butane, isobutane, propane, and ethane. (*Courtesy of Cities Service Co.*)

Cascade refrigeration can also be used to liquefy methane. Three stages of refrigeration are used with propane, ethylene, and methane as the refrigerants. Temperatures required are below $-200\,°F$.

Dry-bed adsorption. Dry-bed adsorption can be used not only to dehydrate natural gas but also to remove some of the heavier hydrocarbons in order to prevent them from condensing in pipelines during transmission. This process can recover 10 to 15 percent of the butane and 50 to 90 percent of the natural gasoline. It is used primarily to ensure that the NGL content of natural gas meets the specifications of the gas sales contract. The natural gas is passed over a bed of solid desiccant, which adsorbs some of the NGLs and separates them from the gas stream.

Fractionation of NGLs

Fractionation, or fractional distillation, is a process used to separate a mixture of NGLs into salable individual products (fig. 7.23). The products can be separated because they have different boiling points. The process, similar to that of crude oil

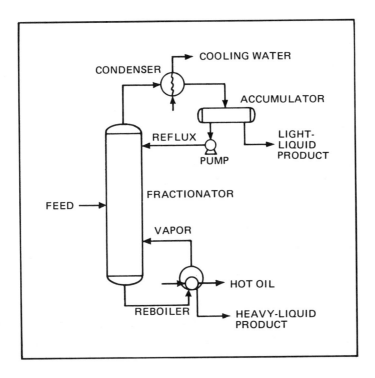

Fig. 7.24. A fractionation system is used to separate NGLs into individual products.

distillation, uses fractionation towers with reboilers and reflux equipment (fig. 7.24). The hydrocarbon components usually separated in a fractionation plant are ethane, propane, isobutane, normal butane, pentane, and a remaining mixture of heavier hydrocarbons (pentane-plus). The NGL products are separated in a series of fractionation towers, the lighter product being removed first. Fractionators are usually named for the main product being separated, known as the overhead or top product because it comes from the top of the fractionator. A deethanizer, for example, produces ethane as the top product, and a depropanizer produces propane.

Because individual products have boiling points that are close together, they are difficult to separate. Getting the necessary vapor-liquid contact requires a large number of trays in the tower, a substantial heat input by the reboiler, and heat removal by the reflux and condenser.

PETROCHEMICALS

Although most petroleum products are used as energy sources, such as gasoline for cars, fuel oil for boats and ships, and natural gas for heating and fuel, petroleum components also provide the raw material for the manufacture of petrochemicals. A *petrochemical* is a chemical substance produced commercially from feedstocks derived from crude oil or natural gas. Consequently, many petrochemical plants are integral parts of large refining complexes and are often subsidiaries of major oil companies.

Types of petrochemicals

The basic job of most petrochemical plants is to turn crude oil fractions or their cracked or processed derivatives into feedstocks that will ultimately be used in the manufacture of a host of other products, from liquid detergent to the plastic bottle containing it. These feedstocks fall into three groups, based on chemical composition and structure—aliphatic, aromatic, and inorganic.

Aliphatic petrochemicals. Aliphatic petrochemicals are straight-chain hydrocarbons, saturated (paraffins) or unsaturated (olefins) with hydrogen. They make up over half the volume of all petrochemicals produced. Two major olefin feedstocks are ethylene and propylene. They are produced by subjecting hydrocarbons to thermal cracking in the presence of steam, a process known as steam cracking. The hydrocarbons used as raw materials range from natural gas to naphtha to gas oil. Ethylene in turn can be chemically converted to several other important feedstocks such as polyethylene, ethylene glycol, vinyl chloride, and styrene. Similarly, propylene can be used to make feedstocks such as polypropylene and isopropyl alcohol. These petrochemicals are ultimately used to make plastics, solvents, synthetic rubbers, and synthetic fibers. Other aliphatic petrochemicals include methanol, made from natural gas; and butadiene, derived from butanes and butylenes (C_4 hydrocarbons). Methanol is used to make resins, polyester fibers, and solvents. Butadiene is used to make rubber, resins, and plastics.

Aromatic petrochemicals. Aromatic petrochemicals are unsaturated hydrocarbons with six carbon atoms in a ring. Major aromatic feedstocks include benzene, toluene, and the xylenes. A large portion of toluene is chemically converted to benzene, the most widely used aromatic feedstock. Aromatic petrochemicals are used to make plastics, resins, fibers, and elastomers.

Inorganic petrochemicals. Inorganic petrochemicals include carbon black, sulfur, and ammonia. These substances do not contain carbon compounds and can be produced from nonhydrocarbon sources. Nevertheless, they are considered to be petrochemicals because the primary raw materials for their manufacture are substances derived from petroleum refining. Carbon black, made from natural gas, is used to manufacture synthetic rubbers, printing ink, and paint. Sulfur, produced from hydrogen sulfide and other sulfur compounds, is used to make sulfuric acid, which in turn is used in the manufacture of steel, fertilizer, paper, and other chemicals. Ammonia, synthesized from natural gas and naphtha, is an important feedstock for the manufacture of fertilizers, fibers, and plastics.

A petrochemical plant

Since each petrochemical plant is designed to convert particular petroleum fractions or other substances into specific petrochemicals, each plant uses different processes, procedures, facilities, and auxiliary operations. To get an insight into a petrochemical plant, consider the operation of Tex-Chem, a hypothetical plant that manufactures two basic plastics—polyethylene and polypropylene.

Situated in the heart of oil country where miles of pipelines crisscross underground and barges ply the waterways, Tex-Chem uses several basic raw materials derived from petroleum or natural gas. From these materials, the plant manufactures polyethylene and polypropylene, which are used to make thousands of products such as appliance and automotive parts, moldings, battery cases, luggage, bottles, plastic toys, transparent food wrappers, plastic containers, and a wide range of construction materials (fig. 7.25).

The plant operates twenty-four hours a day, seven days a week. Although the chemical processes themselves are complex, they involve the application of familiar principles: controlled temperature, pressure, and time, and sometimes the presence of a catalyst and a solvent. All operations take place in a labyrinth of pipes,

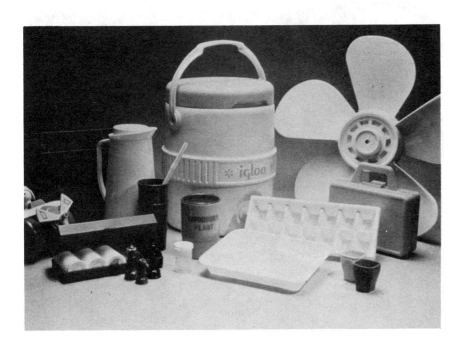

Fig. 7.25. Polyethylene and polypropylene are used to make a host of products.

358

Fig. 7.26. This 20,000-horsepower centrifugal compressor, driven by a high-pressure steam turbine, is an example of the huge yet precise mechanical equipment required to handle process streams in a petrochemical plant. (*Courtesy of ARCO Chemical Co.*)

valves, silos, and towers, powered by heavy equipment (fig. 7.26), controlled by elaborate instruments, and monitored by trained technicians (fig. 7.27). The plant is organized in three systems: the olefin units, the polymer units, and the supporting facilities.

Fig. 7.27. A process technician can tell by checking a meter or chart what temperature or pressure is being applied in any section of a petrochemical unit, adjusting those conditions by touching a switch or turning a dial. Many plants today use computer consoles and keyboards to accomplish the same function.

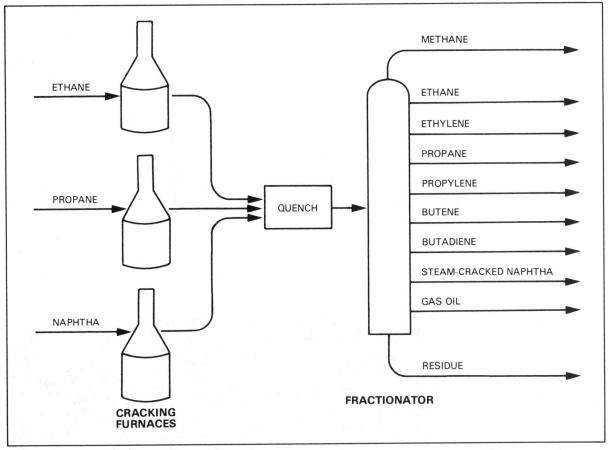

Fig. 7.28. Steam cracking is used to produce ethylene and propylene.

Olefin units. The primary job of the olefin units is to provide ethylene and propylene for the polymer units (polyethylene and polypropylene units), as well as for sale to other chemical or plastic producers. The olefin units, sometimes referred to as a petrochemical refinery because of their wide range of products, also produce butadiene, butylenes, hydrogen, and aromatics for use in the manufacture of other chemicals or gasoline products. These coproducts are sold to other chemical producers or used as feedstocks for other plants operated by the Tex-Chem Company.

The olefin units use liquefied ethane and propane and distillates such as natural gasoline and naphtha as raw materials. These hydrocarbons spend a brief period in a furnace where they are thermally cracked under high temperatures in the presence of steam. The resulting products are sent to a fractionator to be separated (fig. 7.28).

Other processes used in the olefin units include noble metal catalytic conversion of acetylenes, cryogenic separation of hydrogen, and selective adsorption using molecular sieves. (A noble metal is one having high resistance to corrosion and oxidation.) Aromatics are produced by catalytic reforming of virgin naphtha. When the naphtha is passed over a noble metal catalyst such as platinum, it forms ring compounds with no double bonds. These compounds are changed into aromatic compounds by dehydrogenation—removing some of the hydrogen and forming double bonds.

Polymer units. In the plant's polymer units, propylene and ethylene are combined into larger molecules (polymerized) to make polypropylene and polyethylene. The reaction takes place in a liquid solvent in the presence of a catalyst. After the polymers have been formed, they are removed from the reactor and sent to a purification facility to remove the catalyst and the solvent. The powdered product is blended with additives and sent through an extruder to form pellets (fig. 7.29).

Fig. 7.29. Polypropylene is extruded into pellets about the size of BBs in its finished form.

Fig. 7.30. Laboratory analysts run extensive tests on raw materials, process streams, and finished products to assure quality.

Supporting facilities. Extensive laboratory tests are made to maintain quality control during each phase of plant processes (fig. 7.30). Polyethylene and polypropylene pellets are checked for proper content and consistency. If product specifications are met, the pellets are prepared for shipping (fig. 7.31).

Fig. 7.31. Polypropylene pellets are sacked and put in cardboard boxes for transporting to other plants.

Fig. 7.32. In the aeration
basin of the plant's water
treatment facilities, millions of
tiny bacteria (microorganisms)
literally eat the hydrocarbon
waste, purifying the water
before it is reused in the plant
or disposed of.

Plant operations are designed to conserve energy, air, and water. The water used to cool the process units is purified by the plant's water-treatment facilities and reused (fig. 7.32). Only steam is allowed to rise from the plant's cooling towers and units. The steam is odorless, contains no contaminants, and rapidly disperses into the atmosphere. The intermittent discharge of waste hydrocarbon gas from the process units—a potential source of air pollution—is burned in a flare system from tall, derricklike structures. Process units are designed to make full use of the energy consumed, and, in many instances, hydrocarbons created as by-products are recovered and used as fuel.

Companion plant. Linked closely with the operation of this Tex-Chem plant is another plant located nearby. Using feedstocks from plant No. 1, Tex-Chem's plant No. 2 produces specialty chemicals and petrochemical intermediates for the manufacture of other products (table 7.3).

TABLE 7.3
Some Petrochemicals—Their Feedstocks and Final Products

Feedstock	Petrochemical	Typical Final Products
Styrene	Polystyrene	Cups and glasses, phonograph records, radio and TV cabinets, furniture, luggage, telephones, ice chests, lighting fixtures
Paraxylene	Terephthalic acid (TPA), dimethyl terephthalate (DMT)	Polyester fibers used to make wearing apparel, tire cords; polyester film used to make electronic recording tape, photographic film, cooking pouches, specialty packaging items
Metaxylene	Isophthalic acid (IPA)	Fiberglass-reinforced auto bodies, surfboards, snowmobile housings, outboard motor covers, cooling fans, vaulting poles, many paints and coatings
Propylene	Polypropylene	Appliance parts, bottles, safety helmets, disposable syringes, battery cases, construction materials, feed bags, carpet backing

The development of useful products from oil and gas made possible a giant leap in the living conditions of mankind. Used explosively in combination with air, petroleum made twentieth-century man mobile on land, on sea, and in the air. Man is able to heat and cool his environment more effectively with the electricity and gas made available by the industry. Agricultural and manufacturing processes are greatly expanded and improved because of the refinement of produced hydrocarbons.

Petroleum Marketing

Marketing is the necessary interface between the consumer and the petroleum industry. The marketing department of any organization has a dual function: (1) to accurately interpret consumer needs and relay these to its company and (2) to correctly inform the consumer public of the quality, quantity, and kinds of products its company has available. To carry out its functions, the marketing department must manage planning and research, buying and selling, and advertising and public relations so that the right products are in the right place, at the right time, at the right price, and in the right quantity. These five "rights" are the reasons for all marketing functions.

CHANGING MARKET

The detail and depth at which the marketing functions are pursued depend, of course, on the nature of the company being served and the prevailing market. Political pressures, new technology, new discoveries, and different product uses have influenced marketers through the years.

Types of marketers

Basically, two types of marketers operate in the petroleum industry: vertically integrated companies and independent marketers. Naturally, their marketing methods differ.

A *vertically integrated oil company* is one that is large enough to be involved in all aspects of the oil industry. Its marketing practices are integrated within the company's exploration, drilling, production, refining, transporting, and distribution activities. Such a company sells some or all of its automotive products at a service station bearing the company emblem, thereby carrying its petroleum from the reservoir to the retail outlet.

An *independent marketer*, on the other hand, functions at one of several levels in the marketing chain. An independent buys bulk—either the overstock of name-brand refiners or products from independent refineries. He may be a broker who buys and sells mostly on paper (fig. 8.1), or he may arrange to take delivery of his requirements. An independent company often purchases spontaneously in accord with the momentary market rather than according to long-range planning. Or, it may have contractual agreements for specified periods or specified amounts with its suppliers. An independent marketer may also be engaged in storage, transportation, and wholesale or retail distribution.

Development of products market

Petroleum was first valued in the late 1800s for its potential as a lighting fuel (in the form of kerosine). Since then, the petroleum market has grown astronomically, thanks largely to the automobile. Because of the booming demand for internal-combustion engine fuel, less of the crude-oil cut, or fraction, was designated for kerosine and more for the products that would blend into gasoline.

Until 1912, the gasoline sold was only that originally in the crude (straight-run gasoline) or that condensed from natural gas. With the advent of the refining process called *cracking*, more gasoline could be produced per barrel of crude. This gasoline had less tendency to knock than straight-run gasoline from the distilling process. Later high-compression engines required additions such as tetraethyl lead to prevent detonation and engine damage, and most gasoline sold from the 1920s through the 1970s contained

Fig. 8.1. A broker often handles transactions without leaving his office.

lead. Since the mid-1970s, environmental controls required on automobiles manufactured in the United States has increased still further the need for high-octane unleaded gasolines.

As the octane race intensified, a new refinery process was developed: that of *reforming*—both thermal and catalytic reforming. Two other processes, *polymerization* and *alkylation*, helped in the effort over this century to produce a higher percentage of gasoline out of the crude barrel, with an even higher octane rating. In addition, heavy oil cracking units were designed to turn residuals into unleaded gasoline with sufficient octane ratings. Thus, the need for kerosine changed to a need for gasoline, jet fuels, propane, and petrochemicals, creating enormous challenges in the marketing of products from the crude oil barrel.

Use of natural gas

Synthetic gas had an important effect on the eventual development of the natural gas industry, because it was so much cleaner and easier to use than some of the other fuels in common use at the time. Synthetic gas was produced in Europe and America by heating coal, coke, and other solid fuels and then passing steam

368

Fig. 8.2. Gaslights were a nineteenth-century luxury. Note the gas table lamp. (*Courtesy of American Petroleum Institute*)

over them. But because of the high price, manufactured gas was not widely used in the U. S. until the 1920s.

During the nineteenth century, gas was used for lighting but was considered a luxury (fig. 8.2). By the beginning of the twentieth century, natural gas was available as a fuel for homes and industries, but costs were high because the delivery system was still inefficient. The development of gas pipelines played a key role in making natural gas an acceptable and affordable fuel.

RESEARCH AND PLANNING

Analyzing and predicting the petroleum market is a complex task, often involving seemingly contradictory information and trends. For example, just when the planet's way of life was committed to using petroleum at an all-time high, the worldwide supply of that resource was found to be limited. On the other hand, conservation programs were in their heyday about the time an oil "glut" became apparent—a surplus that resulted from the high cost of imported oil and its effect on domestic drilling. The limited supply of petroleum resources is a reality, and an attitude toward conservation has developed on the part of consumers and manufacturers. Users of petroleum and its products have been asked, and in some cases forced, to slow down.

Researchers must prepare for the gradual changeover to other energy sources. Many nations feel that an orderly changeover to other fuels can be accomplished more easily if petroleum is sensibly conserved and therefore used at a slower rate in the coming decades. But predicting this future rate in the face of the expected demand for a continually improving quality of life and the worldwide growth in population is a difficult task for a planner.

Because of its marketing importance, petroleum is often involved in international disputes. A border war in the Middle East can immediately influence the price of oil in Houston. The planner must be flexible enough to take into consideration the unpredictable political implications of the industry. Technology also is changing at such a rapid rate that a petroleum product seemingly irreplaceable today may become obsolete tomorrow. Researchers and planners must keep abreast of technological breakthroughs that, although difficult to predict, may help control the rising cost of energy. The exploration and development of two major oil areas discovered in the 1960s—the North Sea and Alaska—cost ten to twenty times as much as developing similar fields in the Middle East.

International influences

Marketing planners research for three different time periods—the present, the near future, and the more distant future. Policies governing *present* marketing are constantly analyzed for effectiveness. The size of the current market, which brands are selling, and where retail distribution is occurring can all be measured. Seasonal factors must be accounted for: gasoline sells more during the summer vacation months, and heating fuels are needed more during the winter months. As heavier grades of diesel are converted into heating fuels, diesel is not as readily available, thus forcing diesel prices upward in cold months. Research within a company can determine how well its advertising and public relations efforts are working. Consumer behavior must be watched for changing preferences: rejection of, a slight inclination toward, or an overwhelming positive response to a product or service.

The *near future* is usually defined in the marketing world as the next three years. This phase of research usually involves pretesting of new products, services, methods, or publicity. A distinction is made between pretesting and test marketing. *Pretesting* (analysis

Market assessment

of a product) is done under simulated market conditions, and *test marketing* is conducted under real conditions of the ever-changing market. Pretesting is usually used for choosing between alternatives.

Predicting the *more distant future* requires both science and intuition. Clues to the future can be deduced from both the present and the past to predict coming patterns with some accuracy. Demographic indicators such as population numbers, geographical structure, size of families, and educational and financial levels can be fed into computers for predictions of future consumer trends. Marketing research reduces the risk of making decisions for the future; but with the rapid acceleration of technological change, sound judgment will always be needed in petroleum marketing.

Supply and demand

Once the consumer market is assessed, the next and more encompassing role of the planner is to match marketing requirements with resources—in other words, to balance supply and demand. A marketing planner must be able to project whether his company's supplies will be adequate to meet consumer demand without leading to production surpluses. Demand fluctuates for many reasons, some of which are not readily discernible. For example, comparing U. S. individual product demands of June 1985 with those of June 1984, "gasoline was down 2.4 percent, aviation fuel was down 3 percent, and distillates (diesel fuel and heating oil) were up 3.9 percent."[1] This report seems illogical for a summer month, so the marketer must research the "why" of these statistics in order to plan future production.

Traditionally, supply has been able to grow with demand, but a time may be coming in the petroleum market when supply will dictate demand. Along with this problem, a marketing plan must embrace the considerations of possible sources of crude, refinery capacity, and availability of shipping and pipelining. With massive data input, the plan must be able to predict with some accuracy a balance between input and output of an entire system. In a large oil company, meshing together all activities into an efficient and profitable operation becomes quite a task.

Computer evaluations

Sophisticated programs run on computers are used to evaluate the many options available. Various mathematical models generate predictions from computerized data. Supply planners must then decide upon the practicality of the various programs, because the

1. "U. S. Petroleum Situation," *PhilNews*, August 1985, p. 5.

most economical future solution is not always the most desirable in terms of current conditions.

An example of such planning is the mathematical model used by the authors of the National Energy Plan (1977) to predict energy futures. This model, referred to as Project Independence Energy Systems (PIES), was used to answer the kinds of questions a marketing planner might ask. Two types of information were recorded on reels of magnetic tape that could be fed into a computer. One set of tapes held the *program*, describing the expected relationships among factors involved in the energy system. The program sought answers to such questions as "How much more oil would be found in the United States in response to a given effort to find it?" and "What increase in the price of petroleum products would occur if total deregulation were allowed?" The second set of tapes held the actual data from recent years relevant to the relationships set up in the first set of tapes. Examples of these data were the number of drilling rigs used in given years, the amount of oil found annually, recent world crude prices, domestic petroleum products prices, and so forth. These examples comprise only a very few of the relationships and data fed into the PIES mathematical model.

When such tapes are inserted into appropriate computers, the numerical values in the second set of tapes are entered into the mathematical relationships established by the first set of tapes. After the computers perform the calculations, they print out tables of figures that help planners make predictions—in the case of PIES, predictions about energy futures. In the case of marketing models, the predictions would involve pricing, size of markets, inclination of future consumers, and other patterns influencing supply and demand.

Refinery capacity

Planning for future refinery capability is complicated by the fact that it takes about five years from the time of decision to bring a major new refining plant on stream. The estimated profitable output over the life of the refinery must justify the amount of capital required to invest in a new plant. Smaller refining investment projects may include increased dock facilities, environmental protection facilities, or new storage areas.

New government regulations may require extensive restructuring on the part of some refineries. A case in point is the effect of the gasoline lead phase-down ordered by EPA beginning in 1984 (fig. 8.3). The EPA ordered the three-stage phase-down after finding that emissions from leaded fuel are a health hazard. Given a few years leeway, most U. S. refineries will be able to make the transition and still keep their gasoline inventory adequate and

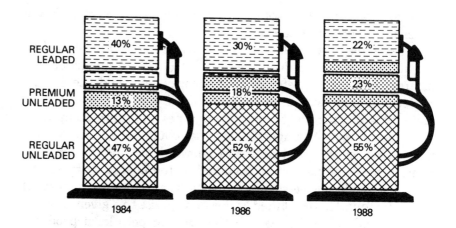

REGULAR LEADED — 40% | 30% | 22%

PREMIUM UNLEADED — 13% | 18% | 23%

REGULAR UNLEADED — 47% | 52% | 55%

1984 1986 1988

Fig. 8.3. Large numbers of American motorists are switching to unleaded gasoline because of EPA's phasing out of lead in gasoline. (*Courtesy of* PhilNews)

SOURCES: Department of Energy; Environmental Protection Agency; "U. S. Petroleum Situation," *PhilNews*, May 1985, p. 5.

their financial heads above water. (One company estimated that it will spend around $140 million to convert its refining processes to produce totally unleaded gasoline by 1988.) Analysts predict some results of this regulation: many small refineries will close their doors; companies may have to rely on imported products in order to supply their customers and outlets during conversion; the oil supply will be depleted sooner because the less lead in gasoline, the more crude oil it takes to produce it; and American motorists will pay higher prices at the pumps. A planner must consider such consequences in his company's marketing strategy.

Transportation scheduling

Having a means of transportation ready to pick up and deliver the proper quantities of feedstock and the right selection of products is another task of planners. To take advantage of the fact that the cost per unit of feedstock usually declines as more volume is carried, the planner must accurately estimate need. A planner, or scheduler, of transportation must select carriers, make sure they are at the proper loading points, monitor crude and product stocks to have them ready for loading, and choose appropriate routings.

Transportation timing and coordination may be affected by market fluctuations, weather, vessel engine trouble, or political upsets. Cargoes can neither be shipped before they are available nor allowed to accrue so much that refinery or process throughput has to be stopped because of lack of storage space. Details such as a ship's capacity, pipeline availability, access to inland waterways, and a myriad of other transportation-related factors must be considered.

The end pricing of petroleum products was greatly affected from 1971 to 1981 by government regulations controlling the wellhead prices of crude oil and natural gas. Among the forces that create prices—availability of reserves, entrepreneurial freedom, and market competition—governmental regulation has been a prime contributing factor. The oil industry has long operated under a variety of governmental regulations and tax burdens. The emphasis after the enactment of President Jimmy Carter's National Energy Plan (NEP) and the Natural Gas Policy Act (NGPA) was toward deregulation. Some wellhead prices for oil were allowed to rise to world levels to encourage further exploration and development, but a large portion of the increased industry profits was taken by the government in the form of the *windfall profits tax.* This excise tax was part of the Energy Security Act of 1980 and continues to be in effect, even though oil prices have dropped considerably since its enactment.

An elaborate set of rules was made to phase out natural gas price controls at the wellhead by 1985. Traditionally, *intrastate gas* (gas that does not cross state lines as it is piped from producer to consumer) has not been federally controlled. *Interstate gas* (gas that travels across state lines) has been regulated by the Federal Energy Regulatory Commission (FERC). The NGPA has sought to equalize gas prices in producing and consuming areas by involving intrastate gas in its phasing out of controls. Gas is priced according to a six-tier system of definitions:

(1) new gas (gas discovered since 1977);
(2) old, interstate gas that is subject to existing contracts;
(3) old, interstate gas that will be made available at the expiration of existing interstate contracts;
(4) the same class of gas formerly sold in intrastate commerce;
(5) specific categories of high-cost gas; and
(6) synthetic natural gas.

As the resulting change in pricing occurs, several marketing problems will arise. Should gas sell at prices based on the actual cost of production? Should different classes of users—industrial and residential—be offered different prices resulting from the different tiers? If gas prices are allowed to rise to compete as a Btu equivalent of world market crude, will that action allow oil-producing countries to set the price of U. S. natural gas as well as oil?

Pricing for crude oil since deregulation has been divided into four tiers:

(1) old oil;
(2) previously discovered oil;

(3) the current world price for newly discovered oil (oil from a well that is either 2.5 miles or more from an existing onshore well as of April 20, 1977, or more than 1,000 feet deeper than any existing well with a 2.5-mile radius of such a well); and

(4) incremental tertiary recovery and stripper oil production that is free of controls.

Interpretation of what constitutes each of these tiers has been in litigation since deregulation began.

Price levels must, of course, be taken into consideration by a petroleum marketing planner. They are part of the challenge of the industry presented daily to planners, middlemen involved in sales, and finally to end consumers.

BUYING AND SELLING

In any marketing concern, sales are the first consideration, although purchases of raw materials and feedstocks, as well as trades, equally affect an operation's profitability. The old adage, "You make your money when you buy," is no less true in the petroleum industry than in any other business. But, unlike most businesses, the petroleum industry's marketing is profoundly affected by federal price regulations.

Crude acquisition For a petroleum marketer, valid considerations concerning crude acquisition include not only the immediate oil needs indicated by marketing research, but also the need to build up or deplete stocks for future periods. The marketing flexibility provided by an oil company's owning storage and transportation facilities becomes crucial at this point. The different grades and quantities of crude that will be available at any one time must also figure into a broad plan. A marketing company may choose to buy from another company or abroad rather than use its own crude, even when the company is also a producing concern. Or it may choose to trade for the various qualities of crude demanded by the dictates of its own predicted market.

The future availability of crude, of course, depends on world-wide reserves and the ability of future technology to extract it at a cost lower than the value of the energy produced. A planning researcher must have reliable worldwide statistics programmed into his machines. He will either keep abreast of current news and scientific journals or employ a consulting service to do so for him.

TABLE 8.1
Typical Marketing Information

Latest week 11/15	4 wk. average	4 wk. avg. year ago*	% change	Year-to-date average*	YTD avg. year ago*	% change
Demand (1,000 b/d)						
Motor gasoline	6,762	6,767	—	6,818	6,702	1.7
Distillate	2,773	2,798	-0.8	2,829	2,843	-0.5
Jet fuel	1,134	1,188	-4.5	1,190	1,169	1.7
Residual	1,268	1,219	3.9	1,159	1,386	-16.3
Other products	3,632	3,648	-0.4	3,529	3,662	-3.6
Total demand	15,569	15,620	-0.3	15,525	15,762	-1.5
Supply (1,000 b/d)						
Crude production	8,903	8,945	-0.4	8,877	8,872	—
NGL production	1,599	1,661	-3.7	1,627	1,626	—
Crude imports	3,473	3,457	0.4	3,004	3,264	-7.9
Product imports	1,715	2,014	-14.8	1,749	2,038	-14.1
Other supply†	699	761	-8.1	675	782	-13.6
Total supply	16,389	16,838	-2.6	15,932	16,582	-3.9
Refining (1,000 b/d)						
Crude runs to stills	12,270	12,047	1.8	11,924	12,107	-1.5
Input to crude stills	12,414	12,207	1.6	12,082	12,237	-1.2
% utilization	79.3	76.8	—	77.3	76.3	—

	Latest week	Previous week*	Change	Same week year ago*	Change	% change
Stocks (1,000 bbl.)						
Crude oil	310,176	314,753	-4,577	350,497	-40,321	-11.5
Motor gasoline	213,467	215,493	-2,026	232,349	-18,882	-8.1
Distillate	128,576	123,817	4,759	155,373	-26,797	-17.2
Jet Fuel	45,708	42,795	2,913	45,696	12	—
Residual	46,591	48,117	-1,526	49,225	-2,634	-5.3
Drilling						
Hughes rotary rig count	1,895	1,864	31	2,597	-702	-27.0

* Based on revised figures. † Includes other hydrocarbons and alcohol, refinery processing gain and unaccounted for crude oil.
Figures are produced from Oil & Gas Journal Energy Database.

SOURCE: *OGJ Newsletter, Oil & Gas Journal,* November 4, 1985.

Commonly used marketing reports are *Platt's Oilgram Price Report* and *The Lundberg Letter.* Such services help not only with availability but also with pricing. Oil companies will frequently have researchers who monitor government legislation affecting availability and pricing. Marketing information, such as that shown in table 8.1, is essential knowledge for a marketing planner trying to construct a valid program of supply and demand.

Three major methods of buying crude today are the traditional *contract* market, the *spot* market where oil is bought and sold daily, and the *futures* market. Since crude oil was defined as a commodity in 1983, the new method of trading—futures marketing— has continued to gain momentum.

Contract market. The contract market is the oldest system for trading oil. Suppliers, such as major oil companies, offer long-term contracts to their customers with the contracts specifying all terms except price, which is fixed at the time of the sale.

Spot market. The spot market is an alternative way to sell oil and products. Prices are set by supply and demand. Trading firms and individuals around the world buy and sell over the phone every day. Usually, in large companies a margin of shipping capacity is left to be acquired by spot buying. Having storage tanks and shipping facilities waiting makes this kind of action possible.

Futures market. In the futures market, prices are determined by open bidding for contracts on the trading floor of a commodities market such as the New York Mercantile Exchange. A futures contract is a commitment to deliver or receive a specific quantity and grade of oil or product during a designated future month at a preset price. Brokerage commissions are charged for trading these contracts. Delivery of the physical product occurs in only a small percentage of cases, with most participants liquidating their positions before the end of trading. Given the current uncertainty in oil prices, petroleum marketers have increasingly resorted to buying and selling petroleum futures in order to *hedge* their price risks and ensure themselves against adverse price moves.

Hedging is a financial insurance device for minimizing losses and protecting profits during the production, storage, processing, and marketing of commodities. A marketer can save a margin of product because he believes prices are going up. But rather than lose value if the prices drop, the marketer can sell the product on the futures market and lock in a price for himself for later delivery. That contract may then continue to be exchanged on the futures market among other marketers with similar concerns and expectations. The futures market provides a way to lessen risk in the volatile pricing situation that is the petroleum market.

Products sales and distribution

The task of distributing petroleum products from processing plants and refineries to the final consumers is complex and often costly. The combined costs of product storage and handling, transportation, and delivery can add up to more than half the total operating costs of a marketing department. The fewer handling or reloading operations done per trip, the lower the total cost will be.

One integrated company found that time and transportation costs have profound effects on the bottom line (fig. 8.4). Its key to successful marketing in spite of refining costs was to shift its logistics; that is, process those raw materials most available and hence most economical; maintain a balanced proximity between supply sources and customers; and produce products that are more profitable and most in demand.

Consumer products, such as gasoline and home heating oil, are usually sent directly from the refinery via pipeline, tanker, barge, or railroad to an installation called a *terminal*. A terminal is a storage and loading area close to a large population center usually owned by a refinery or large oil company. Personnel at a terminal

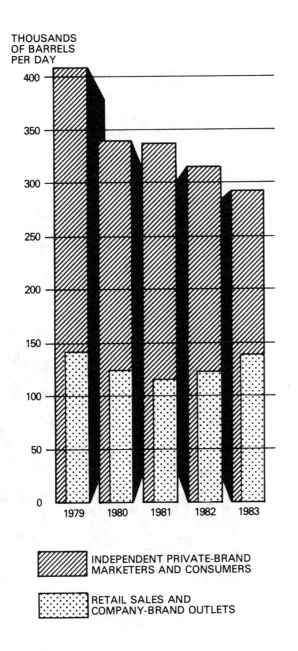

THOUSANDS
OF BARRELS
PER DAY

| INDEPENDENT PRIVATE-BRAND MARKETERS AND CONSUMERS |

| RETAIL SALES AND COMPANY-BRAND OUTLETS |

Fig. 8.4. One integrated company's refined products sales remained stable as the independent market dropped from 1979 to 1983.

SOURCE: "Charting a Course Downstream," *Marathon World*, No. 1, 1984, p. 18.

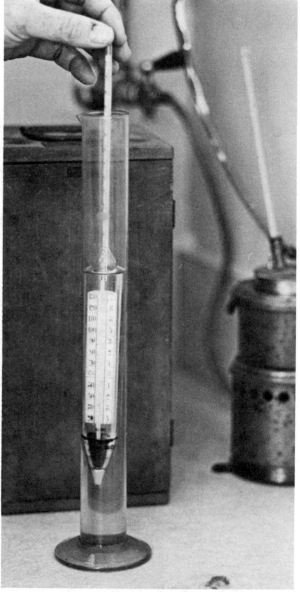

Fig. 8.5. A flash-point indicator is used to verify fuel specifications.

Fig. 8.6. A tester checks gravity of the fuel at the terminal.

are responsible for certain quality checks before they can sell products. These tests may include a flash-point check (fig. 8.5), to verify that each product vaporizes within the specified proper temperature range, and a gravity check (fig. 8.6). Aviation fuel is sent to a laboratory for testing before it can be sold. These tests have to be repeated at the terminal because products may have changed during transport from the refinery.

From the terminal, products are transported to bulk plants serving smaller communities or directly to service stations, airports, homes, or businesses. Although a large volume of bulk products is transported by rail, barge, and coastal tanker, truck transport is the most popular form of final distribution. Large industrial consumers, the exception, are usually supplied by water, rail, or pipeline.

Most terminals will have several truck loading devices, commonly called *loading racks* (figs. 8.7–8.10). The rack is equipped to load the truck compartments with several kinds of products at the same time and to pipe vapors off the top of the truck compartments in vapor recovery efforts. Such a terminal often supplies a number of different kinds of consumers.

A refining company may supply its own service stations and other company-brand outlets. The terminal's owning company may contract with a *jobber* (an independent middleman) to act as a wholesale agent for that brand of product in a given trade area. A jobber may also own a retail outlet where he resells the product from the terminal. Independent service station owners may purchase their fuels from terminals or from jobbers. Jobbers may purchase propane to deliver to rural homes. At a large terminal, the combination of different types of orders, loading instructions, dispatching, and invoicing creates a great deal of physical supervision and paperwork.

Terminal operation is another area of the oil industry that has become automated. Using automatic invoicing, data can be transmitted quickly to a company's centralized computer. Such information also facilitates centralized credit control, sales statistics, vehicle statistics, and stock records. A number of physical functions previously performed manually are now automated. During truck loading, the identity of the driver and vehicle, the amount and description of product exchanged, and product destination can be recorded automatically. Some terminals are equipped with visual display units that enable terminal employees to call up certain computerized files at the head office in order to obtain consumer data. The minicomputer is making possible the storage of customer data and the proper invoicing at a terminal itself.

When a truck is loaded, discharge plans must be designed for optimal offloading performance. Industrial customers and many service stations can take full truckloads, but in the case of farms

Fig. 8.7. A truck operator prepares to load up with several types of products at the terminal.

Fig. 8.8. A truck is loaded by attaching the rack hoses.

and domestic customers, the amount of fuel needed probably will not empty a full truck. Thus, a terminal must cater also to the small-drop market. Computers often are used in this area to plan vehicle loadings and routings. This minimization of distribution time, and thus costs, is of major importance to all marketing companies.

Fig. 8.9. A truck operator uses a card for automatic invoicing.

Fig. 8.10. The terminal meter records type and amount of products purchased.

Petrochemical sales

Marketing petrochemicals is different in several respects from marketing conventional petroleum products. Few petrochemicals are end products themselves. They are usually primary feedstocks or intermediate chemicals that are sold to manufacturers. Factory processes then turn them into such articles as tires, fabrics, plastics, and many other products. Also, some primary petrochemicals are used as intermediates for the manufacture of other petrochemicals.

The volume of petrochemicals is relatively small compared with those of gasolines and fuel oils. And most sales of petrochemicals are by long-term contracts, with little spot selling to influence price fluctuation. Therefore, plant output is easier to predict even before the plant is built. On the other hand, market development is more of a challenge in petrochemicals than in conventional petroleum products. As more and more imaginative end products are developed from petrochemicals, the public must be educated as to their modern usefulness. Television advertising has been very useful in showing the innovation, for instance, of such a product as plastic wrap. The petrochemical market fluctuates with the products made from them. For this reason, many vertically integrated oil companies have stayed out of petrochemical marketing. They may instead sell their primary feedstocks directly to chemical companies or manufacturers.

Natural gas transactions

Natural gas is generally bought at the wellhead by a pipeline company. The pipeline company will transport the gas to processing facilities, if necessary, and then to distribution centers. In the United States, various state regulatory agencies determine how much gas can be gathered from a reservoir on the basis of market demand and a reasonable depletion time for the reservoir.

Gas distribution is almost totally dependent on the availability of pipelines. Transcontinental pipelines carry the marketable product from producing areas to urban consumers. Local distribution companies, mostly privately owned, buy gas from the pipelines and make it available to consumers in homes, apartments, businesses, and sometimes industry. In order to perform this service, local companies operate an underground network of smaller pipelines.

The Natural Gas Policy Act of 1978 started deregulation of the price charged by producers for "new" gas—gas discovered since 1977—at the wellhead. This act was designed to achieve deregulation gradually, with the total release of price controls effective January 1, 1985. Some 60 percent of the gas is now deregulated. As long as "old" gas is subject to price controls, it will probably be sold at prices below the current market clearing price. The effect of this inequity has been the subject of many price discussions, especially since pipelines purchase and sell both new and old gas.

Some pipelines sell their purchased gas at a weighted average price of old and new gas prices; however, not all pipelines have the same "mix," and those in a position to purchase more old gas can charge less than those who rely mostly on new gas.

The FERC has also proposed so-called *block billing* for interstate pipelines, which would necessitate the separation of old and new gas into blocks, or batches, for the purposes of sales and shipping. Oil industry spokesmen have objected to this block billing proposal, stating that it would "adversely affect future gas supplies, encourage fuel switching by large industrial customers, and result in an increase in oil imports. . . . The best solution to the problems addressed by this proposal is the complete deregulation of all gas."[2]

Gas marketing has also been changed by the loosening of FERC regulations controlling the gas pipelines. Starting in 1984, FERC's rulings have made the pipeline industry more competitive. Pipeline companies are being forced to release their customers from long-term contracts, to refrain from requiring a minimum gas volume or a minimum dollar contract from their customers, and to transport third-party gas for a reasonable price.

Many independent natural gas producers are beginning to rely on spot markets for part of their sales. Posted prices and spot market programs offer some advantages to producers: access to a strong market, flexibility of short-term contracts, market-clearing price, transportation assistance, and increased cash flow. Pipelines are also using the spot market as a way to move problem gas. The spot market is a means of getting independent producers' gas into a marketplace heretofore unavailable to them.[3]

PETROLEUM PRODUCT CONSUMPTION

Examining how the crude barrel is broken down into its marketable products is important, since most creature comforts depend on products and services using some form of petroleum (fig. 8.11). Besides power for most modes of transportation, petroleum products play an important role in housing (fuel and electricity), clothing (petroleum-based synthetic fabrics), and food (fertilizers, pesticides, cookware, and packaging). Beyond these basic needs provided by energy-intensive goods are many petroleum-derived products that add to modern man's quality of life.

2. American Petroleum Institute, *Response*, R-338, November 19, 1985, p. 4.

3. Kevin Bumgarner, "Availability of Spot Market Vital to Independents," *The American Oil & Gas Reporter*, October 1985, p. 19.

CHECK LIST OF COMMERCIAL PETROLEUM PRODUCTS
THEIR SOURCE AND END USES

From *Petroleum Products Handbook* by Virgil B. Guthrie. Copyright © 1960 by the McGraw-Hill Book Company, Inc. Used with the permission of McGraw-Hill Book Company.

From Natural Gas

LP-GAS
Dom. and ind. fuels, motor fuel, gas enrichment, solvent, treating and varied ind. uses

PROPANE
Component of LP-Gas, raw material for petrochemicals

BUTANE
Component of LP-Gas and of motor and aviation gasoline, raw material for petrochemicals

AMMONIA
(by syn. from methane)
Fertilizer, petrochemical mfg.

CARBON BLACK
Rubber reinforcing agent, ink mfg., carbon paper, battery cells

LAMP BLACK
Tinting paints and lacquers

ETHANE AND OTHER RAW MATERIALS FOR PETRO-CHEMICAL MFG.

SULFUR
(from hydrogen sulfide)

NATURAL GASOLINE
Component of motor and aviation gasoline, special solvents, refinery processing material

From Refinery Gases

Noncondensable Gases

HYDROGEN
(from cat. reforming)
Chemical and ind. processing

AMMONIA
(from hydrogen)

CARBON BLACK

SULFUR
(from hydrogen sulfide)

Liquefied Gases

REFINERY LP-GAS

MOTOR FUELS

Processed Derivatives Olefins, Diolefins

BUTADIENE
Syn. rubber, petrochemical and plastics mfg.

ETHYLENE
Polyethylene mfg., raw materials for petrochem. mfg. and alcohols

POLYMERS
Lube oil additives

ANTIKNOCK AGENTS
Aviation gasoline blending

ALCOHOLS, ESTERS KETONES, RESINS
Antifreeze agents, drug and lacquer solvents

Syn. Plastic Waxes
Inks and paints

From Light Distillates from Refining Petroleum

Light Naphthas

GAS-MACHINE GASOLINE

SPECIAL GASOLINES

Intermediate Naphthas

AVIATION GASOLINE

MOTOR GASOLINE

MARINE GASOLINE

COMMERCIAL SOLVENTS
Rubber, lacquer and pesticide diluents

BENZENE
High-octane gasoline component, solvents, petrochemical mfg.

TOLUENE
Solvent, high-octane blending agent, chemical intermediate, explosives

XYLENE
High-octane gasoline blending agent, lacquer and enamels mfg., chemicals intermediate

Heavy Naphthas

VM&P NAPHTHA
Thinner for paints, varnishes, lacquers, type cleaner, mfg. cements and adhesives

STODDARD SOLVENT
Special solvent for dry-cleaning trade

MINERAL SPIRITS
Thinner for paints and varnishes, turpentine substitute

Jet Fuel Component

The term "jet fuel" includes fuels for aircraft gas turbine engines of both the jet thrust and propeller type

Fig. 8.11. Commercial petroleum products, their sources, and their end uses (*Courtesy of McGraw-Hill Book Co.*)

385

From Middle Distillates from Refining Petroleum

KEROSINE
Illuminating oil, range oil, stove fuel, tractor fuel, pesticide, diluent
|
JET FUEL

Gas Oil
|
DIESEL FUELS
|
DISTILLATE HEATING OILS
|
WATER GAS CARBURETION OILS
|
METALLURGICAL FUELS
|
NAPHTHENIC ACIDS
Detergents, paint dryers, lubricating oil additives, pesticides

ABSORBER OILS
Used in the recovery of natural gasoline, gasoline, aromatics, and other light hydrocarbon compounds by absorption methods

From Distillate Lubricating Oil Stocks

White Oils (technical)
|
INSECTICIDE AND SPRAY OIL DILUENT
|
BAKERS, FRUIT PACKERS, CANDY MAKERS, EGG PACKERS, SLAB OILS
|
HYDRAULIC OILS
|
RECOIL OILS

White Oils (medicinal)
|
INTERNAL LUBRICANTS
|
SALVES, GREASES, OINTMENTS, COSMETICS

Saturating Oils, Emulsifying, Flotation Oils
|
PAPER, LEATHER, WOOL, TWINE OILS
|
COAL SPRAY OILS
|
DUST-LAYING OILS
|
TEXTILE OILS
|
METAL RECOVERY, QUENCHING OILS

PETROLEUM WAX (PARAFFIN AND MICROCRYSTALLINE)
Paper manufacturing, sanitary containers, waxed wrappers, candle making, match making, drugs, cosmetics, canning, sealing wax, insulation and coatings, candy, chewing-gum manufacture
|
FATTY ACIDS
Lubricating oil pour point depressors, grease and soap lubricant
|
FATTY ALCOHOLS AND SULPHATES
Rubber compounds, detergents, wetting agents

Light Lube Oils
|
TURBINE OILS
|
LIGHT SPINDLE OILS
|
TRANSFORMER OILS
|
HOUSEHOLD LUBRICANTS
|
COMPRESSOR OILS
|
ICE MACHINE OILS
|
METER, DUST-LAYING, TEMPERING OILS

Medium Lube Oils
|
MOTOR OILS
|
AIRCRAFT OILS
|
DIESEL OILS
|
JOURNAL OILS
|
ENGINE OILS
|
RAILROAD OILS
|
CUTTING, METAL WORKING OIL BASE

Heavy Lube Oils
|
LUBRICATING GREASES
|
STEAM CYL. OILS
|
VALVE OILS
|
TRANSMISSION OILS
|
PRINTING OILS
|
BLACK OILS
|
TEMPERING OILS

From Lubricating Oil Stocks Bottoms

MEDICINAL PETROLATUMS
Salves, creams, ointments, petroleum jelly

TECHNICAL PETROLATUMS
Rust-preventing compounds, rubber softeners, wire rope lubricants, cable coating compounds

From Residues from Petroleum Distillation

Residual Fuel Oils
|
HEATING OILS
|
BOILER FUEL
|
BUNKER C FUEL
|
WOOD PRESERVATIVE
|
GAS-MANUFACTURE OILS

Asphalt
|
ASPHALT CEMENT
Paving, surfacing, construction materials, base for paints, lacquers, inks, roofing compounds, undercoatings
|
LIQUID ASPHALT
Cut-back asphalts, road oils, emulsion bases, saturants, roofing compounds
|
BLOWN ASPHALT
Waterproofings, rubber substitutes, insulating asphalts, roof coatings

Petroleum Coke
|
RAW COKE
Ind. and dom. fuel, calcium carbide manufacture, foundry and blast furnace coke
|
CALCINED COKE
Aluminum anodes, furnace electrodes and linings, graphite specialties

Refinery Sludges
|
SULFONIC ACID
Saponification agents, emulsifiers, demulsifying agents, special oils

Fig. 8.11, continued

386

Marketable petroleum products can be divided into broad classes that provide energy for transportation, electrical generation, industry, home, commerce, and agriculture. These general areas are supplied by different *cuts* of crude, including gases, liquefied petroleum gas, gasoline, aviation fuels, kerosine, diesel fuel, lubricating oils, waxes, fuel oils, and miscellaneous products.

Automobiles

Consumers of automotive fuels experience the effects of petroleum marketing most directly at the retail outlet—the service station (8.12). Several transactions occur in the marketing chain of operations as gasoline moves from the terminal to the station pump. However, the enacting of U. S. deregulation in 1981 removed price structuring from product sales as well as sales of crude oil. Therefore, prices are now largely a function of supply and demand.

To the consumer, the most important characteristic of purchased gasoline is probably its antiknock value, or *octane rating*. This number represents a measure of the gasoline's efficiency of combustion. The higher the octane rating, the less likely is the gasoline to cause the engine to knock. Knock happens when the gasoline-air mixture in the engine explodes rather than burns evenly. The knock, or ping, is the sound of small explosions as the gasoline is inefficiently burned by the engine. Knock is harmful, so the octane rating of the gasoline must be matched to the engine's characteristics.

Fig. 8.12. Gasoline is unloaded from a transport truck for underground storage at a service station.

Octane rating is tested and determined under laboratory conditions. A gasoline is marketed according to the average of its octane numbers taken from the research method (r) and motor method (m) (see fig. 7.16), accounting for the formula seen on many gasoline pumps:

$$(r + m)/2 = \text{octane number.}$$

Acceptable octane ratings for gasolines in most countries range from about 86 to 92.

A straight-run fraction of crude at temperatures from 30° C to about 200° C (the gasoline range) would produce violently heavy knock. Therefore, gasoline blends use processed concentrates of those hydrocarbons that have high octane numbers. These include aromatics (obtained from catalytic reforming), olefins (obtained from catalytic cracking or polymerization), and isoparaffins (obtained from isomerization or alkylation). Since the government has required that new cars be manufactured to use only unleaded fuel, oil companies have produced a *super* unleaded gasoline with an octane rating of 90 or more (fig. 8.13). Today's challenge for U. S. refiners and their marketing departments is to produce high-octane, unleaded gasolines that can be sold at competitive prices. Environmental Protection Agency lead phase-down and other federal regulations have put U. S. refiners at a disadvantage in the world products market, since foreign firms are not faced with the same costs.

Fig. 8.13. Super unleaded gasoline is offered by some service stations.

Aircraft   The need for air-transport fuels has developed dramatically in this century, as the size, capacity, and number of airliners have increased. The demands placed on fuel quality for aircraft are, of course, more stringent than those for automotive fuel. Aviation fuel must be free of contaminants (water or dirt), be pumpable at low temperatures, meet regulations for specific gravity and caloric value, and burn cleanly and remain stable when heated to high temperatures.

Two categories of aviation fuel are piston-engine fuel and turbojet fuel. Most passenger-carrying and military aircraft have changed over from piston engines to turbojet engines. However, a need still exists for high-octane aviation fuels for piston engines in older airplanes and smaller private and commercial aircraft (fig. 8.14). Marketing planners try to predict the rate of changeover each year from aviation piston-engine fuel to jet fuel. The most commonly used fuels for jet aircraft are JP 4 and JP 5. JP 5 is kerosine supplied from crude as a straight-run fraction in the range of 150° C to 250° C. A blend of straight-run fractions covering the range of crude from about 30° C to 260° C is made to supply the aviation fuel called JP 4. More specialized fuels are made, such as JP 7 that is used in the SR-71 (strategic reconnaissance aircraft).

As the quality of aviation fuel has gotten better, so also has the design of equipment for its storage and transfer from the refinery to the airplane. Because contamination could cause fatal accidents, aviation fuel movement is monitored very carefully. Laboratory tests are repeated at different stages to guarantee product specifications. Usually, a jobber will buy from a terminal and then store the

Fig. 8.14. Small aircraft, such as this piston-engine plane, uses aviation gasoline for its fuel.

Fig. 8.15. A turbojet commercial airliner is filled with jet fuel.

fuel close to an airport. Different kinds of fuel are color coded for sight identification in loading and servicing. The refueling rate has improved greatly to meet scheduled aircraft turnaround time. Large refueling trucks carry the fuel out to an airplane and pump it under controlled presssure into the aircraft's tanks (fig. 8.15). In a hydrant system, underground pipe carries the fuel from storage tanks to aircraft refueling areas.

Marine vessels

Oil firing, rather than coal burning, became popular for steamships during the first half of the century. Oil allowed greater steaming distance, was safer and easier to use, and took up less storage space. Refueling ships became simpler, and oil burned cleaner. With the development of the marine diesel engine, more new ships with oil-fired engines were built.

Marine fuels include both fuel oil and diesel fuel. Since marine diesel engines run slower than automotive diesel engines, they will accept deeper cuts into crude oil. A marine diesel oil often consists of a blend of gas oil and some residual material. Gas oil is generally defined as being somewhere in the boiling range of 200° C to 350° C. Because of rising prices, ship owners have recently tended to operate diesel engines on high-viscosity residual fuels.

The refueling of ships with oil is referred to as *bunkering.* Bunkering may be achieved through pipelines, from barges, or by road or railcar tanks. When a ship is too large to come alongside a wharf, submarine or floating lines or refueling barges can be used to refuel the ship's tank (bunker) while it is moored at sea. Ships can also be refueled while underway.

Electrical generation

A large portion of the world's natural gas goes into the production of electricity at generating plants (fig. 8.16). Oil and coal are also fuel sources for generating plants (fig. 8.17). Boilers fueled by gas, oil, and coal generate steam; steam turbines drive generators to produce electricity.

Coal-fired plants must have tall stacks, from 450 to 600 feet, in order to release the hot gases well above ground level (fig. 8.18).

Fig. 8.16. Natural gas enters this electricity-generating plant through pipelines.

Fig. 8.17. Oil is stored in tanks for alternate use as fuel for generating electricity.

Gas-burning plants have shorter stacks (fig. 8.19). Stack emissions are carefully monitored for pollutants both at the site and at various distances within the prevailing windstream.

Fig. 8.18. These 600-foot emission stacks of a coal-fired generating plant help protect the atmosphere.

Fig. 8.19. Shorter stacks are sufficient for gas-burning generating plants such as this one.

Industry   The industrial sector consumes more than half of the world's total output of petroleum products. A large portion goes to manufacturers who require energy to run their machines and processes. An example is the steel industry, whose plants consume thousands of tons of fuel oil annually. Besides fuel oil and natural gas, petroleum supplies industry with a wide range of lubricants and process oils.

Products for industry may be marketed directly in bulk to large consumers, such as the armed forces, railways, and major manufacturers, or indirectly through jobbers to smaller consumers. Sales transactions tend to be long and complex.

Home   The conversion from solid fuel to oil for home heating has been widespread in this century. And, since the 1973 rise in crude prices, the use of gas and electricity has intensified. Natural gas and electricity provide the convenience of being piped or wired into the consumer's property, but oil has to be delivered by truck. Natural gas has become the major competitor to oil for heating. Electricity, a secondary energy source, is usually a more expensive alternative for heating but is used extensively for cooling and for operating household appliances, tools, and instruments (fig. 8.20).

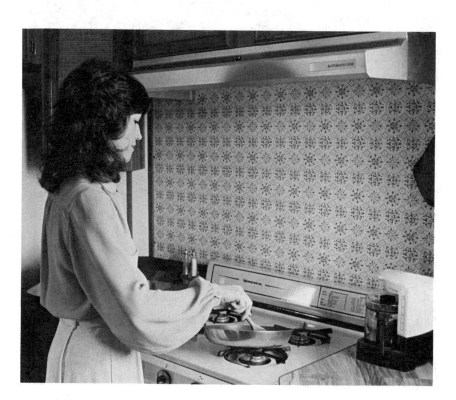

Fig. 8.20. Practically all household appliances are operated by gas or electricity.

Fig. 8.21. An LPG tank furnishes heating fuel for this farm home.

Fig. 8.22. A scrubber cleans gas as it enters an underground network of pipelines at a city gate.

In rural areas and locations not serviced by natural gas pipelines, liquefied petroleum gas (generally propane) is often used for heating. LPG tanks are purchased and maintained by the property owners and filled by local distributors (fig. 8.21). Small, portable propane tanks are often used by owners of recreational vehicles, mobile homes, and remote vacation cabins. These tanks are taken by owners to their local distributors to be refilled. In many rural locations, LPG dealers may keep small filled tanks on hand for sale or rent.

Both natural gas and electricity come under the heading of public utilities. Because gas pipes and electrical wires must be part of a community network, the community contracts as a whole for services (fig. 8.22). Thus, the utility is considered a "public" service, and the two sources of energy are subject to state regulations, which influence their pricing to consumers.

Commerce

Although gas and electricity are used increasingly for residences, heating oil continues to be marketed to heat commercial buildings such as hospitals, schools, and factories. Fuel oils for domestic burners are distilled from the deeper cuts of crude after the lighter products have been taken off. Domestic fuel oils are divided into two grades for marketing—*number 1 oil* and *number 2 oil*—as specified by the American Society for Testing and Materials. Fuel oils are marketed and distributed like gasoline, with the jobber's role being crucial for deliveries and sales to individual customers.

Fig. 8.23. A portable propane
tank furnishes fuel for the
farmer's combine and other
farm implements.

Agriculture

Petroleum products play a vital role in farming today. The internal-combustion engine was largely responsible for transforming the traditional rural farm into a modern industry. Between 1940 and 1983, agricultural productivity increased by 90 percent, a growth attributed to mechanization, fertilizers, pesticides, and irrigation. The development of diesel fuels and LPG for tractors and other mechanized implements, as well as the development of proper lubricants for those machines, revolutionized farming (fig. 8.23).

Other vital petroleum derivatives used for agriculture are the chemicals used for fertilizers. A range of petroleum-based chemicals is also used to protect crops from plant and animal pests. Methods of marketing this wide array of agricultural products vary. Sometimes marketing companies sell directly to the farmer or rancher, sometimes agents or jobbers are used, and sometimes the farmer or rancher buys from a local distributor (fig. 8.24).

Fig. 8.24. Fertilizer is poured into the hopper of a spreader for use on a rancher's grazing land.

ADVERTISING AND PUBLIC RELATIONS

Another aspect of marketing that has evolved over the century is advertising and public relations. During the era of allocations, controls, and short supplies, marketers needed little advertising to sell their products. However, as restraints were removed and competition returned to the petroleum marketplace, promotional activities once again became important to a company's success. These activities may be categorized as institutional advertising, special promotions, or service-oriented public relations.

Leading U. S. oil companies traditionally have used newspaper and magazine ads and sponsored radio and television programs. More recently, they have tried to educate the public on the role of the oil industry in the national and world economies. Philanthropic activities, such as sponsoring research in the humanities

Institutional advertising

and documentary programs on television, bring the company's name before the public without actually advertising its products. A magazine ad may show only a company's logo, just to keep the company name before the public. This type of advertising — known as institutional advertising — establishes a company image in the minds of its viewers. It promotes the company, or institution, rather than its products.

Special promotions

A new company name, a new product, a new retail outlet — these events and others are reasons for special promotional campaigns.

New name. When Esso (originally Humble Oil) changed its name to Exxon, the company conducted an extensive ad campaign to let the public know that Exxon was the same company that "put a tiger in your tank." The company image established as Humble and Esso stayed with Exxon because of its special promotion.

New product. When Shell came out with its new super unleaded gasoline, SU 2000, the company went all-out to tell consumers that its product was a premium one — a cut above its competitors and worth the extra cost. The company decorated its stations, gave away soft drinks, conducted a big advertising campaign on TV and in magazines, held dealer contests, and did all sorts of things to catch the public interest. The advertising campaign stressed *performance*, with TV as its prime medium. The TV campaign centered on sponsoring various sports programs — those watched primarily by men who are most likely to be interested in car performance and maintenance. The radio campaign featured the Shell Answerman, who delivered spot messages with a product plug. The print media campaign included ads in national magazines and periodicals, especially leading automotive magazines. Every retail station displayed signs that advised its customers: "Turn your car into an SU 2000 performer."

New station. Diamond Shamrock had grand openings in 1985 at its newly built and remodeled retail outlets (fig. 8.25). "Store specials, free drinks, and drawings for race cars and sailboats attracted twice the usual number of customers at twenty grand openings. In many instances, normal inside sales were doubled, tripled, and even quadrupled, and substantial increases of gasoline sales were also recorded."[4]

Other, more direct customer appeals have included giving away road maps, dishes, and other premiums at service stations. At one

4. "Grand Openings," *Tattler*, Diamond Shamrock, Spring 1985, p. 14.

Fig. 8.25. Visitors to a new service station are greeted by a "leprechaun" at a St. Patrick's Day grand opening. (*Courtesy of Diamond Shamrock*)

time trading stamps were popular giveaways. The idea behind premiums is to build goodwill and sales volume at the same time. Today, premium merchandising is less prevalent, giving way to service-oriented promotion.

Service-oriented public relations

After the dramatic increase in the cost of crude oil that occurred in 1973, American oil companies began to take measures to cut down on expenses. They closed down unprofitable stations and reduced their marginal services at others. The familiar driveway attendant who checked the water and oil, cleaned the windshield, filled the gas tank, and aired up the tires became practically extinct. Self-service was offered to partly offset the rising price of gasoline

Fig. 8.26. Self-service helps the consumer keep the price of gasoline down.

(fig. 8.26). Price wars, which had been widely advertised in certain areas to draw customers to the lowest priced stations, were no longer valid promotional devices. Credit card accounts were no longer widely solicited. Many service stations supplemented their gasoline business with sales of groceries, snacks, drinks, and ice for the traveler—a service that has come to be expected by the average American on the road today.

In the 1980s, as in the years after World War II, companies are again offering added services to promote their products and enhance their public relations. Mass mailings of credit applications are being made to attract new customers. Many gasoline credit cards can now be used for tires and auto accessories, food and drinks, travelers' insurance, motel accommodations, car rental, and even emergency cash advances. Some companies have long used their credit cards to sell mail-order merchandise of all kinds.

Stations equipped with electronic point-of-sale systems are able to evaluate credit card transactions immediately with sales transmitted directly to the credit card center. Paperwork is reduced, bad credit cards are purged, and companies get their money faster.

While some companies are advocating the use of their credit cards, others are offering cash discounts on their gasoline to attract the motorist who prefers to pay cash, wants to save money, or does not own credit cards. This policy discourages the use of credit cards, thereby saving the company the costs of doing credit business.

One of the oldest services rendered by the service station is its restroom facility, and the condition of the restrooms has long influenced the traveler's choice of service stations. Not only is a poorly kept station bad for business, it is also a large contributor to bad public relations for its parent company.

A service becoming more common in the retail outlet, especially in self-service stations, is the free or "nominal fee" car wash with a fuel fill-up. These automatic, drive-through facilities let busy customers get quick car washes without getting out of their cars.

One company is offering service coupons, which customers can accumulate with purchases and later use toward automotive repair work. The retail outlet offering this service must be capable of mechanical repairs and service. This company is working toward offering a national warranty, which would mean that a customer whose car was previously serviced or repaired at an affiliated station could, in the event of further car trouble, call a toll-free number and get free help from the nearest station or garage associated with the program.

Good public relations comes from customer contact, and that contact is at the retail outlet. To maintain a good overall image, or reputation, dealer-oriented companies conduct marketing seminars for their dealers and jobbers, hold national sales meetings for their sales representatives, and participate in trade shows displaying the latest equipment available for their vendors. The purpose of these activities is to keep up-to-date on industry happenings so that "both dealers and jobbers understand and embrace change and innovation rather than resist them."[5]

In today's petroleum-oriented world economy, countries are defined as producers or consumers. This distinction has helped make the oil industry a complex and fascinating market that vastly influences relationships between countries. Petroleum raw materials and products are among the largest and most influential commodities selling in the world today. An energy source is the basis of human enterprise, and petroleum remains worldwide one of the energy sources most used.

5. James A. Cox, "The Marketers Are Back," *Shell News*, April, 1985, p. 35.

Economics of the Industry

IX

Petroleum is big business. It is the single largest item in the balance of trade between countries. Every day, $2 billion changes hands in petroleum transactions worldwide. Because of its economic clout, and because it is the fuel on which modern industry and the desired standard of living absolutely depend, petroleum is also big politics. Most governments throughout the world have taken nearly complete control of their petroleum resources and industries. Of all oil produced during 1984, three-fourths was done so by government-owned, not private, companies. Regardless of who produces the oil and gas, most governments collect 80 to 90 percent of the net profits made by the petroleum companies of their countries, including the United States, which gets its share through taxes. And the United States is the only country in the world that does not have a government-owned petroleum firm.

Where is this mammoth industry, so closely entwined with national interests and world politics, headed? Any predictions must take into account two realities of the petroleum situation that are likely to overshadow all others in the long term.

First, most experts agree that global oil production will have peaked by the year 2000, if it hasn't peaked already. By the most optimistic estimates, all of the world's remaining oil resources, including those yet to be discovered, will last no more than a hundred years beyond then; some predict less than fifty years. *Reality No. 1: The world will most likely run out of oil within the next century.*

Second, the United States for the most part dominated the petroleum industry from World War I until the mid-1960s. During that time, it controlled most of the world's oil production through development of its own sizable reserves and through favorable concession arrangements between foreign governments and U.S. companies. However, as U.S. reserves began to deteriorate and the Middle East countries banded together and gradually nationalized their massive holdings, the balance of power shifted. *Reality No. 2: The United States is no longer the undisputed leader in the world oil situation.*

The petroleum industry, then, must proceed according to a new set of rules: the resource is running out, and no one player holds all the cards. Neither of these realities happened overnight, but it took the supply disruptions and the price shocks of the 1970s to force them to the attention of the public, and of some industry trackers as well.

Since the mid-1970s, the petroleum industry has been undergoing dramatic changes to meet the challenges of its next century. This discussion examines (1) the key circumstances that led to the current situation, including the emergence of the Middle East as a dominant force in the world oil market; (2) petroleum supply and demand—that is, how much is left, who has it, and how long will it last; and (3) some of the changes now underway or likely to happen in the coming years, such as increased energy conservation and the shift from reliance on petroleum to use of coal and alternative energy sources.

The economics of petroleum are global; that is, understanding trends of oil and gas production and consumption in one country is not possible without taking worldwide patterns into consideration. Hence, the petroleum situation is deliberately viewed here from an international standpoint.

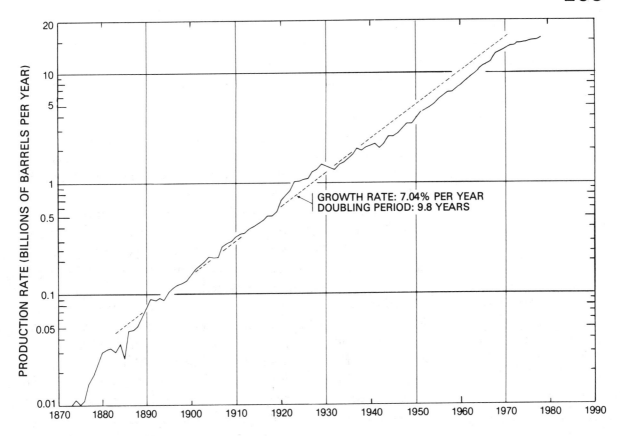

GROWTH RATE: 7.04% PER YEAR
DOUBLING PERIOD: 9.8 YEARS

SOURCE: Adapted from M. K. Hubbert in *U. S. Energy Resources: A Review as of 1972.*

Figure 9.1. Rise in world crude oil production, 1870–1970

THE CHANGING ECONOMIC PICTURE

For most of its history, the petroleum industry experienced tremendous growth. Between 1890 and 1970, crude oil production doubled every ten years (fig. 9.1). Worldwide, more oil was produced in each decade than the total amount that had been produced in all the years before then. Demand by the industrializing world for this cheap, clean, efficient fuel appeared insatiable—to power growing industries, to fire electrical plants, and, particularly in the United States, to fuel automobiles. By the early 1970s, however, it had become clear that the world could, and indeed would, run out of oil. That realization has colored the industry's direction since.

404

Era of U. S. dominance: 1915 to 1945

The United States convincingly dominated the world petroleum market between World War I and World War II. Even at the turn of the century, it was producing more than half of the world's oil, including a quarter of that needed outside of the United States. The United States remained the largest petroleum producer in the world and a net exporter into the 1950s.

The first major U. S. discovery was Spindletop in 1901, which alone increased world oil production by more than 20 percent. Producing 84,000 barrels of oil per day, Spindletop launched the petroleum age in Texas. During the 1920s, industry watchers began to warn that the United States might run out of oil. Then in 1930, just as talk of shortages had become serious, the mammoth East Texas field was discovered (fig. 9.2). It contained 6 billion barrels of oil—more than that found in the next twenty largest fields in the United States combined.

During this era, seven huge petroleum companies—called the *majors*—controlled almost all development, production, refining, transport, and marketing of the international industry. The original seven majors were Exxon, Chevron, Mobil Oil, Gulf Oil, Texaco, Royal Dutch/Shell, and British Petroleum. Chevron and Gulf merged in 1984. Five of the original majors were based in the United States. Three of these emerged from the breakup of Standard Oil in 1911. A sixth major—Royal Dutch/Shell—has a large percentage of its holdings in the United States.

Figure 9.2. C. M. "Dad" Joiner, third from left, is congratulated by geologist A. D. Lloyd in front of the Daisy Bradford No. 3 discovery well in the East Texas field. Discovery of the 6-billion-barrel field in 1930 quieted early warnings that the United States might run out of oil. (*Courtesy of Texas Mid-Continent Oil & Gas Association and American Petroleum Institute*)

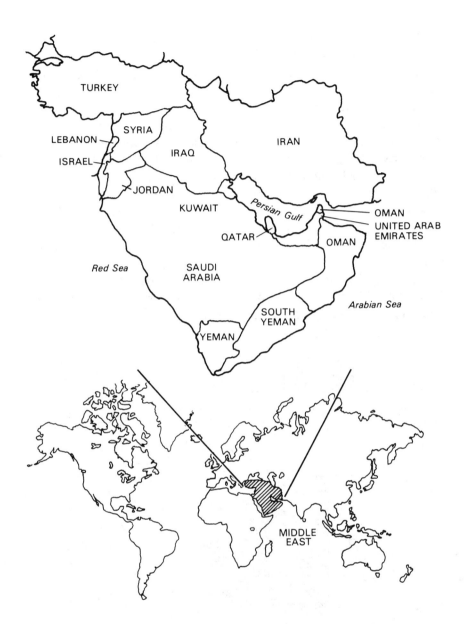

Figure 9.3. The countries surrounding the Persian Gulf are commonly referred to as the Middle East.

After World War II, worldwide demand for energy, oil in particular, rose dramatically. Consumption was greatly accelerated in Western Europe and Japan, for example, as they strove to rebuild war-shattered economies. As it became clear that the United States would no longer be able to meet its own needs for oil, much less export to other countries, the non-Communist world logically turned to the known massive oilfields and expected great potential of the countries surrounding the Persian Gulf, which are collectively called the Middle East (fig. 9.3).

**Turn to the
Middle East:
1945 to 1970**

Oil companies scrambled to explore unclaimed Middle East territory and to produce fields that had already been found. Joining the majors in the search were many smaller U. S. and European companies—now called the *internationals*. Examples of U. S. internationals are Amoco, Conoco, ARCO, Sun Oil, Phillips Petroleum, and Marathon Oil. European internationals include Compagnie Française des Pétroles of France, the Belgian Petrofina Company, and Ente Nazionale Idrocarburi of Italy.

The rate of Middle East oil discovery peaked between 1945 and 1950, owing particularly to new finds in Saudi Arabia (fig. 9.4). The 150-mile-long Ghawar field, discovered in Saudi Arabia in 1948, contained 60 billion barrels of oil—ten times as much oil as was found in the East Texas field. Middle East production also rose sharply during the postwar years, jumping from a half million barrels per day in 1945 to more than a million barrels in 1948. By 1952, the Middle East was producing 2 million barrels per day, accounting for one-sixth of world production.

Figure 9.4. Rates of discovery of world oil reserves, inside and outside the Middle East, 1930–1983

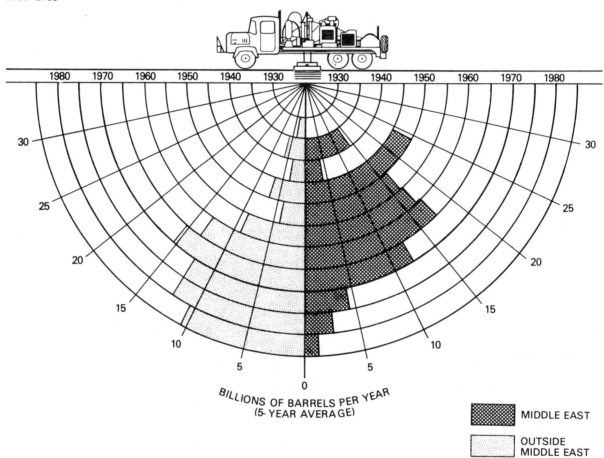

SOURCE: Adapted from Exxon Background Series, *Middle East Oil and Gas*, 1984, p. 8.

Figure 9.5. The supertanker *Cactus Queen* engages in single-point mooring at the Fateh offshore loading terminal at Dubai in the Arabian Gulf. Supertankers helped to open the world market to Middle Eastern oil in the 1950s. (*Courtesy of Dubai Petroleum Company and American Petroleum Institute*)

Several other developments contributed to the increased interest in Middle East oil after the war. One was that by the early 1950s, the price of oil had begun to rise, after a prewar low of less than $1 a barrel, making Middle East production more economically feasible. Technological improvements had also begun to lower the cost of exploration and production in the Middle East, where the harsh climate and primitive living conditions had previously discouraged investment.

The introduction of the supertanker in the mid-1950s also encouraged Middle East production (fig. 9.5). Producing companies now had an economical way to get oil from the Persian Gulf to the industrial centers in Western Europe, the United States, Japan, and other high-demand areas. That breakthrough allowed Middle East oil to compete with U. S. and other oil on the world market.

As production rose and the world became more reliant on Middle East oil, the relationships between the governments of the producing countries and the oil companies operating within their borders — mainly the majors and the U. S. and European internationals — began to change. These changes were the first signal that a shift in power was underway.

Under concession agreements set up in the 1920s and 1930s between the oil companies and the governments of the producing countries, the companies paid the governments rent for use of the land and a set amount for each barrel of oil produced. The governments were paid this flat fee, regardless of the oil's posted price (the price at which the oil companies officially agreed to sell the oil).

However, as the oil companies began to raise posted prices and as the costs of production dropped, yielding higher profits for the oil companies, the host countries began to levy income taxes. By the early 1950s, the governments of most Middle East producing countries shared profits 50/50 with the oil companies. Because of the income tax, the producing countries now had a stake in what the oil sold for: if the companies lowered their posted prices, and therefore their profits per barrel, the governments' "take" also dropped.

That's just what happened in the late 1950s, when overproduction led to a surplus of oil, and the companies, in response to market pressures, lowered their prices. Discontent over the decrease in revenues per barrel was a major force leading the governments of five countries to form the Organization of Petroleum Exporting Countries (OPEC) in 1960. Four of OPEC's original members were in the Middle East: Saudi Arabia, Iran, Iraq, and Kuwait. The fifth was Venezuela.

During its first ten years, OPEC had fairly little impact on the world petroleum industry. Even when the Middle East surpassed the United States as the world's largest oil producer in 1965, OPEC made no drastic moves. The international oil companies still controlled how much OPEC oil was produced and, for the most part, what it sold for. OPEC's main goal at that time was to keep its members' revenue per barrel from falling. It did this by pressuring the oil companies to keep their posted prices stable and by increasing the taxes on their profits.

OPEC asserts itself: 1970 to present

By the early 1970s, the world supply and demand situation had placed OPEC in a more prominent role. Rising world consumption had greatly increased the demand for OPEC oil. In Western Europe, for example, oil made up only 10 percent of all energy consumed in 1950; by 1970, that percentage had jumped to 55 percent. In Japan, oil's share of the energy pie rose from 9 percent to 33 percent over the same time period. Even in the United States, oil use as a percentage of all energy consumed increased from 40 percent in 1950 to 44 percent in 1970. More significantly, one-third of the increased demand in the United States was met by imported oil.

OPEC had grown by the early 1970s to thirteen members, six of which were in the Middle East. (Those six members continue to dominate OPEC, accounting for 60 to 70 percent of its total production.) Joining the original five were Qatar, Indonesia, Libya, Abu Dhabi (later part of the United Arab Emirates), Algeria, Nigeria, Ecuador, and Gabon. With the weight of their expanded ranks, the OPEC countries began to demand a larger share of the profits and a greater say in what their oil would sell for.

The tight supply situation gave OPEC a valuable new tool to use in seeing that those demands were met—namely, the ability to restrict, or even halt, production that the rest of the world, United States included, could not replace. Although the United States in 1968 had made another great discovery—the 10-billion-barrel Prudhoe Bay field in Alaska—it would be ten years before transport of Alaskan oil to the lower forty-eight would be feasible.

In 1970, Libya was the first OPEC country to impose production restrictions and to demand a higher tax rate and higher posted prices from the international oil companies. Under the threat that Libya would suspend all production, the companies agreed to the new terms. Over the next two years, the remaining OPEC countries followed Libya's lead, successfully negotiating new contracts with the oil companies under the threat that supply would be halted to any companies that did not meet their demands.

Buoyed by its early successes at raising prices and profits, OPEC in 1973 took complete control over setting posted prices. In October of that year, it unilaterally raised the posted price of its crude from $3.01 per barrel to $5.12. The selling price, or the price at which OPEC would sell crude to the companies, was raised to $3.65. By December, the posted price, that is, the price that the oil companies charged their customers, had shot to $11.28 per barrel.

A second major development of 1973: OPEC for the first time followed through on its threat to cut off supply. In the fall of that year, the Arab members of OPEC (referred to as OAPEC, the Organization of Arab Petroleum Exporting Countries) embargoed all of its oil shipments to the United States and the Netherlands in retaliation for their support of Israel in the Arab-Israeli War (fig. 9.6).

Figure 9.6. U. S. consumers endured long lines at the gas stations during the Arab oil embargo of 1973 and the shortage scare of 1979–1980. (Courtesy of Standard Oil Company of California and American Petroleum Institute)

As OPEC was gaining control of pricing and supply in the early 1970s, its members were also beginning to demand greater "participation" in, meaning ownership of, the oil industries in their countries. (Iran, the exception, had nationalized its oil industry in 1951.)

Agreements made in the 1970s between OPEC members and the international oil companies operating in their countries allowed for a gradual increase in government ownership, with some compensation to the companies for unrecovered investments. By 1980, more than 90 percent of Middle East oil was produced by government-owned national oil companies (fig. 9.7).

Figure 9.7. International oil companies' equity interest in Middle East crude oil production, 1965–1982

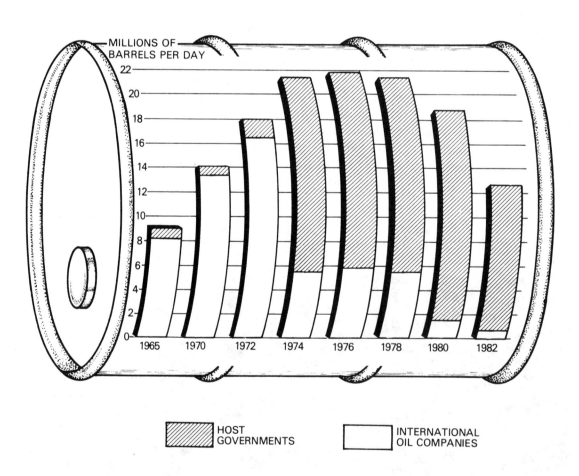

HOST GOVERNMENTS

INTERNATIONAL OIL COMPANIES

SOURCES: *Annual Statistical Bulletin*, Organization of Petroleum Exporting Countries; Adapted from Exxon Background Series, *Middle East Oil and Gas*, 1984, p. 35.

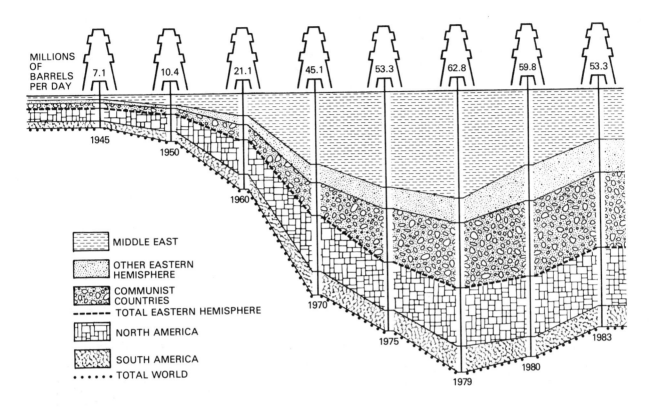

MILLIONS OF BARRELS PER DAY

7.1   10.4   21.1   45.1   53.3   62.8   59.8   53.3

1945   1950   1960   1970   1975   1979   1980   1983

MIDDLE EAST
OTHER EASTERN HEMISPHERE
COMMUNIST COUNTRIES
------ TOTAL EASTERN HEMISPHERE
NORTH AMERICA
SOUTH AMERICA
• • • • • TOTAL WORLD

SOURCES: *Oil & Gas Journal,* 1984; Adapted from Exxon Background Series, *Middle East Oil and Gas,* 1984, p. 32.

Figure 9.8. World crude oil production by region, 1945–1983

The once all-powerful international oil companies were, then, essentially reduced to buying and reselling oil, having lost ownership of production. They did, however, continue to staff and advise the government-owned companies. Many producing countries later began to market the oil themselves, bypassing the majors and internationals altogether. By the early 1980s, the U. S. majors' share of OPEC exports had dropped to 45 percent, from 90 percent in the early 1970s.

After the escalations of 1973–1974, oil prices rose steadily but not drastically for several years. Middle East production even declined slightly in 1975 in response to lowered demand brought on by worldwide recession, conservation, and a partial shift from oil to other energy sources. By the late 1970s, however, Middle East production had risen to an all-time high of 22.5 million barrels a day (fig. 9.8).

A second major "price shock" came in late 1979 and 1980, when OPEC increased selling prices from $18 to $32 per barrel (fig. 9.9). (Posted prices, by then, were no longer significant. Because the OPEC members had acquired nearly complete ownership of their countries' production, government selling prices became the price

Figure 9.9. OPEC crude oil selling prices, 1970–1983

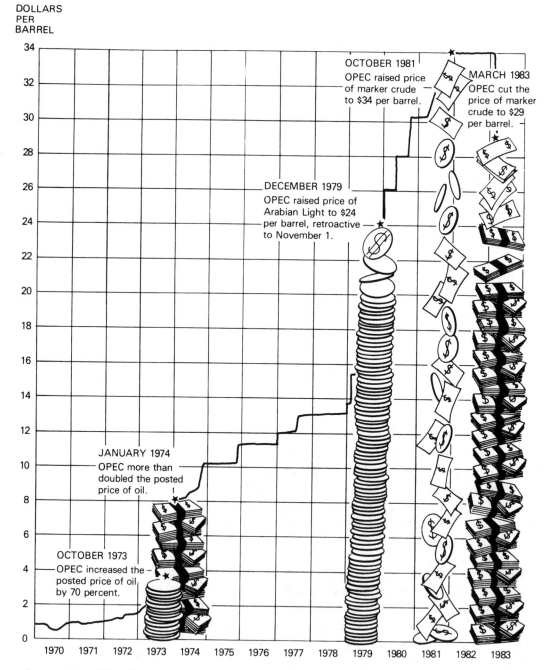

DOLLARS PER BARREL

OCTOBER 1981
OPEC raised price of marker crude to $34 per barrel.

MARCH 1983
OPEC cut the price of marker crude to $29 per barrel.

DECEMBER 1979
OPEC raised price of Arabian Light to $24 per barrel, retroactive to November 1.

JANUARY 1974
OPEC more than doubled the posted price of oil.

OCTOBER 1973
OPEC increased the posted price of oil by 70 percent.

SOURCE: Adapted from Exxon Background Series, *Middle East Oil and Gas*, 1984, p. 28.

at which crude was sold to companies or third parties.) The escalations of 1979–80 resulted in part from the fear of shortages brought about by the halt of all production in Iran, owing to the revolution and war in that country. Saudi Arabia also increased prices in an attempt to unify OPEC's pricing structure, which had begun to break down.

Worldwide response to the 1979-1980 price escalations was dramatic, setting into motion a series of events that would challenge OPEC's ability to retain control of supply and demand. Reduced economic growth spurred by a recession even more severe than that of 1974, stronger conservation measures, and an accelerated shift to alternative energy sources greatly depressed world demand for oil. And demand for OPEC oil in particular was depressed even further, as importing countries began to search in earnest for non-OPEC oil supplies in areas such as Mexico and the North Sea.

That search was encouraged in the United States in 1980, when President Jimmy Carter began the phased decontrol of oil prices, making new exploration more profitable for U. S. companies. OPEC's boost in prices, along with improved technology and completion of the Trans-Alaskan Pipeline, had also made it economical to begin production of the huge Prudhoe Bay field in Alaska. One measure of the success of U. S. conservation and price decontrol was that by 1982, U. S. imports of crude oil and petroleum products had dropped to about 4 million barrels per day, from more than 7 million per day in 1979 (fig. 9.10).

Figure 9.10. Changes in the oil imports of the United States, Europe, and Japan, 1970–1982

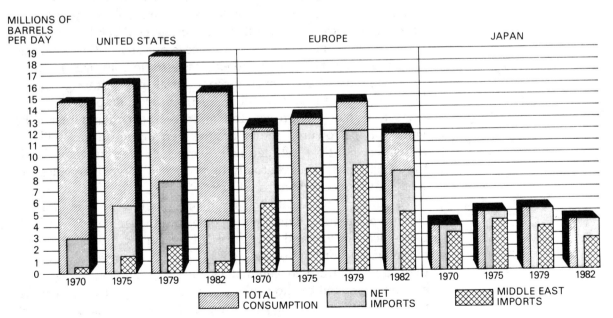

SOURCE: Adapted from Exxon Background Series, *Middle East Oil and Gas,* 1984, p. 7.

Faced with decreased demand for its oil, OPEC began to lower the quotas that its members were allowed to produce. But some members continued to produce beyond their official limits and also began to discount prices. In 1983, Middle East production was down to 12 million barrels a day, and OPEC was forced to make its first official price cut, from $34 to $29 per barrel.

By the mid-1980s, with additional price cuts and falling demand, OPEC's united front appeared to be eroding further. OPEC had become unable to keep its members within their production limits and therefore unable to prevent further price drops. From the discussion of current supply and demand that follows, it is clear that dwindling oil reserves and increased demand will lead to a tighter supply situation within the next decade. And that situation will undoubtedly put the Middle East producing countries into a more favorable bargaining position. But whether OPEC can ever regain dominance of the world petroleum situation remains an open question.

SUPPLY AND DEMAND

Estimating how much oil and gas remain to be produced worldwide is no simple proposition. First, a large part of the petroleum that is left has not yet been discovered. And even the petroleum that has been found but not yet produced is difficult to measure. Oil and gas are trapped underground in porous rock, similar to the way water is held in a sponge. Only about one-fourth to one-third of the oil in an average field can be recovered using conventional methods. To squeeze out any of the remaining three-fourths, sophisticated and expensive techniques must be used. Recovery rates for conventional natural gas, although much higher than those for oil, still range from 80 to 95 percent.

As technology improves and as increased demand raises the value of petroleum, more of the remaining target will be recovered. For now, estimating how much of the petroleum that is thought to exist will eventually be produced depends on guessing what future recovery rates will be.

To further complicate the issue, many different classification systems are used to describe the petroleum that remains. The U.S. Geological Survey and Bureau of Mines, for example, categorize oil and gas according to degree of geologic assurance (how certain they are that the petroleum does, in fact, exist) and degree of economic feasibility (how likely it is that the petroleum can be recovered at a profit using existing technology and under current economic conditions).

The two categories of classifying petroleum that are most commonly used in the industry are ultimately recoverable resources and proved reserves. *Ultimately recoverable resources* are all of the petroleum that remains and is expected to be producible, including what has not yet been discovered. *Proved reserves*, often called simply reserves, are that portion of ultimately recoverable resources that has been discovered and is considered commercially producible under current economic conditions. Estimates of proved reserves are adjusted constantly as new discoveries are made, as existing reserves are produced, and as the prices of oil and gas fluctuate (making production of previously uneconomic resources feasible or production of once-economic resources no longer profitable).

Because of the difficulties inherent in trying to calculate petroleum resources and reserves, estimates of both vary greatly. Different countries and different petroleum companies have come up with very different numbers. This variation is especially true of resource estimates, which rely heavily on future technology and economic conditions.

Estimates of ultimately recoverable oil resources given by twenty-nine experts at the World Energy Conference of 1978 ranged from 1.2 to 3 trillion barrels (those estimates included the 1 trillion barrels that had already been discovered). At predicted rates of consumption (also subject to great change), the high estimate would last a bit more than a hundred years, the low estimate less than half that.

In its 1982 *World Energy Outlook*, the International Energy Agency of the Organization for Economic Cooperation and Development (OECD) presented two scenarios for the lifespan of conventional natural gas resources (conventional resources exclude gas produced from sources such as tight sands, Devonian shale, or coal). These scenarios estimate that remaining conventional gas resources total 5,900 to 9,200 trillion cubic feet. According to predicted rates of consumption, the high estimate would last around one hundred years, the low estimate about sixty-five years.

Because proved reserves have been discovered and evaluated, they can be estimated with greater precision than can resources. Estimates of proved world oil reserves for selected years are given in table 9.1, proved gas reserves in table 9.2.

Figure 9.11 shows the distribution of proved oil reserves among the principal producing countries of the world as of January 1985. OPEC, its Middle East members in particular, clearly dominates, underscoring its continued economic importance in the world petroleum situation. The Soviet Union leads in proved reserves of natural gas, with more than 40 percent of the world total (fig. 9.12). As oil reserves are depleted and natural gas becomes a

TABLE 9.1
ESTIMATED PROVED WORLD RESERVES OF CRUDE OIL
(Thousands of barrels)

Year	United States	Canada	Latin America	Middle East	Africa	Asia	Western Europe	Communist Nations	Total World
1966	31,352,391	6,711,237	25,170,500	215,360,000	23,049,000	11,008,750	2,085,000	33,377,000	348,118,878
1967	31,452,127	7,791,751	27,100,300	235,614,600	32,355,500	11,808,472	1,924,500	33,838,000	381,885,250
1968	31,376,670	8,168,924	26,907,250	249,209,000	42,285,250	11,815,945	2,017,000	35,773,000	407,553,039
1969	30,707,117	8,381,613	28,782,775	270,760,000	44,568,800	13,720,200	1,937,000	55,877,000	407,553,039
1970	29,631,862	8,619,805	29,179,750	333,506,000	54,679,500	13,138,150	1,779,000	60,000,000	530,534,067
1971	39,001,335[1]	8,558,980	26,185,250	344,574,900	74,757,520	14,408,648	3,708,500	100,000,000	611,195,133
1972	38,062,957[1]	8,333,087	31,558,750	367,386,000	58,886,200	15,604,500	14,222,000	98,500,000	632,553,494
1973	36,339,408[1]	8,020,141	32,601,750	355,852,000	106,402,000	14,922,260	12,082,000	98,000,000	664,219,559
1974	35,299,839[1]	7,674,150	31,640,250	350,162,500	67,303,750	15,635,040	15,990,950	103,000,000	626,706,479
1975	34,249,956[1]	7,171,229	40,578,000	403,858,200	68,299,450	21,047,700	25,814,000	111,400,000	712,418,535
1976	32,682,127[1]	6,653,002	35,368,000	368,410,570	65,085,220	21,234,230	25,487,700	103,000,000	657,920,849
1977	30,942,166	6,247,082	29,608,950	367,681,220	60,570,200	19,391,140	24,538,810	101,100,000	640,089,568
1978	29,486,402	5,970,872	40,270,000	366,166,000	59,200,150	19,749,270	26,862,500	98,000,000	645,805,194
1979	27,803,760	6,860,000	41,246,500	369,996,000	57,892,125	20,007,200	23,966,000	94,000,000	641,771,585
1980	27,051,289	6,800,000	56,472,500	361,947,300	57,072,100	19,355,200	23,476,400	90,000,000	642,174,789
1981	29,805,000	6,400,000	69,489,837	362,071,000	55,148,375	19,630,500	23,085,000	86,300,000	651,929,712
1982	29,426,000	7,300,000	84,982,270	362,839,950	56,171,630	19,150,800	24,634,500	85,845,000	670,350,150
1983	27,858,000	7,020,000	78,482,066	369,285,893	57,821,690	19,756,077	22,923,680	85,115,000	668,262,406
1984	27,735,000	6,730,000	81,675,900	370,100,800	56,907,020	18,969,400	23,019,480	84,600,000	669,737,600

(1) Figures include 9.6 billion barrels in Prudhoe Bay, Permo-Triassic Reservoir, Alaska (discovered in 1968) not yet available for production due to lack of transportation facilities.

SOURCE: 1948–1980: United States—American Petroleum Institute, Committee on Reserves and Productive Capacity
1981–1984: United States—U.S. Energy Information Administration
Canada—Canadian Petroleum Association (1952–1980)
Rest of the World—*Oil and Gas Journal*, "Worldwide Oil" Issues

NOTE: Reprinted from *Basic Petroleum Data Book* with permission from the American Petroleum Institute

TABLE 9.2
ESTIMATED PROVED WORLD RESERVES OF NATURAL GAS
(Billions of cubic feet)

Year	United States	Canada	Latin America	Middle East	Africa	Asia	Western Europe	Communist Nations	Total World
1966	31,352,391	6,711,237	25,170,500	215,360,000	23,049,000	11,008,750	2,085,000	33,377,000	348,118,878
1967	31,452,127	7,791,751	27,100,300	235,614,600	32,355,500	11,808,472	1,924,500	33,838,000	381,885,250
1968	31,376,670	8,168,924	26,907,250	249,209,000	42,285,250	11,815,945	2,017,000	35,773,000	407,553,039
1969	30,707,117	8,381,613	28,782,775	270,760,000	44,568,800	13,720,200	1,937,000	55,877,000	454,734,505
1970	29,631,862	8,619,805	29,179,750	333,506,000	54,679,500	13,138,150	1,779,000	60,000,000	530,534,067
1971	39,001,335[1]	8,558,980	26,185,250	344,574,900	74,757,520	14,408,648	3,708,500	100,000,000	611,195,133
1972	38,062,957[1]	8,333,087	31,558,750	367,386,000	58,886,200	15,604,500	14,222,000	98,500,000	632,553,494
1973	36,339,408[1]	8,020,141	32,601,750	355,852,000	106,402,000	14,922,260	12,082,000	98,000,000	664,219,559
1974	35,299,839[1]	7,674,150	31,640,250	350,162,500	67,303,750	15,635,040	15,990,950	103,000,000	626,706,479
1975	34,249,956[1]	7,171,229	40,578,000	403,858,200	68,299,450	21,047,700	25,814,000	111,400,000	712,418,535
1976	32,682,127[1]	6,653,002	35,368,000	368,410,570	65,085,220	21,234,230	25,487,700	103,000,000	657,920,849
1977	30,942,166	6,247,082	29,608,950	367,681,220	60,570,200	19,391,140	24,538,810	101,100,000	640,089,568
1978	29,486,402	5,970,872	40,270,000	366,166,000	59,200,150	19,749,270	26,862,500	98,000,000	645,805,194
1979	27,803,760	6,860,000	41,246,500	369,996,000	57,892,125	20,007,200	23,966,000	94,000,000	641,771,585
1980	27,051,289	6,800,000	56,472,500	361,947,300	57,072,100	19,355,200	23,476,400	90,000,000	642,174,789
1981	29,805,000	6,400,000	69,489,837	362,071,000	55,148,375	19,630,500	23,085,000	86,300,000	651,929,712
1982	29,426,000	7,300,000	84,982,270	362,839,950	56,171,630	19,150,800	24,634,500	85,845,000	670,350,150
1983	27,858,000	7,020,000	78,482,066	369,285,893	57,821,690	19,756,077	22,923,680	85,115,000	668,262,406
1984	27,735,000	6,730,000	81,675,900	370,100,800	56,907,020	18,969,400	23,019,480	84,600,000	669,737,600

(r) Revised

(1) Figures include 26 trillion cubic feet in Prudhoe Bay, Alaska (discovered in 1968) for which transportation facilities are not yet available.

SOURCE: 1967–1980: United States—American Gas Association, Committee on Natural Gas Reserves
1981–1984: United States—Department of Energy
Rest of the world—*Oil and Gas Journal*, "Worldwide Report" Issues

NOTE: Reprinted from *Basic Petroleum Data Book* with permission from the American Petroleum Institute

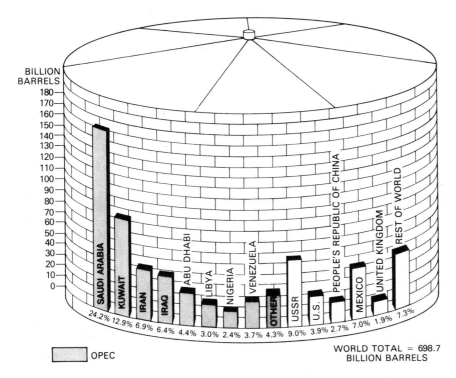

Figure 9.11. Estimated proved world reserves of oil as of January 1, 1985

SOURCE: Adapted from *International Petroleum Encyclopedia*, PennWell Publishing Co., 1985.

Figure 9.12. Estimated proved world reserves of natural gas as of January 1, 1985

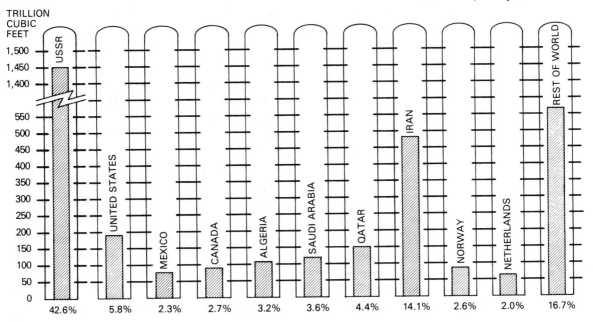

WORLD TOTAL = 3,402 Tcf

SOURCE: Adapted from *International Petroleum Encyclopedia*, PennWell Publishing Co., 1985.

more valued commodity, the Soviet Union will probably play a more important role in the world market.

A scenario of supply and demand: 1984 to 2000

Just as it is difficult to estimate how much petroleum remains, it is hard to predict in what manner it will be used up. One study released recently by Conoco, a subsidiary of the DuPont Company, projects non-Communist world supply and demand for all energy sources, including petroleum, from 1984 to 2000. Its projections rely on certain basic assumptions about the primary influences that will affect the energy situation during those years. If any of these assumptions proves wrong, the situation could change drastically. With that caution in mind, this scenario is more a model for understanding the many factors involved in supply and demand than a precise forecast.

The assumptions on which the study is based include (1) that oil prices will continue to decline during the 1980s but will begin to increase faster than inflation by the 1990s, (2) that OPEC will remain intact and that there will be no major disruptions of Middle East oil supplies, (3) that economic growth of the non-Communist world will average about 3 percent a year during the rest of the 1980s and then slow somewhat during the 1990s, (4) that natural gas will remain competitive with oil product prices, and (5) that coal will retain its price advantage over other fuels for electricity generation but will lose its advantage for other uses.

The Conoco study predicts that non-Communist world demand for all types of energy will grow roughly 2 percent a year to 1990 and then 1.5 percent a year to 2000 (fig. 9.13). Demand for oil will grow at a slower rate than that for total energy sources—about 1 percent a year to 2000—as coal, nuclear, and other alternative energy sources provide an increasing share of the world supply, particularly in developed countries (fig. 9.14).

The developing countries of the Third World (which the study defines as those non-Communist countries that are not members of OECD) will account for nearly half of the predicted *additional*, or incremental, demand for all energy, fully two-thirds of the increased demand for oil. As the Third World continues to develop, it will need a growing energy supply, oil in particular, to produce electricity and to power the energy-consuming cars and appliances that more of its population will be able to afford. Because of large external debts, the Third World will be less able than the developed world to take advantage of oil substitutes such as nuclear power and coal, which have high start-up costs.

Conoco forecasts that despite gains made by alternative energy sources, oil and gas will still supply close to three-fifths of the world's total energy by the year 2000. Of the *additional* energy

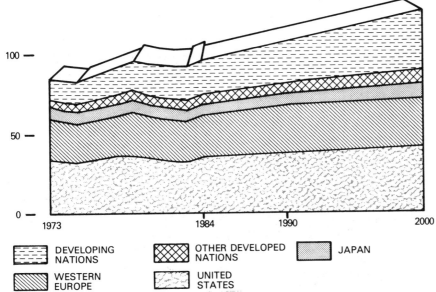

MILLION BARRELS
PER DAY OIL
EQUIVALENT

Figure 9.13. Actual and
predicted world energy
demand by region,
1973–2000 (*Courtesy of
Conoco*)

Source: Conoco, *World Energy Outlook Through 2000*, 1985, p. 3.

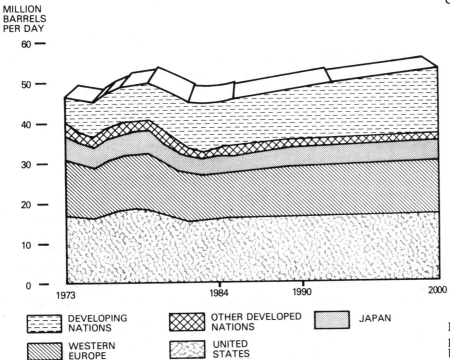

MILLION
BARRELS
PER DAY

Figure 9.14. Actual and
predicted world oil demand
by region, 1973–2000
(*Courtesy of Conoco*)

Source: Conoco, *World Energy Outlook Through 2000*, 1985, p. 5.

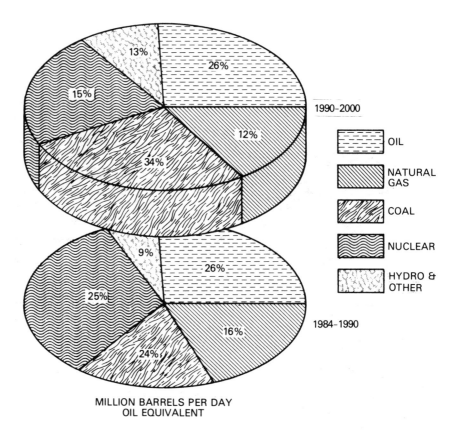

Figure 9.15. Predicted shares of incremental world energy supply, 1984–1990 and 1990–2000 (*Courtesy of Conoco*)

MILLION BARRELS PER DAY
OIL EQUIVALENT

Source: Conoco, *World Energy Outlook Through 2000*, 1985, p. 4.

needed to the year 2000, beyond consumption levels of 1984, oil will consistently supply one-fourth; however, the share supplied by natural gas will decline from 16 percent through 1990 to 12 percent by the year 2000 (fig. 9.15). Natural gas production had increased steadily from World War II through the mid-1970s. This long-term drop is predicted to occur because a large amount of gas is used for purposes that can also be served by substitute fuels, such as coal or nuclear power for electricity generation.

At the rate of increase projected by the Conoco study, oil demand will rise from 46 million barrels per day in 1984 to 49 million in 1990 and 54 million in the year 2000. This increased demand will almost certainly need to be met by OPEC, according to the study. OPEC production of crude oil and natural gas liquids is expected to rise from 19 million barrels per day in 1984 to 28 million by the year 2000, accounting for nearly all of the increased demand anticipated. Non-OPEC oil production, in contrast, is expected to peak by the late 1980s (U.S. production peaked in 1970 at 9.6 million barrels per day and isn't expected to reach that level again). If the study's predictions hold true, OPEC will supply more than half of the non-Communist world's oil by the year 2000 (fig. 9.16).

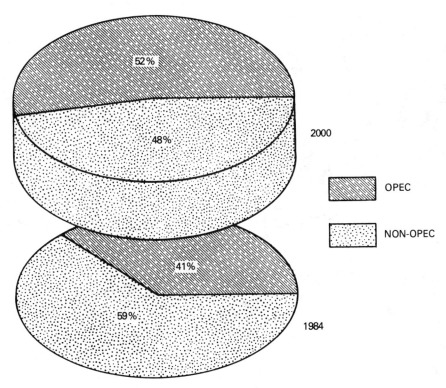

2000

OPEC

NON-OPEC

1984

Figure 9.16. Present and predicted sources of world oil supply, 1984 and 2000 (*Courtesy of Conoco*)

SOURCE: Conoco, *World Energy Outlook Through 2000*, 1985, p. 2.

FUTURE TRENDS

Many changes are now underway to prepare for the inevitable cutback in world oil and gas supplies. Within the petroleum industry, the trend is toward exploration for and development of more difficult targets and improved recovery techniques to produce more of known resources. Response by governments and the general population is expected to focus on increased conservation of the petroleum that is left and switching to alternative energy sources.

Disruptions of oil shipments from the Persian Gulf in 1973–74 and 1979–80 spurred the United States, Western Europe, and other regions dependent on OPEC imports to search for more secure supplies. The rapid price increases during the 1970s and early 1980s also made exploration for and development of resources in more difficult environments, particularly offshore, economically feasible. Although falling oil prices in the mid-1980s slowed some of these exploration efforts, they are likely to pick up considerably by the year 2000 as world dependence on OPEC and the price of

Drilling hot spots for the 1990s

Figure 9.17. World drilling
hot spots through the 1990s
are expected to include
Alaska, the North Sea, the
Pacific Basin, Mexico, and
South America.

oil again increase. Expected hot spots for worldwide exploration
and production through the 1990s include Alaska, the North Sea,
the Pacific Basin, Mexico, and South America (fig 9.17).

Alaska. In 1984, the National Petroleum Council estimated that
Alaska's total oil resource base is 82 billion barrels — that accounts
for half of all the oil yet to be discovered in the United States.
Much of that resource is thought to lie offshore (fig 9.18). Unlike
that of the lower forty-eight states, Alaska's oil and gas potential
has barely been tapped; little exploration has been done outside of
Cook Inlet and the Prudhoe Bay area. Frigid, hostile drilling conditions pose serious technological challenges to Alaskan exploration, but the appeal of a secure domestic supply makes it an almost
certain target for future U. S. activity.

North Sea. Since the late 1970s, the North Sea has been the most
prolific center of oil production in Western Europe. Producing

Figure 9.18. Ice dwarfs workers on an artificial gravel island off the tip of Cross Island in the Beaufort Sea north of Alaska.

nearly 3 million barrels of oil a day during 1983, it followed only the Middle East and Latin America in total offshore production. Between 1985 and 1990, more than $9 billion is expected to be spent developing fields already found in the area. Exploration is also likely to continue to increase in the coming decade, particularly in the United Kingdom and Norwegian sectors of the region.

Pacific Basin. China, Japan, the countries of Southeast Asia, Australia, New Zealand, and the Pacific island states—collectively referred to as the Pacific Basin—are expected to emerge as a major center of petroleum exploration and production by the end of this century. With proved oil reserves of 18.5 billion barrels in 1985, the People's Republic of China has the greatest potential of the region (fig. 9.19). The Chinese government plans to double the country's crude oil production from 2 million barrels per day in 1984 to 4 million barrels by the year 2000. Much of that added production may come from the South China Sea, although exploration had proved disappointing by the early 1980s. The Tianjin area of northern China, the Taklimakan Desert, and the Sichuan Province show promise for onshore production. China will most likely export a large part of the additional oil it will produce to buy the technology it needs for its modernization program.

Figure 9.19. This fixed drilling rig platform in China's Bohai Bay is designed for depths of 13,000 to 20,000 feet and can handle up to twelve producing wells. (*Courtesy of National Supply Company and American Petroleum Institute*)

Figure 9.20. Seismograph workers hack their way through dense undergrowth in the jungles of Indonesia. (*Courtesy of Conoco and American Petroleum Institute*)

Indonesia, the only Asian member of OPEC, is expected to continue to be a major exporter through this century (fig. 9.20). Its oil reserves in 1985 totaled 8.6 billion barrels, and only about half of the country's prospective sedimentary basins had been explored. Australia, another important resource base in the region, had oil reserves of more than 1.4 billion barrels in 1985. The Australian government has estimated that an additional 1.9 billion barrels will be found on the continent.

Mexico. With crude oil reserves of more than 48 billion barrels in 1985 and exports of 1.5 million barrels a day during 1984, Mexico will remain among the world's top producers and exporters during the next decade. Mexico accounted for 18 percent of the United States' total oil imports during the first quarter of 1985, more than any other country, including OPEC members. Although offshore exploration was down in Mexico during 1984 because of the drop in oil prices and the country's huge foreign debt, it was expected to recover quickly. Future offshore targets of Petroleos

426

Mexicanos (Pemex), the country's nationalized oil company, include the Bay of Campeche, the Sea of Cortez, and the Gulf of Mexico off Veracruz.

South America. Although enormous foreign debts had begun to discourage exploration and production in most of South America by the mid-1980s, activity will probably pick up by the 1990s. Brazil shows the most promise, particularly in production from the Campos Basin and other offshore targets (fig. 9.21). Colombia also is expected to do very well during the coming years. In 1985, Occidental Petroleum confirmed the discovery of a 1-billion-barrel oilfield in the Llanos Basin. That find was expected to more than double Colombia's oil production by mid-1986.

Figure 9.21. A jackup rig drills in Brazilian waters. (*Courtesy of Gulf Oil Corporation and American Petroleum Institute*)

Figure 9.22. Engineer-in-charge at the site of an improved recovery project in West Texas. The workover rig in the background is preparing wells for carbon dioxide injection. (*Courtesy of Conoco and American Petroleum Institute*)

Improved recovery methods

As the petroleum industry searches for new sources of oil and gas during the remainder of this century, it will also attempt to recover a higher percentage of the resource that lies in the world's known fields. Improved recovery is a catch-all term to describe the many methods being tested and used to increase the amount of petroleum that can be produced from a reservoir. Using heat, carbon dioxide, chemicals, and other treatments, improved recovery techniques can more than double the yield from a typical reservoir, from about 30 percent to more than 50 percent of the original oil in place (fig. 9.22).

By the early 1980s, improved recovery techniques were producing an additional 1 million barrels of oil per day in non-Communist countries. By the year 2000, the total is predicted to rise to 5 million additional barrels of oil per day. One drawback of improved recovery is cost. However, as oil prices begin to rise again, improved recovery methods will account for a larger share of total production.

Conservation

As oil and gas supplies become more scarce, energy conservation by industry and private consumers is expected to play an increasingly important role in extending the life of those finite resources. In the non-Communist world, the free market forces of supply and demand are the greatest inducement to conserve: as the price of energy goes up, demand drops. Many economists had once thought that oil was fairly immune to market forces, but the sharply reduced demand after the 1973–74 and 1979–80 price hikes proved otherwise.

The ratio of total energy use to economic output is the measure most commonly used to judge the overall energy efficiency of an economy. This tells the amount of energy consumed to produce a given amount of goods and services. Between 1973 and 1980, that ratio declined in non-Communist, developed countries (members of OECD) by 12.1 percent. Although declining oil prices in the mid-1980s somewhat slowed the trend toward increased conservation, energy efficiency will become more important when prices pick up again by the 1990s. OECD predicts that between 1980 and 2000, the ratio of total energy use to economic output in its member countries will decline an additional 16 to 20 percent.

The three major energy-consuming sectors of the economy are industry, residential/commercial, and transportation. The International Energy Agency of OECD has identified many ways to conserve energy in each of these areas by the year 2000.

OECD predicts that between 1980 and 2000, total industrial energy consumption could be reduced up to 30 percent, primarily by switching from oil to alternative fuels in the high-consumption industries that produce iron, steel, chemicals, petrochemicals, cement, brick, and nonferrous metals.

Space heating and cooling offer the greatest potential for energy conservation among residential and commercial users. Measures such as retrofitting older buildings, requiring higher thermal efficiency for new buildings, and increasing the energy efficiency of major appliances could cut energy use in the residential and commercial sector from 30 to 70 percent by the year 2000, according to OECD estimates (fig. 9.23).

Oil accounts for 99 percent of all energy sources used for transportation. That percentage is likely to remain high through

429

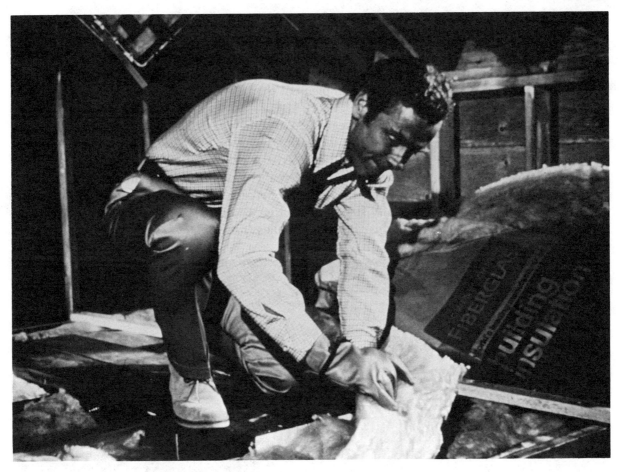

Figure 9.23. Increasing attic insulation and other forms of retrofitting can improve the energy efficiency of an older home. (*Courtesy of Owens/Corning Fiberglas Corporation and American Petroleum Institute*)

the 1990s because of the lack of economical substitutes. However, motor gasoline consumption has declined steadily since 1978, and is expected to continue to do so. OECD predicts that gasoline consumption could be cut by as much as 25 percent in Europe, even more in the United States, by the year 2000. Among the conservation measures that will continue to contribute to energy savings are improved fuel efficiency of new cars, reduction in the average annual distances traveled by cars, lowered driving speeds, better car maintenance, and greater use of car pooling and public transportation.

Fuel switching

In addition to conservation, a second major trend that is expected to stretch remaining petroleum supplies over as long a period as possible is fuel switching, that is, substituting other fuels such as coal, nuclear power, and new or renewable sources for oil and gas.

430

Governments of countries throughout the world have begun to play a major role in fuel switching. For example, in 1980 Canada introduced its off-oil conversion program, which set oil-reduction targets for all sectors of the economy and provided financial assistance to implement them.

Coal. Between 1984 and the year 2000, coal is expected to supply a steadily increasing share of the additional, or incremental, energy that will be needed worldwide. Conoco predicts that coal will account for 24 percent of all worldwide incremental energy from 1984 to 1990, 34 percent from 1990 to 2000. Coal is an especially strong alternative to oil or gas for electricity generation and industrial heat supplies (fig. 9.24).

Figure 9.24. The Charles P. Crane power plant in Baltimore City, Maryland, is being converted to burn coal instead of oil. The conversion will save 3 million barrels of oil per year. (*Courtesy of Baltimore Gas & Electric Company and American Petroleum Institute*)

Figure 9.25. The Susquehanna two-unit nuclear power plant in Berwick, Pennsylvania, produces electricity for nearly 1 million consumers. (*Courtesy of Bechtel Power Corporation and American Petroleum Institute*)

Weak oil prices and the increased costs of environmental safeguards required at coal-burning facilities began to erode coal's price advantage over oil and gas during the early 1980s, but coal is expected to regain that advantage when oil prices rise. And improved methods of burning coal, such as fluidized bed combustion — which removes sulfur in the combustion chamber and thereby reduces the amount of pollutants released into the atmosphere — have begun to make coal environmentally acceptable for a broader range of applications.

Nuclear power. Use of nuclear power is expected to increase at a faster rate than that of any other energy form between 1984 and 2000 (fig. 9.25). However, most of that growth will occur during the 1980s, as the reactors under construction are completed. Public concern over nuclear reactor performance and waste disposal safety, particularly in the United States, has led to considerable uncertainty about the future of nuclear power, making long-term predictions difficult. According to a Conoco study, nuclear's share of additional energy supplies will drop from 25 percent between 1984 and 1990 to 15 percent between 1990 and 2000. If the reactor safety and waste disposal problems are resolved, nuclear power could play a much greater role in meeting future energy demands.

432

Synthetic fuels. Often referred to as synfuels, synthetic fuels are high-energy liquid or gaseous fuels made from coal or other substances and can be used in many of the same ways as gasoline or natural gas. Because oil and gas prices are expected to remain fairly low to the year 2000, synthetic fuels are not likely to be economical before then. By the year 2000, however, pioneer plants could contribute the equivalent of as much as 372 million barrels of oil to OECD energy supplies.

Two synfuel processes now being developed and improved are coal gasification and liquefaction. Gasification produces synthetic natural gas (SNG) from coal (fig. 9.26). Although some SNG projects were cancelled or put on hold during the mid-1980s because of dropping oil and gas prices, four to six commercial gasification

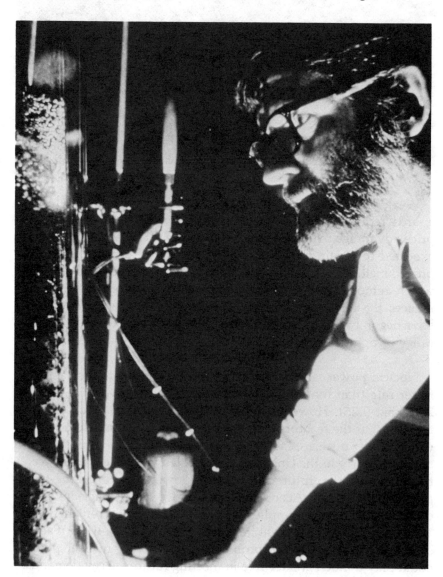

Figure 9.26. A researcher at the Lawrence Livermore Laboratory investigates the process of underground coal gasification. (*Courtesy of American Petroleum Institute*)

plants are expected to be in operation in OECD countries by the year 2000.

Liquefaction, which converts coal into a liquid fuel similar to gasoline, likewise holds great promise for the future. By the mid-1980s, South Africa was the only country in the world that was producing liquid synfuel, but three commercial plants are expected to be in operation in OECD countries by the year 2000.

Shale oil and tar sands. Two other sources of future energy supplies involve extracting oil from shale and tar sands. Both processes require mining or the application of heat, and the recovered oil must be cleaned of mineral and metal contaminants before it can be processed in a refinery.

Many small-scale shale oil projects were begun during the 1970s, and several commercial-size projects were being developed by the mid-1980s. One technique being explored involves in situ extraction, which would avoid the problems of waste disposal and land reclamation associated with shale-oil mining. Since 1967, tar sands have been commercially exploited in Canada, where they are plentiful (fig. 9.27). Most new tar sand projects, however, had been put on hold by the mid-1980s because of falling oil prices.

Figure 9.27. Workers discuss operations at a tar sands project at Fort McMurray, Alberta, in western Canada. The giant bucket wheel excavator is used to mine the tar sands. (*Courtesy of Sun Company, Inc., and American Petroleum Institute*)

Renewable energy resources. Renewables encompass all the nondepletable energy resources, including hydropower; solar energy; geothermal energy; wind, wave, and tidal power; and biomass. Apart from hydropower, which is already used extensively in some parts of world for electricity generation, renewables are expected to make only a marginal contribution to world energy supplies by the year 2000. They will, however, continue to be used in many small-scale local or regional applications and will become more important in the coming years.

Solar power shows great promise for domestic heating, especially in the United States, the Pacific Basin, and Southern Europe (fig. 9.28). Geothermal energy likewise may be used for heating and also for electricity production. Geothermal resources are plentiful in the United States, Italy, Japan, New Zealand, and Turkey. Wind power, which has been used for centuries for pumping and other applications, now is being used to produce electricity. Biomass energy is derived from agricultural or forestry products, animal wastes, marine culture, and even municipal wastes. New techniques are being developed to produce methane and alcohol fuels economically from biomass. Canada, Ireland, and Sweden are the most likely candidates for greater biomass energy production.

Figure 9.28. This 3,100-square-foot solar demonstration house in Carlisle, Massachusetts, features a 7.3-kilowatt photovoltaic array. (*Courtesy of Solarex Company and American Petroleum Institute*)

Industry planners must take into account the many complex and highly volatile aspects of petroleum economics in preparing for the future. Social or political changes in one part of the world can drastically affect petroleum supply and demand in the rest. Among the trends facing the industry between now and the year 2000, two are evident: despite increased conservation and switching to alternative fuel sources, worldwide oil demand will continue to rise; and despite stepped-up exploration for petroleum outside the Middle East, dependence of the non-Communist world on that region's supplies will become greater.

To meet the energy needs of the future, the petroleum industry must be ready for change. One example of such change already well underway is the diversification of many oil and gas companies into coal, nuclear, and other energy forms. The *petroleum* industry is becoming the *energy* industry. If this industry can find, develop, and market the vast energy supplies to be found in alternative sources with even a fraction of the success it demonstrated with oil and gas, then fear of energy shortages will become a thing of the past. That's the challenge of the 1990s and beyond.

References

Abbott, John P. "Barging." *Shell News*, February 1985.

Abbott, John P. "We Just about Cover the Map." *Shell News*, February 1985.

Allvine, Fred C., and James Patterson. *Competition, LTD.: The Marketing of Gasoline.* Bloomington, Ind.: Indiana University Press, 1972.

American Petroleum Institute. *Introduction to Oil & Gas Production.* Dallas: Production Department, American Petroleum Institute, 1983.

Amoco Chemicals Corporation. "Amoco Chemicals Corporation, Texas City Plant." Houston: Amoco Public and Government Affairs Department, 1974.

Amoco Chemicals Corporation. "Chocolate Bayou Plant, Alvin, Texas." Houston: Amoco Public and Government Affairs Department, 1974.

Anderson, Robert O. *Fundamentals of the Petroleum Industry.* Norman: University of Oklahoma Press, 1984.

Australian Institute of Petroleum. "Economic Power Moves from the Atlantic to Pacific." *Petroleum Gazette*, Summer 1984.

Baker, Ron. *A Primer of Oilwell Drilling,* 4th ed. Austin: Petroleum Extension Service, The University of Texas at Austin, 1979.

Baker, Ron. *A Primer of Offshore Operations.* Austin: Petroleum Extension Service, The University of Texas at Austin, 1985.

Baker, Ron. *Oil and Gas: The Production Story.* Austin: Petroleum Extension Service, The University of Texas at Austin, 1983.

Bell, Harold Sill, ed. *Petroleum Transportation Handbook.* New York: McGraw-Hill, 1963.

Berger, Bill D., and Kenneth E. Anderson. *Modern Petroleum.* Tulsa: Petroleum Publishing Co., 1978.

Berkhout, A. J. "It Is Time to Aim for the Best." *World Oil*, December 1984.

Bland, William F., and Robert L. Davidson, eds. *Petroleum Processing Handbook.* New York: McGraw-Hill, 1967.

British Petroleum Company Limited. *Our Industry Petroleum.* London: British Petroleum Co., 1977.

Bumgarner, Kevin. "Availability of Spot Market Vital to Independents." *The American Oil & Gas Reporter*, October 1985.

Burton, Dudley J. *The Governance of Energy: Problems, Prospects, and Underlying Issues.* New York: Praeger Publishers, 1980.

Chevron Corporation. "Chevron Shipping—Battened Down and Making Way through Rough Seas." *Chevron World*, Summer 1985.

Conoco Inc. "Conoco's First Sealift Headed North." *Conoco World*, July 1985.

Conoco Inc. *World Energy Outlook through 2000.* Wilmington, Del.: Coordinating and Planning Department, Conoco, 1985.

Cox, James A. "The Marketers Are Back!" *Shell News*, April 1985.

Denny, Sharon. "The Hunt for Offshore Oil Moves into Deeper Waters." *The Oil Daily*, May 6, 1985.

438

Diamond Shamrock Refining and Marketing Company. "Grand Openings." *Tattler*, Spring 1985.

Diamond Shamrock Refining and Marketing Company. *The Refining and Marketing Resource*. San Antonio: Diamond Shamrock, n.d.

Dobrin, Milton Burnett. *Introduction to Geophysical Prospecting*, 3rd ed. New York: McGraw-Hill, 1976.

Eubank, Judith. *Land and Leasing*. Austin: Petroleum Extension Service, The University of Texas at Austin, 1984.

Exxon Corporation. *Middle East Oil and Gas*. New York: Public Affairs Department, Exxon, 1984.

Fischer, Joel D., and Constantine D. Fliakos. *British Petroleum: A Basic Report*. New York: Drexel Burnham Lambert, Inc., 1978.

Galloway, W. E.; T. E. Ewing; C. M. Garrett; N. Tyler; and D. G. Bebout. *Atlas of Major Texas Oil Reservoirs*. Austin: Bureau of Economic Geology, The University of Texas at Austin, 1983.

GATX TankTrain. Chicago: General American Transportation Corp., n.d.

Guthrie, Virgil B., ed. *Petroleum Products Handbook*. New York: McGraw-Hill, 1960.

Hayes, C.; M. M. Thacker; and J. H. Viellenave. "Exploration Techniques Aided by Computer Tiering." *World Oil*, December 1984.

Hinds, Bill. "Competition Comes to the Gas Industry." *Venture*, American Natural Resources Co., Summer 1985.

Hosmanek, Max. *Pipeline Construction*. Austin: Petroleum Extension Service, The University of Texas at Austin, 1984.

"Industry Scoreboard 11/25." *OGJ Newsletter, Oil & Gas Journal*, November 25, 1985.

International Energy Agency. *World Energy Outlook*. Paris: Organization for Economic Cooperation and Development, 1982.

International Petroleum Encyclopedia. Tulsa, Okla.: PennWell Publishing Company, 1985.

Levander, Kai; Torsten Heideman; Henrik Segercrantz; and Gustav Lindqvist. *Ice 2000: Ice-Going Ships for the Next 20 Years*. Helsinki, Finland: Wärtsilä Shipbuilding Division, 1985.

Leecraft, Jodie, ed. *A Dictionary of Petroleum Terms*, 3rd ed. Austin: Petroleum Extension Service, The University of Texas at Austin, 1983.

Makinen, Eero, and Raimo Roos. "Ice Navigation Capabilities of Lunni-Class Icebreaking Tankers." Paper presented at the meeting of Eastern Canadian Section of SNAME in Montreal, November 22, 1977, and Quebec City, November 23, 1977.

Mangan, Frank. *The Pipeliners*. El Paso: Guynes Press, 1977.

Marathon Oil Co. "The Right Stuff." *Marathon World*, No. 1, 1984.

Megill, Robert E. "Exploration Outlook: Is the Past a Key to the Future?" *World Oil*, June 1984.

Moody, Graham B., ed. *Petroleum Exploration Handbook*. New York: McGraw-Hill, 1961.

Morris, Jeff; Annes McCann Baker; and Richard House. *Practical Petroleum Geology*. Edited by Jodie Leecraft. Austin: Petroleum Extension Service, The University of Texas at Austin, 1985.

Moses, Leslie. "From Lease to Release." *AAPL Guide for Landmen*. Fort Worth, Tex.: American Association of Petroleum Landmen, 1970.

Oil and Gas Pocket Reference. Houston: National Supply Co., 1984.

"Oil on the Move," *Marathon World*, No. 1, 1985.

Oil Pipelines in the United States: Progress and Outlook. Washington, D. C.: Association of Oil Pipe Lines, 1983.

Our Magnificent Earth: A Rand McNally Atlas of Earth Resources. New York: Rand McNally & Co., 1979.

Parshall, Joel S. "The Moving Force." *Marathon World*, No. 1, 1984.

Petroleum Extension Service. *Field Handling of Natural Gas*, 3rd ed. Austin: Petroleum Extension Service, The University of Texas at Austin, 1972.

Petty, O. Scott. *Seismic Reflections.* Houston: Geosource Inc., 1976.

Phillips Petroleum Company. "How to Sail in an Oil Surplus" *Shield* Vol. X, No. 2, 1985.

Phillips Petroleum Company. "Pasadena Terminal Operation Makes Fuel Loading Easier." *PhilNews*, July 1985.

Phillips Petroleum Company. "Remote Self-Service Is First." *PhilNews*, September 1985.

Pinkard, Tommie. "Gulf Intracoastal Waterway: Texas' Big Ditch." *Texas Highways*, August 1984.

Pressure Transport, Inc. *Pressure Transport.* Austin: Texas Gas Transport Co., n.d.

The Rail Tank Car Industry. Alexandria, Va.: Railway Progress Institute, n.d.

Royal Dutch/Shell Group of Companies. *The Petroleum Handbook*, 6th ed. New York: Elsevier Science Publishing Co., 1983.

Sheppard, Nora, ed. *Introduction to the Oil Pipeline Industry*, 3rd ed. Austin: Petroleum Extension Service, The University of Texas at Austin, 1984.

Simson, Stephen F., and H. Roice Nelson, Jr. "Seismic Stratigraphy Moves toward Interactive Analysis." *World Oil*, December 1984.

Sklarewitz, Norman. "Shell Goes for the Guinness Book of Records with the GATX TankTrain® System." *Shell News.* Offprint by General American Transportation Corporation, n.d.

Tank Cars. *The Car and Locomotive Cyclopedia*, 4th ed. New York: Simmons Boardman Publishing Corp., 1980.

Telford, W. M., et al. *Applied Geophysics.* Cambridge and New York: Cambridge University Press, 1976.

Thomasson, M. Ray; John Sandy; Robert G. Reeves; James R. Lucas; Bill Ehni; and Benjamin Prokop. "Non-seismic Methods: A Key to Integrated Exploration." *World Oil*, December 1984.

Wallace, Britann E. "FERC Rule Changes Gas Marketing." *The American Oil & Gas Reporter*, October 1985.

White, D. A. "Assessing Oil and Gas Plays in Facies-Cycle Wedges." *AAPG Bulletin*, Vol. 64, No. 8, 1980

Williamson, Harold F.; Ralph L. Andreano; Arnold R. Daum; and Gilbert C. Klose. *The American Petroleum Industry: The Age of Energy, 1899–1959.* Evanston: Northwestern University Press, 1963.

Williamson, Harold F., and Arnold R. Daum. *The American Petroleum Industry: The Age of Illumination, 1859–1899.* Evanston: Northwestern University Press, 1959.

Wood, Robert A. "Farm Power." *Marathon World*, No. 3, 1985.

Wood, Robert A. "Charting a Course Downstream." *Marathon World*, No. 1, 1984.

Wood, Robert A. "The Right Stuff." *Marathon World*, No. 1, 1984.

Index